INTRODUCTORY ENGINEERING MODELING EMPHASIZING DIFFERENTIAL MODELS AND COMPUTER SIMULATIONS

INTRODUCTORY ENGINEERING MODELING EMPHASIZING DIFFERENTIAL MODELS AND COMPUTER SIMULATIONS

William G. Rieder
North Dakota State University

Henry R. Busby.
The Ohio State University

JOHN WILEY & SONS
New York   Chichester   Brisbane   Toronto   Singapore

Library of Congress Cataloging in Publication Data:
Rieder, William G.
  Introductory engineering modeling emphasizing
differential models and computer simulations.

  Includes index.
  1. Engineering models.   2. Computer simulation.
  I. Busby, Henry R.   II. Title.
TA177.R54   1986   620'.00724   85-6295
ISBN 0-471-89537-7

Printed in the United States of America

10 9 8 7 6 5 4 3 2 1

PREFACE

This book presents introductory material on how to set up and use applied mathematical models containing derivatives. It reflects more than 15 years of experience in conveying the elements of this art to undergraduate engineering students. Since the emphasis is on the applied rather than the theoretical, classical mathematical details are minimized to get at simulation methodology, including some finite-element approaches, as directly as possible. Thus this is not a text book on "mathematical modeling," but rather a text book on "applied modeling and simulation." Although intended primarily as a textbook, it can be used effectively as a reference and review source by engineering practitioners who want to improve their skills in applying a computer to technical problems. The overall thrust is toward using the digital computer as an intrinsic part of engineering modeling; but it is emphasized that successful computer-generated simulations must be built on *both* model-setup skills and knowledge of how to choose and implement a simulation method.

Traditionally, engineering curricula have placed a strong emphasis on classical mathematics and physics courses for the first two years. Following this, the student is subjected to discipline-specific engineering courses that provide skills in the handling of certain types of engineering problems. However, it is widely recognized that the textbook examples and homework problems in the latter courses tend to be overly simplistic relative to "real" problems. In recent years, some textbooks have attempted to overcome this deficiency by including special homework problems (usually at the end of the list in each chapter) that require the use of the computer. Unfortunately, the extra preparation effort required of both the instructor and student to handle these problems usually detracts from the course objectives. This book is directed toward this preparation problem.

The course level at which the material in this book fits best into the undergraduate engineering curriculum has been one of our major concerns. Ideally, the course should come precisely between the classical preparatory courses (mathematics, physics, chemistry) and the discipline-specific engineering courses.

This presumes that some exposure to differential equations is included in the mathematics and that some background in *applied* computer programming is present. In fact, the structures of most accredited curricula do not permit such a precise placement. In consequence, many compromises in content have been made to allow use in a course placed anywhere from mid-sophomore to mid-senior level. It should be noted, however, that most of the material has been used successfully at the lower-sophomore level, and for short courses.

This book contains more material than can be covered in a three-credit, one-semester, or four-credit, one-quarter course. Typically, for an upper-sophomore or junior-level course, most of Part One can be covered in about 10 hours. About 2 hours are sufficient for Chapter 4, and about 12 hours are required for Chapter 5. Most of Chapter 6 requires another 10 hours. Selected material from Chapters 7 and 8, selections from sections previously skipped, exams, and contingencies easily absorb the remaining 6 hours. If convenient, one of the course credits should be allocated to the designated "L" exercises and case studies. A 2 or 3 hour laboratory associated with this credit provides the appropriate question and consultation time. Although not absolutely necessary (because review material is given in Appendix A2), some background in general-purpose numerical procedures is helpful. In this regard, the laboratory time can be useful in reviewing the Appendix A2 material.

Several alternatives to the above are feasible. For example, if the course is offered at the upper-junior level or higher, more time can be spent on Chapters 3 and 7. At any level, Part One can be treated as a review, with concentration being placed on Part Two, or Appendixes A1 and A2 can be studied in place of Part One. A particularly successful method that has been used at the sophomore level is to go through Part Two (excluding some sections and Chapter 7), reviewing Appendix A2 as necessary. Next, Part One is covered with the emphasis on only one or two sections from each chapter, and the Appendix A1 is reviewed in the context of Chapter 8.

Some comment is necessary in regard to the deemphasis of units in the examples and exercises. Early teaching experiences with the foregoing materials, including the units, has showed that the student invariably becomes mired down in the units and becomes unable to concentrate on the wide-ranging modeling topics in the allocated time. Furthermore, much professional modeling uses dimensionless approaches anyway. Thus units are disregarded. However, this does not mean that they are unimportant, and this should always be pointed out to the student. If case studies are used, the problems associated with using units can be incorporated.

Finally, we acknowledge with thanks the contributions of countless individuals over many years. We would be remiss if we didn't single out the mechanical and civil engineering students who suffered through the many pedagogical versions of this material, and who always responded with constructive advice. We are especially grateful to our wives and children, to whom this book is dedicated.

William G. Rieder
Henry R. Busby

CONTENTS

LIST OF SYMBOLS

An attempt has been made to use commonly recognized engineering symbols in this book. As a consequence, because of the multidisciplinary nature of many of the topics, some symbols have more than one meaning. The appropriate meaning will have to be determined from the context. However, the multiple meanings are listed here wherever practical.

A, A	area; matrix of nodal coordinates
B	transfer parameter $= 1/Fo$; matrix relating displacements to strains
C	capacitance or storage capability; damper constant; concentration; $\cos\theta$; elasticity matrix (plane stress; with prime, plane strain)
C_u	capacitance per unit volume
CE	characteristic equation
CM	center of mass
CV	control volume
D	diameter; mathematical operator representing a derivative; plate stiffness parameter
DF	degree of freedom
E	modulus of elasticity (Young's modulus)
F	force, $\Delta F = F$ difference
FB	free body
Fo	transfer parameter $= \alpha\,\Delta t/(\Delta x)^2$
G	shear modulus
H	liquid level
I	area moment of inertia
J	polar mass moment of inertia
K	stiffness or flow resistance general matrix, permeability
L	length (see also l)
M	external bending moment

N	shape function matrix
P	pressure; external load; component of force; vector of nodal pressures
Q	quantity flowing or stored during an accounting period; beam shear force
Q_i	Q into a control volume
Q_e	Q out of a control volume
Q_s	Q stored or lost from a control volume
R	flow resistance; reaction force, external nodal force
R_{i-j}	flow resistance from cell i to cell j
R_c	combined flow resistance; convective flow resistance
R_u	flow resistance per unit length of flow path
S	stress; flow conductance $= 1/R$; $\sin \theta$
S_{i-j}	length of an element edge between node i and j
T	temperature, $\Delta T = T$ difference; temperature vector
T_c	convective transfer environment temperature (see also T_∞)
T_∞	ambient (surroundings) temperature
V	velocity; volume
VV	volumetric flow rate; vector of nodal flow rates
W	weight; external load; beam nodal displacement vector
$a_1, a_2, \ldots$	recursion values for Runge–Kutta formulas; coefficients for u displacements; polynomial coefficients
$b_1, b_2, \ldots$	recursion values for Runge–Kutta formulas; coefficients for v displacements
f	vector of element or global forces
$f(x, y)$	a function of x and y
f_x	x-direction force
f_y	y-direction force
f_i	$f(x_i, y_i)$; force on node i
g	acceleration owing to gravity
g_c	gravitational conversion constant
h	heat transfer coefficient at a boundary; distance to neutral axis on a beam cross section
i	grid index in x direction; iteration numbers; cell, element, or node number
j	grid index in y direction; cell, element, or node number
k	time index; spring rate; thermal conductivity; element or node number; element stiffness matrix
l	length (also see L)
m	mass; internal beam moment
$\dot{m}$	mass flow rate
p	pressure; distributed force
q	quantity flow rate; heat transfer rate vector or element; displacement vector element
q_{i-j}	flow rate from cell i to cell j

q_i	inlet flow rate to a control volume $= q_{xi}$ in the x direction
q_e	exit flow rate from a control volume $= q_{xe}$ in the x direction radius
r	of curvature
s	movement coordinate
$\dot{s}$	velocity along s
$\ddot{s}$	acceleration along s
t	time, $\Delta t = t$ difference or time span; element thickness
u	x direction displacement
v	y direction displacement
w	unit load; element of displacement vector in $x - y$ plane
x, y	rectangular coordinates, $\Delta x = x$ difference, $\Delta y = y$ difference; displacement vector
$\dot{x}$	velocity along x
$\ddot{x}$	acceleration along x

Greek Letters

α	flow diffusivity (units of length^2/time)
$\alpha_1, \alpha_2, \alpha_3$	geometric grouping for shape function
$\beta_1, \beta_2, \beta_3$	geometric grouping for shape function
$\gamma_1, \gamma_2, \gamma_3$	geometric grouping for shape function
γ	specific weight
γ_{xy}	local shear strain in x-y plane
Δ	general displacement; represents a "difference of" or an "increment" when a prefix
Λ	matrix relating stresses to joint forces
ε	convergence parameter for iterative solution methods; local strain vector
ε_x	local normal strain in x direction
ε_y	local normal strain in y direction
θ	slope; angular displacement
λ	parameter dependent on v
v, μ	Poisson's ratio; viscosity
ξ	dimensionless position $= x/l$
ρ	density
σ	stress vector
σ_x	local normal stress in x direction
σ_y	local normal stress in y direction
τ_{xy}	local shear stress in $x - y$ plane

Subscripts

i, j	cell number; node number; position in x, y directions, respectively; element number; inlet flow station
k	time index; element number

c	convection environment at boundary
e	exist flow station
x, y	x and y directions
0	central value; initial value; zero position
s	stored value, shear stress
u	unit volume, length, etc., value
∞	surrounding (ambient) conditions
1, 2, etc.	node number; element number; cell number
r, l, t, b	indicates right, left, top, and bottom

Superscripts

c	corrector value
p	predictor value
o	old value, or value from preceding step
$*$	new value
i, j, k	node number; element number
x, y	x and y directions
1, 2, etc.	node number; element number
$'$	single prime, an alternate value; first derivative
$''$	double prime; second derivative

Matrix Symbols

$[-]$	multiple-column matrix
$\{-\}$	column matrix, or vector
$[-]^{-1}$	inverse matrix
$[-]^{T}$	transpose of matrix (row to column interchange)

PART ONE
MODEL SETUPS

1
Some Engineering Approaches to Model Setups and Simulations

For most engineers the term *modeling* implies the use of one system's behavior to simulate (to allow prediction of) performance of a different system. However, the entity used as the "model" can vary widely. In some instances carefully fabricated miniature aircraft, or automobiles, are placed in wind tunnels and subjected to controlled tests. In other instances mathematical equations are manipulated by classical techniques, or through numerical methods using calculators or computers, to give the same simulation results. We are interested in the latter cases in this book. In these cases the equations themselves are viewed as the "models"; and the "simulations" are the manipulation processes that generate the performance numbers. This rather restricted perspective is taken here to avoid the slight confusion that has been a traditional part of much of the technical jargon surrounding the use of the word *modeling*.

In this book we define models to be the mathematical equations that represent physical (engineering) systems. Furthermore, we view these equations to be the kind that are set up by the typical engineer analyst. This means that much empirical information is mixed together with a few basic principles and concepts to develop some new equations to predict behaviors. In other words our models can always be constructed using relatively few basic principles. These classical principles, although they may be in equation form themselves, cannot be derived from other equations. Often they are recognized as simple algebraic or transcendental formulas from elementary physics. Occasionally, one will have been refined into differential equation form, and these can sometimes be incorporated directly into a model setup without change. We are particularly interested in models that contain derivatives because they tend to represent highly complicated physical systems in a condensed fashion and provide by far the most realistic simulations of behavior over wide ranges of conditions. This is important in engineering work because of the diversity of physical systems that may be found coupled together.

Simulation results will always be sets of numbers rather than single values.

3

Typically, matched pairs of independent and dependent variable values will occur. When more than one independent variable is involved, subgroups of numbers for each independent variable will be identifiable within the simulation results. Occasionally, a classical mathematics solution of a model will be found. Such solutions may become an intermediate step in a simulation; they typically come from relatively simple algebraic, or transcendental, equations. Students will often ask why a computation of a single stress, or temperature, value from some elegantly simple algebraic formula found in a handbook is not really a simulation by our definition. Important as such quantities might be, they generally do not give the broad overall picture of behavior that sets of numbers in a complete simulation will give. Furthermore, the model from which the formula originates is usually a restricted model in concession to classical mathematics, and thus provides only limited realism. So even if we generate a complete simulation by computing multiple values, the practical usefulness of the results may be limited by the simplistic starting model. On the other hand, these classical solution simulations can be very valuable for verifying numerical methods as benchmark standards.

The present engineering value associated with using the mathematical equation model, that is, the "soft model," as opposed to an actual physical model, the "hard model," is clearly related to the wide availability of large-capacity digital computers. The digital computer can generate numbers cheaply. Of course, part of the penalty paid for these "simulations" may be hidden in the engineering quality of the simulation numbers. Many professional engineers prefer the hard-model approach for critical problems and have a certain amount of distrust (probably justified) of the computer numbers. Part of the problem has been that early over-enthusiasm about applying the computer to everything has led to simulations based on faulty models. In at least some cases engineering realities have been ignored by the people setting up the simulations. An adequate "feel" for the real problem was not there. Consequently, the engineering student is warned not to place too much confidence in simulation results simply because they were generated by a computer. In almost all respects, the setup of the model is more important than the computer simulation process.

The basic concepts and principles necessary to start the setup of a model can come from many sources. The common situation is for the engineers to expend great effort in visualizing what is happening physically. They may then search the professional literature and handbooks, or they may even get a little information directly from the laboratory or field to identify basic relationships that can be used to start building their models. The diversity and creative complexity of the foregoing defies a decent description, and we will relegate what little can be said to later examples and case studies. We will short-circuit the process slightly in this book by sticking with some relatively uncomplicated cases for which the basic concepts can be found in commonly used engineering textbooks and handbooks. However, a significant appeal is made to the student to also use his personal observations of the world about him. We will then proceed into the details of constructing the mathematical model. A final comment is necessary at this point. Some individuals seem to be able to utilize a combination of meticulous observations and "tuning" of a computer program (i.e., running successive trial simulations) to eventually give an acceptable engineering simulation, apparently without much effort, if any, in setting up any mathematical models. Of course, the success depends in part on where the program originated. In the authors' opinion, such simulations have to be suspect until complete independent verification can be shown. Nevertheless, it is often surprising how small an arsenal of basic formulas is needed to allow model setups to proceed.

The next few chapters illustrate some of the procedures used by engineers in setting up some common differential models. The examples shown have been picked because they represent fairly large classes of engineering problems. They include flow systems, moving object systems, and elementary structure systems. However, it should be recognized that many other system classifications exist in engineering, for example, management and economic systems, and certain electronic systems, and that the details of setting up models may differ from those described here.

Finally, it should be observed that in most of the material that follows, units are ignored. This is deliberate. This is not a book for practice in handling units; rather, attention should be focused on the broader concerns of model development and simulation methodology. However, if the user so desires, he or she can usually substitute a consistent set of units wherever an example or exercise so designates.

2
Setups Involving Ordinary Derivatives

FLOW SYSTEMS

Central to the setup of models for flow systems, regardless of what may be flowing, is the concept of *control volume*. We will use the abbreviation "CV" to stand for "control volume" in much of the following. Simply a spatial region upon which attention can be focused for an acceptable accounting period, a CV can be very large, such as, a complete city, or very small, such as an infinitesimal segment of the thin oil film space in a journal bearing. At least some portions of the CV boundary (sometimes referred to as the "control surface") are open or permeable to the flowing material. Sometimes the flowing material may be something to which we can attach mass units, and other times it may be something we cannot see and to which we may attach things like energy units. Although not necessary, it is usually convenient to place the analyst's spatial reference stationary with respect to some easily recognized feature of the control volume. It must be recognized that the boundaries of a CV seldom coincide with actual physical boundaries of the system under consideration; in fact, they are strictly imaginary.

How to choose CV boundaries to simplify model setup might be considered to be part of the engineering analyst's art. A common practice is to choose several different groupings of control volumes at various stages in the model setup for a single system. The obvious usefulness of the CV can be associated with the fact that just a few formulations (of conservation and flow principles) can be applied repeatedly in completing a model setup. Skill at visualizing the physical situation in terms of those control volumes which minimize model setup difficulties is a talent that most engineering students can learn at least to some degree.

The most basic principle needed to set up flow system models is the conservation principle. This principle can be posed in simple algebraic form, as shown in Equation 2.1.

$$Q_i - Q_e = Q_s \tag{2.1}$$

where Q_i = quantity flowing into the CV during an accounting period

Q_e = quantity flowing out of the CV during the same accounting period

Q_s = quantity stored, or lost from, the CV during the same accounting period

If we divide both sides of Equation 2.1 by the accounting period, and then visualize this accounting period as being infinitesimally small, we get our first differential (or derivative-containing) model—the flow rate equation for a single control volume, Equation 2.2.

$$q_i - q_e = \frac{dQ}{dt} \tag{2.2}$$

where q_i = quantity flow rate into the CV

q_e = quantity flow rate out of the CV

Q = quantity existing in the CV at any instant

Of course, both q_i and q_e must account for all the separate flows if more than one inlet and/or exit are involved. Also, we have implicitly ignored any quantity conversions that might have occurred within the CV. The typical effect is to see another term on the left-hand side of Equations 2.1 and 2.2, which if shown with a negative sign is called a "sink," and if shown with a positive sign is called a "source," Although we mention source/sink concepts later, in this book, we will concentrate on systems that do not exhibit source/sink behavior.

A special form of the conservation principle occurs if no change in the quantity stored is observed (i.e., the derivative in Equation 2.2 goes to zero). Then Equation 2.2 reduces to Equation 2.3.

$$q_i - q_e = 0 \tag{2.3}$$

We digress slightly here and divert our attention to how the foregoing can be related to a common situation observed by most students. We look at liquid flowing steadily through a horizontal pipe. A reasonably short segment of the pipe is chosen as our CV. It is apparent that we can find an average velocity at any cross section. Then, for mass flow, we can restate Equation 2.3 as Equation 2.4.

$$(\rho A V)_i = (\rho A V)_e \tag{2.4}$$

where ρ = density of the fluid at CV inlet i or exit e

A = flow area normal to the average velocity vector

V = average flow velocity

A further simplification can be made if the fluid does not exhibit much of a density change as it flows through the CV. Equation 2.4 can be reduced to Equation 2.5.

$$V V_i = V V_e \tag{2.5}$$

where $V V_i = A_i V_i$ = volumetric flow rate into the CV

$V V_e = A_e V_e$ = volumetric flow rate out of the CV

Equations 2.4 or 2.5 are often referred to as the "continuity principle." However, it should be noted that in the broadest interpretation "continuity principle" and "conservation principle" are one and the same.

An extremely useful principle for setting up flow system models, which relates the flows to the forces driving the flows, can be applied to most control volumes for which Equation 2.3 applies, even if only instantaneously. This principle, which is sometimes referred to as the "universal flow principle," is encompassed in Equation 2.6.

$$q = \frac{\Delta F}{R} \tag{2.6}$$

where q = instantaneous flow rate through the CV governed by Equation 2.3
ΔF = instantaneous force difference causing the flow to occur
R = instantaneous resistance of the CV to the flow

For fluid flow systems, ΔF is simply the pressure difference between the inlet and exit. For heat flow systems, ΔF is simply the temperature difference between the hot side and cold side of the CV; for electrical systems, ΔF is simply the voltage difference between the high and low terminals.

All the well-known basic laws governing flows can be re-posed in the form of Equation 2.6. The implication is that a flow computation is just a matter of looking up the appropriate value for the resistance R and plugging it into Equation 2.6. In fact, this can be done in only a few cases, usually those shown in textbooks. The R is seldom a constant and is often a strong, complicated function of the flow rate itself. This is illustrated in Example 2.1. Significant engineering research may be necessary to get good relationships for the R. Nevertheless, the usefulness of Equation 2.6 cannot be denied, and almost all flow-model setups depend on it.

EXAMPLE 2.1

The following formulas are common in engineering handbooks and are presented for computing (1) radiant heat transfer, (2) pressure loss in a liquid pipeline, (3) the diffusion rate of gas through a membrane, and (4) the voltage drop in a resistor element. Re-pose each of the formulas into the universal-flow form of Equation 2.6 and identify the flow resistances.

1. $q_r = \sigma FEA(T_h^4 - T_c^4)$, where q_r is the heat transfer rate, σ is the Stefan Boltzmann constant, F is a shape factor, E is the emmissivity, A is the surface area, T_h is the hot-surface temperature, and T_c is the cold-surface temperature.

2. $\Delta P = KV^2\gamma/(2g)$, where ΔP is the pressure drop in the direction of the flow, K is an empirical loss coefficient, V is the flow velocity, γ is the liquid specific weight, and g is the acceleration owing to gravity.

3. $N = DA(C_i - C_o)/L$, where N is the mole diffusion rate, D is the diffusion coefficient, C_i is the molar concentration at the inside surface, C_o is the molar concentration at the outside surface, and L is the thickness of the membrane.

4. $E_1 - E_2 = IR$, where E_1 and E_2 are the high and low voltages across the resistor element, I is the current, and R is the resistance of the resistor element.

▶ **1.** $q_r = (T_h - T_c)/R_t$, where $R_t = (T_h - T_c)/[\sigma A F E(T_h^4 - T_c^4)]$.

2. $\Delta P = K\gamma(VV/A)^2/(2g)$ from substitution with Equation 2.5. $VV = [2gA^2 \Delta P/(K\gamma)]^{0.5} = \Delta P/R_f$, where $R_f = (\Delta P \, K\gamma)/[2gA^2]^{0.5}$.

3. $N = (C_i - C_o)/R_m$, where $R_m = L/(DA)$.

4. $I = (E_1 - E_2)/R_e$, where $R_e = R$.

Almost any flow system can be broken down into subsystems to ease the task of choosing control volumes. These control volumes can be recombined later to get the combinations of equations needed for the model. We are interested at this point in how the control volumes can be combined. By far the simplest combining techniques can be used for control volumes, where Equations 2.3 through 2.6 are valid and where the control volumes are grouped together in an end-to-end fashion. We say the latter is a series-connected group. A well-known resistance combinatorial principle can be used to replace the group with a larger equivalent control volume. This principle is embodied in Equation 2.7, which is a concise representation of the statement that for n steady flow resistances connected in series, the equivalent resistance R_c is the sum of the separate individual resistances R_j,

$$R_c = R_1 + R_2 + \cdots + R_n \qquad (2.7)$$

If we should find any of the subsystem control volumes grouped together in a side-by-side fashion, with a common inlet and exit, we say they are parallel-connected and use another simple principle incorporated in Equation 2.8. The resulting equivalent resistance R_c for this larger equivalent control volume is the reciprocal of the sum of the separate reciprocal resistances $1/R_j$,

$$R_c = \frac{1}{1/R_1 + 1/R_2 + \cdots + 1/R_n} \qquad (2.8)$$

Most readers will recognize the foregoing procedures to be exactly analogous to combining electrical resistances and should note that analogy concepts are an intrinsic part of model setups and system simulations. This will become more evident later. Once combinations of the type described above have been made, the process can be repeated for any similar groupings, which might contain coincidently an element or two made up of equivalent control volumes combined as above. Thus a complete system setup can be built up gradually. Use of the above techniques is illustrated in Examples 2.2 and 2.3. These examples also show how schematic sketches, using symbols borrowed from some specific discipline (in this case, electrical engineering) can be employed to simplify the representation of the actual system.

EXAMPLE 2.2

A hydraulic flow system is shown in Figure 2.1. This is a steadily operating system consisting of a number of flow resistance zones as well as a pump and motor. The pressure rise through the pump is a function of the volumetric flow rate through the pump, $\Delta P_p = f_p(VV_p)$; and the pressure drop through the motor is a function of the volumetric flow rate through the motor, $\Delta P_m = f_m(VV_m)$. Ignoring any energy-conversion effects, do the following setup tasks.

1. Sketch a schematic of the system using analogous electrical symbols for the resistances and show with broken-line boundaries a group of control volumes corresponding to subsystems. Label the control volumes and flow directions.

2. Develop an expression for the combined resistance of the shutoff valve line, the filter and bypass lines, and the control-valve line.

3. Develop an expression for computing the volumetric flow rate through the system.

4. Give an equation for computing the pressure at location A, assuming that the system flow rate has already been computed.

5. Give an equation for computing the flow rate through the filter, assuming that the system flow rate has already been computed.

▶ **1.** (See Figure 2.2)

    2. We observe that the control volumes shown in Figure 2.2 do not correspond to the most elementary breakdown we could have used. What is shown is the result of some preliminary intellectual leaping, where some elementary control

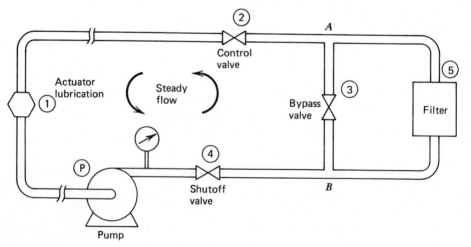

Figure 2.1 Steady flow system, Example 2.2.

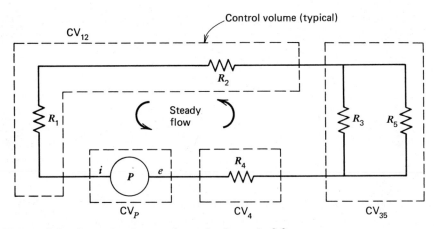

Figure 2.2 Control volume schematic, Example 2.2.

volumes have already been combined (e.g., the filter and bypass have been precombined). Most engineer analysts use this approach to varying degrees, but it can be confusing to the uninitiated. We establish how the equivalent resistance for the bypass-filter CV is obtained before proceeding.

This is a parallel connection, and we use Equation 2.8 to give R_{35},

$$R_{35} = \frac{1}{1/R_3 + 1/R_5}$$

The control valve line, the filter-bypass group, and the shutoff valve line make up a series connection, and we use Equation 2.7 to give the desired R_c.

$$R_c = R_4 + R_{35} + R_2$$

3. To get at the volumetric flow rate through the whole system, we have to invoke another well-known principle, which says that the algebraic sum of all the driving force differences for the control volumes making up a closed flow loop must be zero. This can be recognized as the analog of Kirchoff's voltage law from electrical disciplines.

For a constant density fluid, which is a reasonable assumption for most hydraulic fluids, the continuity principle gives $VV = VV_m = VV_p$. Using Equation 2.6 and the foregoing information, we get an implicit equation for the desired VV.

$$f_p(VV) - VV \times R_c - f_m(VV) - VV \times R_1 = 0$$

4. Assuming that we know the pressure at the pump exit, P_e, the equivalent resistance of the series-connected shutoff valve line and filter/bypass control volumes is found. This in turn is used with Equation 2.6 and continuity to give the desired P_A.

$$R_{435} = R_4 + R_{35}$$

$$P_A = P_e - VV \times R_{435}$$

5. First we have to find P_B. Then we can use Equation 2.6 again and P_A from above to give the desired VV_5.

$$P_B = P_e - VV \times R_4$$

$$VV_5 = \frac{P_B - P_A}{R_5}$$

EXAMPLE 2.3

A cross section of a portion of a building wall is shown in Figure 2.3. The different materials and surface air films are labeled with numbers and exhibit different resistances to heat flow. Assume that each resistance is based on the appropriate area normal to the heat flow, that only the inner and outer air temperatures are known, and that steady flow exists with no storage, and complete the following.

1. Sketch a schematic of the flow system using the analogs of electrical resistances; then using broken-line enclosures, show subsystem control volumes. Label the control volumes and resistances with subscripted CVs and R's.

2. Determine an equation for computing the flow resistance of the complete system. Label this resistance R_c.

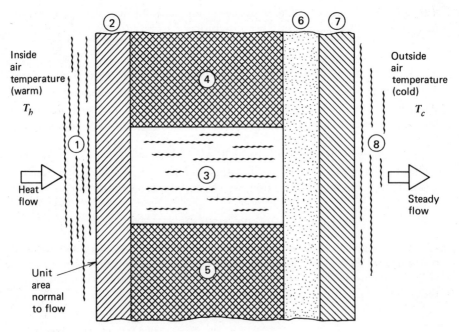

Figure 2.3 Wall thermal flow system, Example 2.3.

3. Develop an equation for computing the flow (heat transfer) rate through the overall system.

4. Develop equations for finding the temperatures between layers 6 and 7, T_{6-7}, and at the inside surface, T_{1-2}.

▶ **1.** The parallel and series resistance groupings are obvious from the description, so only three combined control volumes are shown in Figure 2.4.

2. Equations 2.7 and 2.8 are used to give the equivalent resistances for the three control volumes. The results in turn are combined to give the overall equivalent resistance to heat flow, R_c.

$$R_{12} = R_1 + R_2$$

$$R_{345} = \frac{1}{1/R_3 + 1/R_4 + 1/R_5}$$

$$R_{678} = R_6 + R_7 + R_8$$

$$R_c = R_{12} + R_{345} + R_{678}$$

3. Equation 2.6 is used to get the overall heat flow through the system.

$$q = \frac{T_h - T_c}{R_c}$$

4. Continuity says that q through the subsystem made up of layers 7 and 8 must be the same as the overall system q. Likewise q through layer 1 must equal the system q. Equation 2.6 can then be substituted and algebraic juggling done to get the desired temperatures.

$$q_{78} = q$$

$$R_{78} = R_7 + R_8$$

$$\frac{T_{6-7} - T_c}{R_{78}} = \frac{T_h - T_c}{R_c}$$

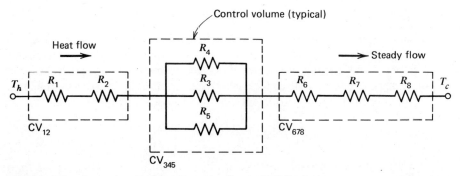

Figure 2.4 Control volume schematic, Example 2.3.

$$T_{6-7} = \frac{(T_h - T_c)R_{78}}{R_c} + T_c$$

$$q_{12} = q$$

$$\frac{T_h - T_{1-2}}{R_1} = \frac{T_h - T_c}{R_c}$$

$$T_{1-2} = T_h - \frac{(T_h - T_c)}{R_c} R_1$$

When subsystem connections become more complicated than the series and parallel arrangements illustrated in Examples 2.2 and 2.3, then special control volumes must be placed around each junction where more than two regular control volumes are connected. However, the only thing special about these control volumes is that they are so small that they have only *one* driving force value associated with them, and consequently have no flow resistance. This visualization is consistent with a scheme used later, called lumped analysis. Obviously, Equation 2.6 does not apply, but Equation 2.3 must be used. Unfortunately, students often become confused over which flow paths should be taken as inlets, and which should be taken as exits in this case of multiple paths. A trick used by many analysts is recommended. To avoid confusion, q_e is set to zero arbitrarily, and Equation 2.3 is re-posed as Equation 2.9.

$$q_{i1} + q_{i2} + \cdots + q_{in} = 0 \qquad (2.9)$$

In other words, all n flow paths (control volumes) are viewed as inlets for the junction special control volume. Then careful attention is focused on the inherent directional nature of Equation 2.6. Since flows are known to proceed only in the direction from large to small values of the driving forces [for those doubting their own observations about this, the perusal of a textbook explanation of the second law of thermodynamics might be worthwhile[1,2]], the flow rates through all the CVs connected to the special control volume are written as shown in Equation 2.10.

$$q = \frac{F_k - F_j}{R_k} \qquad (2.10)$$

Thus the single driving force F_j in the junction control volume is always assumed to be smaller than the upstream force F_k of each connecting control volume having resistance R_k. Of course, this means that a set of simultaneous equations in the F_j will always have to be written for each system containing more than one junction control volume. Each equation will look like the one that follows.

$$\frac{F_{i1} - F_j}{R_1} + \frac{F_{i2} - F_j}{R_2} + \cdots + \frac{F_{in} - F_j}{R_n} = 0$$

An alternative approach has already been illustrated in Example 2.2 (part 3). We identify all complete flow loops in the system and place a driving force symbol

F_j at each multiple-path junction. Then we set the algebraic sum of all driving force differences around each complete loop to zero. In electrical work such a procedure is known as Kirchoff's voltage law. Again the result is a set of simultaneous equations in the unknowns F_j.

Where the system has some control volumes exhibiting storage, then Equation 2.2 must be used; time derivatives of stored quantities show up along with the driving-force unknowns. The trick then is to find how to relate the driving forces to the stored quantities. Various "capacity" relationships, with a variety of equations, can be found, depending upon the discipline with which one is dealing. Basically, all these equations can be shown to be of the form given in Equation 2.11. The capacitance, C, is a measure of the ability of the control volume to store. This idea leads to another extremely useful technique for setting up flow system models, which we call *lumped analysis*.[3,4]

$$C = \frac{dQ}{dF} \qquad (2.11)$$

Setups of models for flow systems whose behaviors vary with time require the use of Equation 2.2 and the technique called lumped analysis. The dQ/dt in Equation 2.2 can be replaced by $C\, dF/dt$, using Equation 2.11, provided that we can identify the subsystem where storage takes place and assign to it a single driving force F. Obviously, if we do this, we cannot use the steady state Equation 2.6 in any direct fashion. However, we can argue that the storage should be imagined as occurring at a single point or "lump." If it becomes necessary to consider the flow resistance through this control volume, we break it down into instantaneously steady inlet and/or exit control volumes attached to a single lumped control volume having no resistance, only storage capacity. Alternately, we can break it down into two capacitance-only control volumes lumped at the inlet and exit, with an instantaneously steady, resistance-only control volume between them. Just as we chose the symbol for resistance from electrical disciplines, we will also choose the electrical capacitance symbol to represent our lumped control volume. This lumped analysis approach permits us to set up models for complicated flow systems containing multiple control volumes with and without storage. The resulting equations will contain time derivatives of driving forces, as well as just the driving forces, as dependent variables. Some applications of the foregoing schemes are illustrated in Examples 2.4 and 2.5. We will use lumped analysis again when we discuss the setup of flow system models having partial derivatives.

The tasks involved in setting up a model for a flow system can vary in accordance with the complexity of the system. It is obvious that a steady system will be less likely to require the use of lumped analysis techniques, and some systems may contain only one or two subsystem control volumes. If only series and/or parallel groupings of control volumes are found, tasks may become minimal. Although many of the setup tasks interact and overlap, depending partly on the analyst's preferences, a task list that has worked well for many engineers is summarized in Table 2.1.

Table 2.1 Task List For Setup Of Flow System Models

Task Number	Description
1	Sketch system flow paths using electrical analog symbols.
2	Label driving forces, flow directions and branch points, flows, resistances, storage points, and capacitances. Show known values.
3	Choose and sketch appropriate subsystem control volumes. Tasks 1 and 2 may have to be repeated. Continuous flow paths can be broken into several control volumes. Some subsystems might be grouped into a single control volume to save effort. Place "lumped" control volumes around storage zones. Show inlet and exit control volumes if necessary.
4	Identify dependent variables, usually the driving-force values. However, flow rates may be sought, and Equation 2.6 will have to become a direct part of the model. Establish starting values and simulation spans.
5	Write rate-type equations (Equation 2.2) for each lumped control volume. If no lumped control volumes are present, make sure Equation 2.3 is satisfied everywhere. Remember, if no branch or storage is present, the exit stream of one control volume must be the inlet stream of the adjacent control volume by continuity arguments.
6	Use Equations 2.6 through 2.11 as applicable to get the number of unknowns reduced. Significant effort is usually required to identify capacitance values when applying Equation 2.11.
7	Simplify and re-pose results into standard equation-set forms.

EXAMPLE 2.4

A flow system consists of two open-top tanks, one draining into the other. The top tank is a vertical cylindrical tank (diameter $= 5$; height $= 15$), which is drained by gravity through a valved pipe in its bottom having a flow resistance of $R = 0.5(H)^{0.5}$, where H is the instantaneous liquid level in the tank. The drain volumetric flow from this tank, VV, is directed into the top of a funnel tank (top diameter $= 10$; bottom diameter $= 2$; symmetrical cone height $= 10$), which in turn drains by gravity through a line in its bottom having a flow resistance of $R_f = 8.2(H_f)^{0.5}$, where H_f is the instantaneous liquid level in the funnel. The top tank is supplied with liquid by a pump outside the system at a volumetric rate of $VV_i = 5 \sin 0.2t + 6$, where t is time from start of any simulation. Liquid levels in both tanks are initially at 0.1. Provide a schematic of the system showing lumped analysis control volumes, give an equation for the flow rate from the top tank, and set up an appropriate model for simulating the liquid levels in both tanks and the flow rate from the funnel tank.

▶ 1.

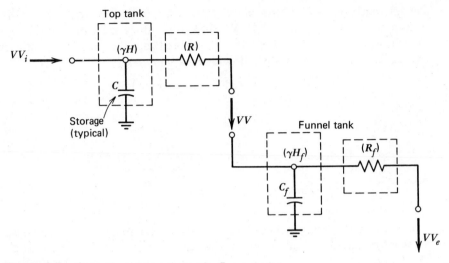

Figure 2.5   Control volume schematic, Example 2.4.

2.   Treating the drain line as an instantaneously steady control volume as indicated in the schematic and applying Equation 2.6 gives an equation in terms of the upstream pressure P_i. We then have to introduce the manometric principle, which says that $P_i = \gamma H$, where γ is the specific weight of the liquid in the top tank. The exit pressure from the drain line is atmospheric, taken zero in this case.

$$VV = \frac{P_i - 0}{R} = \frac{(\gamma H)}{[0.5(H)^{0.5}]}$$

$$VV = 2\gamma(H)^{0.5}$$

3.   Equation 2.2 is applied to both lumped control volumes, where liquid volume is the flowing quantity.

$$VV_i - VV = \frac{d(\text{top-tank liquid volume})}{dt}$$

$$VV - VV_e = \frac{d(\text{bottom-tank liquid volume})}{dt}$$

We observe that the exit flows from both tanks are implicitly taken as the inlet flows for the drain-line control volumes. Of course, by problem description the exit flow from the top tank drain line becomes the inlet flow to the funnel tank. The VV has already been developed in part (2), and the same procedure can be used to get VV_e.

$$VV_e = \frac{\gamma(H_f)^{0.5}}{8.2}$$

The differential volumes in the time-derivative terms must correspond to $C\,dH$ terms according to Equation 2.11. We look at dH thick slices of liquid storage in each tank to determine these. It is discovered that the capacitance in each case is simply a function of the following tank geometry.

$$d(\text{top-tank stored volume}) = \pi(2.5)^2 \, dH$$

or

$$C = \pi(2.5)^2$$

$$d(\text{bottom-tank stored volume}) = \pi(1 + 0.4H_f)^2 \, dH_f$$

or

$$C = \pi(1 + 0.4H_f)^2$$

Substituting back, bringing in the VV_i expression, and doing some algebra gives us the set of equations making up the desired model.

$$\frac{dH}{dt} + \frac{0.32\gamma}{\pi} H^{0.5} = \frac{0.8 \sin 0.2t + 0.96}{\pi}$$

with $H = 0.1$ at $t = 0$.

$$\frac{dH_f}{dt} + \frac{\gamma H_f^{0.5}}{8.2\pi(1 + 0.4H_f)^2} - \frac{2\gamma H^{0.5}}{\pi(1 + 0.4H_f)^2} = 0$$

with $H_f = 0.1$ at $t = 0$.

$$VV_e = \frac{\gamma H_f^{0.5}}{8.2}$$

From this model it is evident that the liquid levels in both tanks have to be involved in any simulation of the system's exit flow.

EXAMPLE 2.5

The building wall and conditions given in Example 2.3 are the same except that the outside temperature is varying with time ($T_c = T_h + A \sin 0.262t + Bt$, where A and B are constants) and thermal energy storage occurs in layer 7. We are interested in simulating the heat flow and temperature at the inner surface of layer 2. Initially, all temperatures are at a uniform T_h. Set up an appropriate model including a sketch of a schematic.

▶ After several preliminary schematics we realize that the storage in layer 7 will control both flow and temperature at the inner surface. Consequently, we

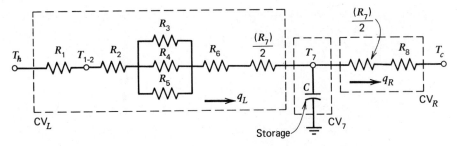

Figure 2.6 Unsteady thermal flow control volumes, Example 2.5

concentrate first of all on a lumped analysis scheme and choose the simplest "lump" we can for layer 7, placing it at the midplane of the layer with two half-layer resistances on either side. The simplest combinations of the other subsystems are then used to complete the working schematic, as shown in Figure 2.6. Preliminary applications of Equations 2.7 and 2.8 give the equivalent resistances of the control volume on the left (R_L), and to the right (R_R), of the lumped volume.

$$R_L = R_1 + R_2 + \frac{1}{1/R_3 + 1/R_4 + 1/R_5} + R_6 + \frac{R_7}{2}$$

$$R_R = \frac{R_7}{2} + R_8$$

Equations 2.2 and 2.11 are applied to CV_7. It is noted that the capacitance C in this case will be the mass of the layer 7 section times the specific heat of the material; however, in this problem we avoid extra terms by sticking with C.

$$q_{7i} - q_{7e} = C \frac{dT_7}{dt}$$

The exit flow from the instantaneously steady control volume on the left is the same as the inlet flow to the lumped control volume by continuity arguments. Similarly, the inlet flow to the steady control volume on the right is the same as the exit flow from the lumped control volume. Equations 2.3 and 2.6 can be applied to the two steady control volumes.

$$q_{7i} = q_L = \frac{T_h - T_7}{R_L}$$

$$q_{7e} = q_R = \frac{T_7 - T_c}{R_R}$$

or

$$\frac{T_h - T_7}{R_L} - \frac{T_7 - T_c}{R_R} = C \frac{dT_7}{dt}$$

Obviously, the heat flow at the inner surface is the same as q_L. With this known, a control volume around layer 1 can be identified so that Equation 2.6 can be applied again to get at the temperature between layers 1 and 2, T_{1-2}. Substituting and re-posing gives the desired four-equation model.

$$\frac{dT_7}{dt} + \frac{(1/R_L + 1/R_R)}{C} T_7 = \frac{T_h}{CR_L} + \frac{T_c}{CR_R}$$

with $T_7 = T_h$ at $t = 0$.

$$q_L = \frac{T_h - T_7}{R_L}$$

$$T_{1-2} = T_h - \left(\frac{R_1}{R_L}\right)(T_h - T_7)$$

$$T_c = T_h + A \sin 0.262t + Bt$$

MOVING-OBJECT SYSTEMS

Here, in contrast to the flow system models just discussed, the system is perceived to be moving relative to the analyst, and accountings of all external forces or moments causing movement have to be done in order to set up a model. Obviously, some systems can exhibit movements simultaneously in several different directions, or subsystems may exhibit different kinds of simultaneous movements. As a result we speak of each coordinate associated with movement as a "degree of freedom," abbreviated in some of the following descriptions as DF. In general, a separate equation, with one independent variable per equation has to be set up for each DF; the resulting set of coupled equations is the model for the system. In this book we restrict our attention to relatively uncomplicated translational and rotational systems and subsystems. By this we mean that we are interested in relatively simple systems, or subsystems, which move along straight lines and/or rotate about some clearly identifiable axes. Subsystems in this context can move relative to each other as well as to the observer.

The engineer working only with certain classes of moving-object systems will invariably develop his or her own streamlined procedures for setting up models. These short-cut approaches are reflected in much of the professional literature; in some cases they lead to procedures where direct setup of equation models is not even necessary. However, for the student trying to introduce himself to the subject, such approaches have some serious shortcomings, not the least of which is the lack of clear connections between the physical problem and the classical mathematics foundations used in the simulations. Consequently, we take a well-tested, although maybe not as cost-effective, approach to model setup methodology and consider a moderately long list of tasks. As the student gains experience, his or her ability to incorporate the shortcuts develops naturally.

What the control volume is for flow system models, the "free body" is for moving-object models. The free body (FB) is a moving object that has been isolated in imagination from its surroundings and its motion frozen. The free-body diagram is a symbolic sketch of the frozen motion object with all its external forces identified by labeled arrows and a movement coordinate designated. In trying to get free-body diagrams generated, there appears to be no substitute for the ability to visualize physical behaviors clearly. This ability can be developed beyond that typically found from everyday observations, and that is the objective of most engineering curriculums.

Once free-body diagrams have been generated, variations on Newton's second law can be used to proceed with model setup. One variation that seems to have some advantages when dealing with complicated systems is given in Equation 2.12, a vector equation.

$$\text{sum of all external forces} + \text{inertia force} = 0 \qquad (2.12)$$

If an external force acts on the object in the same direction as the positive direction of the coordinate system shown on the free-body diagram, it is entered as a positive value in Equation 2.12; if it acts in the opposite direction, it is entered as a negative value. The inertia force, sometimes called the D'Alembert force, will always be negative in this equation and can be determined from Equations 2.13 or 2.14.

$$\text{inertia force} = -\left(\frac{W}{g}\right)(\text{acceleration}) \qquad (2.13)$$

$$\text{inertia force} = \frac{-d(\text{mass} \times V/g_c)}{dt} = \frac{-d(WV/g)}{dt} \qquad (2.14)$$

where W = weight of the object
g = acceleration owing to gravity
g_c = gravitational conversion constant
V = velocity (positive in the positive coordinate direction)

For rotational systems, the counterpart of Equation 2.12 is Equation 2.15.

$$\text{sum of all external moments} + \text{inertia moment} = 0 \qquad (2.15)$$

If an external moment acts on the FB in the same direction as the positive direction shown on the free-body diagram coordinate, it is placed into Equation 2.15 as a positive value; if it acts in the opposite direction, it is placed as a negative value. The inertia moment will always be negative in this equation and can be found from Equation 2.16.

$$\text{inertia moment} = -J(\text{angular acceleration}) \qquad (2.16)$$

where J = the polar mass moment of inertia taken with respect to the rotational axis shown on the free-body diagram.

The angular acceleration (or second time derivative of angular displacement) in Equation 2.16 is always around the rotational axis shown on the FB diagram and is always positive in the positive-coordinate direction. In general, the moment of inertia J will have to be looked up in an engineering handbook; for a complicated geometry object, finding a value may require extensive effort using numerical integration. We will use relatively simple geometries in this book and will furnish values for J.

Second derivatives of linear or angular displacement with respect to time are always present in moving-object system models. This should be contrasted with the flow system models, where first derivatives of driving forces with respect to time dominate. Moreover, the multiple degrees of freedom that always seem to be present in even the simplest real problems, tend to make the moving-object system models look more complicated. The problem of coupling between displacements in different degrees of freedom is probably the most difficult aspect with which the student has to cope. Many times a student will want to know why the simulation of an everyday simple thing like a thrown baseball is so difficult. Even with very nice computer packages, things just do not seem to work out. Basically, the reason can always be traced back to two areas in the setup of the model. Probably, decoupling of displacements was used to some extent, to make the model readable, and various external force parameters were oversimplified because good data could not be found. The extent of engineering effort required to treat these areas correctly is generally underestimated. Unfortunately, within the scope of such a book as this, we cannot treat these areas as correctly as we might desire. The student is urged to consult the literature[5,6, etc.] and think in terms of advanced engineering courses to get some of the answers.

Moving-object systems can also be categorized according to how the object is connected with its surroundings. For example, the external force may be relatively independent of displacement, like weight, an air drag, or a magnetically coupled driving force. On the other hand, the external force can show a springlike effect and be very dependent on displacement. In another category, the external force can be dependent on velocity or acceleration, like dampers, or some bearing frictions. Typically, any driving forces or moments are positive because of the common practice of choosing the FB positive coordinate the same. In Examples 2.6 and 2.7 we treat single-degree-of-freedom systems, which illustrate some of the above categories.

EXAMPLE 2.6

An automobile is moving down a constant-slope hill (slope $= \theta$). The driver is trying to pick up speed against a strong headwind and is using the engine to give positive tractive effort to the vehicle. Provide a free-body diagram and set up an appropriate model to simulate vehicle position (s) versus time from the top of the hill, where speed is V_0. The pertinent parameters for this problem follow: the weight of the vehicle $= W$; tractive force provided by the engine $= F_e(t)$; air (and wind) drag on the vehicle $= F_a = C_1(ds/dt)^n$, where C_1 and n are constants; rolling and road resistance drag on the vehicle $= F_r = C_2 W(ds/dt)^m \cos\theta$, where C_2 and m are constants; and gravitational acceleration $= g$.

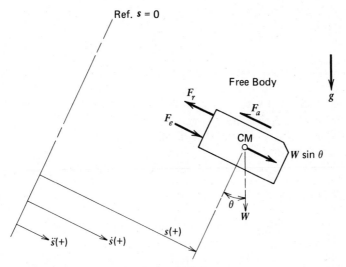

Figure 2.7 Free-body schematic, Example 2.6

▶ The symbols $\dot{s}$ and $\ddot{s}$ are used to represent velocity and acceleration, respectively, and CM is used to represent the center of mass in the free-body diagram shown in Figure 2.7.
We identify the inertia force first by using Equation 2.13. We use Equation 2.13 instead of Equation 2.14 because we have no change in mass occurring.

$$\text{inertia force} = -\frac{W\ddot{s}}{g}$$

Next we substitute all the external forces and the inertia force directly into Equation 2.12, using the FB diagram to determine the appropriate signs.

$$W \sin \theta + F_e - C_1(\dot{s})^n - C_2 W(\dot{s})^m \cos \theta - \left(\frac{W}{g}\right)\ddot{s} = 0$$

Applying some algebra and showing the derivatives, we re-pose the equation into a standard form for our desired model.

$$\frac{d^2s}{dt^2} + \left(\frac{gC_1}{W}\right)\left(\frac{ds}{dt}\right)^n + gC_2\left(\frac{ds}{dt}\right)^m \cos \theta = g \sin \theta + \left(\frac{g}{W}\right)F_e$$

with
$$s = 0 \quad \text{at } t = 0$$
$$\dot{s} = V_0 \quad \text{at } t = 0$$

We observe that if θ should become a function of s, that is, the slope of the hill is not a constant, then additional degrees of freedom enter and the model becomes much more complicated.

EXAMPLE 2.7

A large energy-storage flywheel is mounted vertically in a gas-filled chamber. Its dislike shape is symmetrical around the vertical axis of rotation, and it has a polar mass moment of inertia around this axis of J. The gas drag moment on the flywheel has been studied and is given by $M_a = G(\dot{\theta})^{1.6}$, where G is a constant and θ is angular velocity. The bearing drag moment is given by $M_b = B(\dot{\theta})^{0.4}$, where B is a constant. When connected, the drive motor applies a torque to the flywheel given by $M_d = C_1 - C_2 \cos 0.05t$, where C_1 and C_2 are constants. Give a sketch of a free-body diagram (Figure 2.8) and provide appropriate models for the following situations.

1. Simulation of angular velocity (θ) versus time from power cutoff during a coastdown test of the bearings. The initial θ is θ_0.

2. Simulation of the angular displacement θ versus time during the time a drive torque M_d is applied to the flywheel. Initial conditions for both displacement and velocity are zero.

▶ **1.** We recognize that $d\dot{\theta}/dt$ is angular acceleration, use Equation 2.16, and substitute values directly into Equation 2.15, getting the signs from the sketch.

$$-G(\dot{\theta})^{1.6} - B(\dot{\theta})^{0.4} - J\frac{d\dot{\theta}}{dt} = 0$$

which re-posed gives us our model.

$$\frac{d\dot{\theta}}{dt} + \left(\frac{G}{J}\right)(\dot{\theta})^{1.6} + \left(\frac{B}{J}\right)(\dot{\theta})^{0.4} = 0$$

with
$$\dot{\theta}(0) = \dot{\theta}_0$$

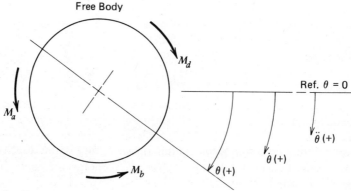

Figure 2.8 Free-body schematic, Example 2.7.

2. Again we use Equation 2.16 and the free-body diagram to give us the correct signs and substitute into Equation 2.16.

$$C_1 - C_2 \cos 0.05t - G(\dot{\theta})^{1.6} - B(\dot{\theta})^{0.4} - J \frac{d\dot{\theta}}{dt} = 0$$

which put into a standard form for our model.

$$\frac{d^2\theta}{dt^2} + \left(\frac{G}{J}\right)\left(\frac{d\theta}{dt}\right)^{1.6} + \left(\frac{B}{J}\right)\left(\frac{d\theta}{dt}\right)^{0.4} = \left(\frac{C_1}{J}\right) - \left(\frac{C_2}{J}\right) \cos 0.05t$$

with

$$\theta(0) = 0$$

$$\dot{\theta}(0) = 0$$

A commonly used notation has been used to show the conditions above.

$$\theta(0) = 0 \text{ means the same as } \theta = 0 \qquad \text{at } t = 0, \text{ and so on}$$

In neither Example 2.6 or 2.7 were the external forces or moments the result of springs or dampers. Both springs and dampers can give reversal of forces depending on displacements and/or velocities. Because real springs often have a slight dead space at their neutral point, complicating the issue tremendously, we are going to focus our attention on dampers for the moment. Basically, most fluid-type dampers exert a force opposite to the sense of the velocity. Obviously, if the velocity is likely to reverse sign during a simulation, the damper force must also reverse direction. This is no problem for the analyst setting up the model if the damper is linear, that is exerts a force directly proportional to the velocity. However, most dampers are nonlinear and exert a force proportional to the velocity raised to some noninteger power. Consequently, with a negative velocity we end up with an undefined mathematical operation. The trick commonly used to overcome this problem is to save the sense by always using the velocity raised to the first power as a multiplier times the absolute value of the velocity raised to whatever noninteger power is necessary to make the expression equivalent to the original. This same scheme can also be applied to a damper force expressed as a multiterm polynomial in the velocity. The procedure is illustrated in Example 2.8, where both a damper and a spring are involved. Two traditional parameters of linear systems, the spring constant with units of force/unit displacement, and the damper coefficient with units of force/unit velocity, may not be directly usable where the more realistic simulations are desired. Although more complicated equations are often used in such problems, we will use only one parameter in each case, with the units assumed to be correct for whatever nonlinearity is given. Consequently, when we give a value for the spring constant, it may have units of force/(unit displacement)^n, and when we give a damper coefficient, it may have units of force/(unit velocity)^m.

EXAMPLE 2.8

A portable weighing device consists of a coil spring fastened to a hook at both its top and bottom, and a piston-cylinder damper mounted inside the spring coil. The upper end of the piston rod is mounted to the upper hook; the lower end of the cylinder is mounted to the lower hook. The cylinder rod passes through a dynamic seal at the upper end of the cylinder, which is filled with oil around the loosely fitting piston. In use the upper hook is fastened to a firm overhead support, and the unknown weight is placed on the lower hook. A calibrated transducer mounted on the outside of the hooks permits a reading signal to be fed to a data-acquisition system located elsewhere. Because this device is to be used in a process situation where weights will be placed on it rapidly in succession, we are interested in simulating its dynamic behavior before use. Some theoretical, as well as some empirical, data lead us to believe that the spring force will be given by $F_s = k$(displacement), where k is a spring constant with units of force/unit displacement, and that the damper force will be given by $F_d = C$(velocity)$^{0.8}$, where C is a constant with units of force/(unit velocity)$^{0.8}$. Give a sketch of a free-body diagram and set up an appropriate model (Figure 2.9).

▶ We choose our reference position $(X=0)$ to be the conceptual equilibrium position with the weight, W, on the hook. We note that the spring force is now $F_s = W + kX$. Also, since some oscillation is expected, we have to worry about velocity sign reversals. As a consequence we make a major modification on the damper force equation, which then reads $F_d = C(dX/dt)(|dX/dt|)^{-0.2}$. Using Equation 2.13, everything is substituted back into Equation 2.12 to give the following.

$$W - F_s - F_d - \left(\frac{W}{g}\right)\frac{d^2 X}{dt^2} = 0$$

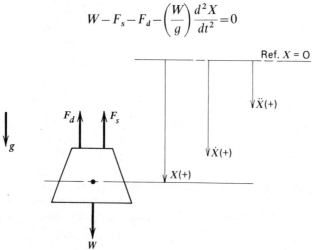

Figure 2.9 Free-body schematic, Example 2.8.

or

$$\frac{d^2X}{dt^2} + \left(\frac{gC}{W}\right)\left|\frac{dX}{dt}\right|^{-0.2}\frac{dX}{dt} + \left(\frac{gk}{W}\right)X = 0$$

with

$$X = \frac{-W}{k} \quad \text{and} \quad \dot{X} = 0 \quad \text{at } t = 0$$

It should be apparent that we could have chosen our reference position to be the unweighted position just as easily as the weighted position in Example 2.8. When more than one degree of freedom is involved with subsystems interconnected by spring and damper elements, the choice of reference positions can become a major point of confusion for the student. Although no single procedure seems to be the most comfortable for everyone, many analysts simply go through the separate free-body diagrams and arbitrarily place constraint conditions in relation to any adjacent free-body displacements. Then it becomes a matter of meticulous consistency in identifying the appropriate relative displacements from one free body to the next. Of course, it helps to choose reasonably visualized constraints initially, but it is not absolutely necessary. This procedure allows the analyst to show an arbitrary local reference for each free body at the start of the setup process. Illustrations of the procedure are given in the two-degree-of-freedom Examples 2.9 and 2.10.

EXAMPLE 2.9

A support system for some heavy equipment is to be modeled as a two mass system. The upper mass (weight $= W_1$) is connected to the lower mass (weight $= W_2$) by the equivalent of one damper [damper force $= F_d = C$ (velocity)^n] and one compression spring [spring force $= F_{k1} = k_1$ (displacement)]. The lower mass is supported from below by the equivalent of one compression spring [spring force $= F_{k2} = k_2$ (displacement)] and in turn supports the upper mass as described above. Assume that two degrees of freedom are sufficient to describe the movements of this system, furnish free-body sketches, and set up a model that can be used to simulate the up and down movement of both the masses after a disturbance.

▶ In the free-body diagrams, Figure 2.10, the coordinate labeled y has been chosen to represent the displacements of the upper mass, and the coordinate labeled x the lower mass. We arbitrarily place the constraint $y > x$ on both free-body sketches; this allows us to identify the connecting forces from the upper spring and damper. However, it is noted that we could also choose the constraint $y < x$, and the model at the finish would be the same. Since we have chosen

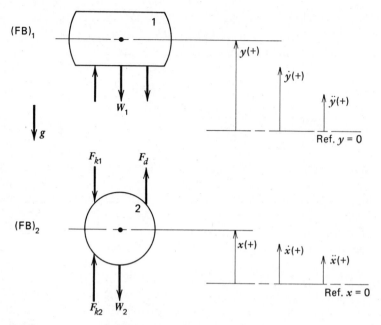

Figure 2.10 Free-body schematic, Example 2.9.

$y > x$, then dy/dt must also be taken greater than dx/dt, to be kinematically consistent, and $d(y-x)/dt$ will be treated as positive for setup.

Treating the upper mass first, we determine the spring and damper forces by using the above constraint. The equilibrium position is used as the global (overall) reference.

$$F_{k1} = W_1 - k_1(y - x)$$

$$F_d = C\left(\frac{d(y-x)}{dt}\right)^n$$

The fact that we have chosen our coordinates positive upward, along with the constraint $y > x$, has forced us to use the negative k_1 term in F_{k1} in order to be consistent. Also note that we are not including any air drag forces in this setup; in most setups we would have to include this for simulation accuracy. Going to the lower mass, we develop the lower spring force.

$$F_{k2} = W_1 + W_2 - k_2 x$$

Finally, using Equations 2.12 and 2.13 for both degrees of freedom, we get the following.

$$-W_1 + (W_1 - k_1(y-x)) - C\frac{d(y-x)}{dt}\left|\frac{d(y-x)}{dt}\right|^{n-1} - \left(\frac{W_1}{g}\right)\frac{d^2y}{dt^2} = 0$$

$$-W_2 + (W_1 + W_2 - k_2 x) - (W_1 - k_1(y-x))$$

$$+ C\frac{d(y-x)}{dt}\left|\frac{d(y-x)}{dt}\right|^{n-1} - \left(\frac{W_2}{g}\right)\frac{d^2x}{dt^2} = 0$$

which can be simplified and re-posed to give the desired model.

$$\frac{d^2y}{dt^2} + \left(\frac{gC}{W_1}\right)\frac{d(y-x)}{dt}\left|\frac{d(y-x)}{dt}\right|^{n-1} + \left(\frac{gk_1}{W_1}\right)(y-x) = 0$$

with $\quad\quad\quad\quad\quad y = y_0 \quad$ and $\quad \dot{y} = 0 \quad$ at $t = 0$

$$\frac{d^2x}{dt^2} - \left(\frac{gC}{W_2}\right)\frac{d(y-x)}{dt}\left|\frac{d(y-x)}{dt}\right|^{n-1} + \left(\frac{gk_2}{W_2}\right)x - \left(\frac{gk_1}{W_2}\right)(y-x) = 0$$

with $\quad\quad\quad\quad\quad x = x_0 \quad$ and $\quad \dot{x} = 0 \quad$ at $t = 0$

EXAMPLE 2.10

A crude approximation of a motorcycle (weight $= W$) as an object supported by two compression springs is shown in the top sketch of Figure 2.11. Both springs are linear, and the one in the front (on the left in the sketch) has a constant of k_1; the rear has a constant k_2. The center of mass is located L_1 from the front and L_2 from the rear. Use two degrees of freedom, X for the up and down motion and θ for the rotational (pitching) motion about the center of mass, and set up an appropriate model for simulating movement. Assume that the polar mass moment of inertia about the center of mass is J.

▶ A free-body sketch is given in Figure 2.11. We will assume that θ is small enough to keep the springs working properly, and we arbitrarily take as our reference the equilibrium position of the motorcycle.
We choose θ positive clockwise, and X positive downward, but we could have chosen three other combinations of positive directions without affecting the final model. We simply have to be consistent in our evaluation of the forces and moments. The spring forces are developed first.

$$F_{k1} = W_1 + k_1[X - L_1 \sin\theta]$$

$$F_{k2} = W_2 + k_2[X + L_2 \sin\theta]$$

where
$$W_1 = \frac{WL_2}{L_1 + L_2}$$

$$W_2 = W - W_1$$

Observing that $\sin\theta$ for a small θ is almost equivalent to θ, we use the forces to develop the moments, M_1 and M_2.

$$M_1 = F_{k1}L_1$$

$$M_2 = -F_{k2}L_2$$

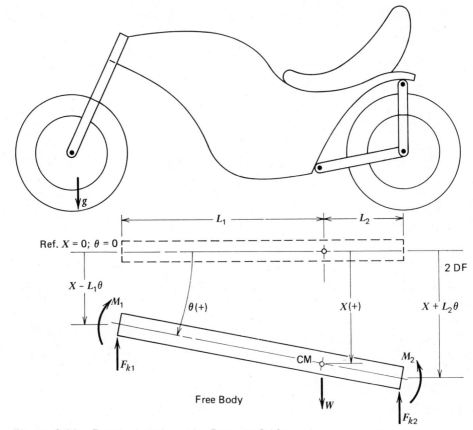

Figure 2.11 Free-body schematic, Example 2.10.

Using Equations 2.12, 2.13, 2.15, and 2.16, and substituting, we get

$$W - \left[\frac{WL_2}{L_1+L_2} + k_1(X-L_1\theta)\right] - \left[\frac{WL_1}{L_1+L_2} + k_2(X+L_2\theta)\right] - \left(\frac{W}{g}\right)\frac{d^2X}{dt^2} = 0$$

$$\left[\frac{WL_2}{L_1+L_2} + k_1(X-L_1\theta)\right]L_1 - \left[\frac{WL_1}{L_1+L_2} + k_2(X+L_2\theta)\right]L_2 - J\frac{d^2\theta}{dt^2} = 0$$

These can be simplified and put into a standard matrix form popular among some analysts. We exclude conditions here, but in actual simulation they would be given.

$$\begin{bmatrix} \dfrac{W}{g} & 0 \\ 0 & J \end{bmatrix} \begin{Bmatrix} \ddot{X} \\ \ddot{\theta} \end{Bmatrix} + \begin{bmatrix} (k_1+k_2) & -(k_1L_1-k_2L_2) \\ -(k_1L_1-k_2L_2) & (k_1L_1^2+k_2L_2^2) \end{bmatrix} \begin{Bmatrix} X \\ \theta \end{Bmatrix} = \begin{Bmatrix} 0 \\ 0 \end{Bmatrix}$$

Our last example is a two-degree-of-freedom problem, which we use to show the characteristics of setting up a model that has a spring-coupled driving force, Example 2.11. It should be observed that all the springs and dampers we treat are massless, a gross approximation in many cases.

EXAMPLE 2.11

A large valve (weight $=W$) is restrained from above by a linear spring (spring constant $=k_1$) and linear damper (damper coefficient $=C$), as in Figure 2.12. The valve stem behaves like a small stiff spring with a spring constant of k_2. The stem, on the underside of the valve, is driven by a rotating cam. Assume that the stem never leaves the cam surface and setup an appropriate model for simulating movement of the valve.

▶ The y displacements give us the cam profile effects. Here we have no clear-cut equilibrium position to use as a coordinate reference. Furthermore, we do not know what preloading, if any, may exist on the springs. However, from all the other translational examples, it is apparent that spring preloadings (weight in the previous examples) will always cancel out after the forces are substituted into Equation 2.12.
Consequently, we are going to write all the spring forces as though they had no preloading and ignore the weight of the object; we will worry only about relative displacements. The spring and damper forces are

$$F_{k1} = k_1 x$$

$$F_{k2} = k_2(y - x)$$

$$F_d = C \frac{dx}{dt}$$

Then, using Equation 2.13 and substituting into Equation 2.12, we get

$$k_2(y - x) - k_1 x - C \frac{dx}{dt} - \left(\frac{W}{g}\right)\frac{d^2 x}{dt^2} = 0$$

This can be re-posed into the following standard form:

$$\frac{d^2 x}{dt^2} + \left(\frac{gC}{W}\right)\frac{dx}{dt} + \frac{g(k_1 + k_2)}{W}x = \left(\frac{gk_2}{W}\right)y$$

with $\qquad x = x_0 \qquad$ and $\qquad \dot{x} = \dot{x}_0 \qquad$ at $t = 0$

The reader may wish to verify that this same model can be set up by assuming an arbitrary reference position that gives a preload on the upper spring, showing

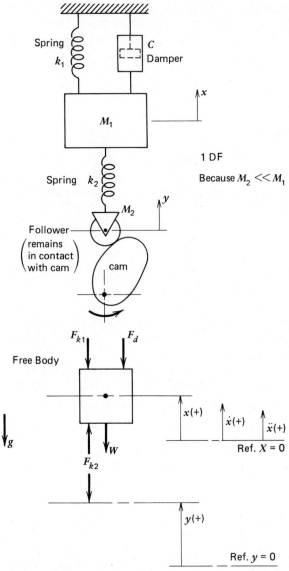

Figure 2.12 Free-body schematic, Example 2.11.

the W force on the free body and substituting as before. Obviously, in all problems of this type, we can save setup time if we recognize that we can cancel the spring preloads as we write the force terms.

As moving-object systems become more complicated, displaying more subsystems and degrees of freedom, the model setup can appear to be an impossible

Table 2.2 Task List For Setup Of Moving-Object System Models

Task Number	Description
1	Clearly identify system and subsystems. Visualize system's motion frozen and sketch free bodies. Estimate mass centers (CM) and rotation centers (CR). Start identifying degrees of freedom.
2	Sketch force field and/or moment field arrows. For many systems these will be the weight arrows (show the gravity field arrow also).
3	Sketch coordinates of perceived movements and establish local references. Identify degrees of freedom if not done already. Introduce time function and geometric influences as necessary. Task 2 may have to be repeated.
4	Sketch motion arrows on the free body and choose positive direction(s) for each degree of freedom. Remember that positive direction for a displacement is also the positive direction for the velocity and the acceleration.
5	Sketch force and/or moment arrows for each degree of freedom. Identify and label all external forces that might influence movement.
6	Determine all force and/or moment expressions (or values) associated with the arrows from task 5. Resolve components for each degree of freedom. The use of basic laws, observations of similar systems, and test-laboratory data will probably be necessary here. Task 5 may have to be repeated.
7	Sum all external force and/or moment components aligned with the previously chosen coordinates (taken positive when in the previously chosen positive direction). Determine the inertia force from Equations 2.13 or 2.14, or the inertia moment from Equation 2.16. Substitute into Equation 2.12 and/or 2.15. Use Newton's second law as necessary.
8	Simplify, re-pose into the desired standard form, and add conditions for each degree of freedom.

job. However, we know that if certain procedures are followed, task by task, the complete model can be constructed in stages, just as one would build a house or bridge. In the foregoing examples we have demonstrated most of these tasks. It should have become evident that certain tasks can be shortened, or eliminated, as one gets the experience with certain types of systems. But for the individual who gets involved in the setup of moving system models only occasionally, remembering all the tasks may be impossible. For these people primarily, we present a summary table of the tasks discussed previously, Table 2.2.

STRUCTURAL SYSTEMS

Structural systems differ from flow systems and moving-object systems in that nothing moves relative to the observer. Most structural systems consist of subsystems made up of beams, braces, gussets, fasteners, plate segments, shell segments, and so on. The simulations typically seek stress levels and deformations within subsystems and overall (global) deformations under prescribed overall loads.[7] Any global model has to be able to project loads from the boundaries of the system to the interior subsystems. But such aspects of a model are the precise topics upon which traditional structures books concentrate. Furthermore, we treat this more as a simulation topic in Chapter 7. Consequently, we direct our attention in this chapter to the local or subsystem part of the model setup. Here we are interested in the part of the model that deals with local deformations and stresses. Of course, in some cases the system may be taken so simple (e.g., a single beam) that our part of the model becomes essentially the whole or global model.

Since the most common subsystem is the beam, we will concentrate our efforts on it here. The typical model setup starts with a visualization of a small portion of the beam cut out and isolated from the rest of the beam. We will refer to this portion as the "cutout" in the following discussion. Then the beam is imagined to be loaded, and the resulting shape of the cutout is sketched, along with all of its external forces and moments labeled, in a manner similar to the free-body diagram of moving-object systems. The deformations and displacements of this cutout are then meticulously related to the external forces and moments, which in turn are related back to the overall loadings on the beam. Of course, the resultants of all forces and moments on the cutout have to be zero, but this can be set up in different ways, as illustrated in examples later. Typically, in these setups several basic principles from the area of strength of materials have to be used. One of these is the definition of stress, which is given by Equation 2.17.

$$S = \frac{F}{A} \tag{2.17}$$

where
$S =$ stress (force units/unit area)
$F =$ force
$A =$ area upon which the force is acting

If the force is directed normal into the area, the stress is known as *compressive stress*; if the force is directed normal out of the area, the stress is called *tensile stress*; and if the force is parallel with the area, we have *shear stress*. The deformation of the material when stressed depends on how great the stress is and on the microscopic characteristics of the material itself. For most materials of interest in structures, the deformation will be directly proportional to the stress and will exhibit no permanent deformation as the stress is removed.[8] This is known as elastic behavior and is described by Hooke's law, given by Equations 2.18 and 2.19.

$$\varepsilon = \frac{S}{E} \tag{2.18}$$

$$\varepsilon_s = \frac{S_s}{G} \tag{2.19}$$

where $S =$ tensile or compressive stress
$S_s =$ shear stress
$E =$ elastic (Young's) modulus; a different constant for each material
$G =$ shear (rigidity) modulus; a different constant for each material;
  E and G are related through Poisson's ratio, μ

$\varepsilon =$ strain (stretch or compression/original length)

$\varepsilon_s =$ shear strain (deformation/original length)

Additional basic principles are taken directly from classical mathematics. However, the developed forms (models themselves) are the most useful here, and development is left as exercises. In an $x-y$ rectangular Cartesian coordinate system, the radius of curvature, r, of any curve is given by Equation 2.20.

$$r = \frac{[1+(dy/dx)^2]^{1.5}}{d^2y/dx^2} \qquad (2.20)$$

The area moment of inertia, I, is given by Equation 2.21.

$$I = \int\int_A h^2 \, dA \qquad (2.21)$$

where $I =$ area moment of inertia about some specified axis (length^4 units)

$h =$ distance from an infinitesimal area to the specified axis

$dA =$ infinitesimal area

Another important principle has no mathematical form and is just stated. An imaginary surface exists in the cutout, on which both tensile and compressive stresses are zero. This longitudinal surface is called the "neutral surface," or "neutral axis" in an elevation view; the longitudinal material layers on one side of this surface are compressed, while on the other side they are stretched. In other words, the neutral axis exhibits neither stretching nor compression. As with all model setups, astute assumptions have to be made also. One that is particularly important for the beam model setup is that the "cut" surfaces of the cutout remain plane regardless of the loading on the beam. Other setup principles and procedures are shown in Examples 2.12 and 2.13.

EXAMPLE 2.12

A simple cantilever beam has a uniform rectangular cross section with an area A. The beam base is at $x=0$; it has a concentrated load W at the free end, where $x=L$, and its deflection is measured in the y plane. Assume that the deflection is measured from the unloaded neutral axis to the loaded neutral axis. The modulus of elasticity for the beam material is E. Give a sketch of a cutout and set up an appropriate model for simulating beam deflection (y) versus longitudinal position along the beam (x), as shown in Figure 2.13.

▶ The stress in the material layer at dA is expressed in terms of the strain by using Equation 2.18. The strain, ε, is then found using the arc length $r \, d\theta$.

$$\varepsilon = \frac{(r+h)\,d\theta - r\,d\theta}{r\,d\theta}$$

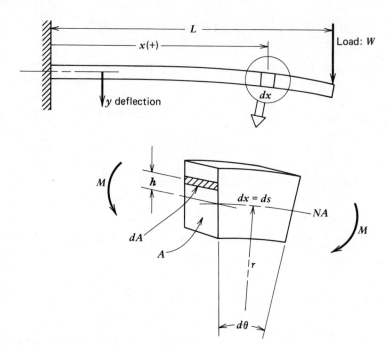

Figure 2.13 Beam cutout, Example 2.12.

Substituting into Equation 2.18 gives a new equation for stress.

$$S = \frac{Eh}{r}$$

The differential moment around the neutral surface line caused by the stress force acting on dA is dM.

$$dM = (S\, dA)(h)$$

The total bending moment for the cross section A is found by integrating the dM expression over the area A. But substitution of the S expression into the integral gives a well-known group (see Equation 2.21).

$$M = \left(\frac{E}{r}\right) \iint_A h^2\, dA$$

or

$$M = \frac{EI}{r}$$

Substituting for r from Equation 2.20 and re-posing, gives the general model.

$$\frac{d^2y}{dx^2} - \frac{M[1 + (dy/dx)^2]^{1.5}}{(EI)} = 0$$

Now we have to relate the M back to the overall loading on the beam. Choosing the same neutral surface line on the cutout as used before for the moment pivot, and recognizing that the W load at the end of the beam will be acting clockwise, we get the following for the beam moment:

$$M = -W(L-x)$$

Substituting back into the general model and adding conditions gives us the desired model.

$$\frac{d^2y}{dx^2} + \frac{W(L-x)[1+(dy/dx)^2]^{1.5}}{(EI)} = 0$$

with
$$y = 0 \qquad \text{at } x = 0$$
$$y' = 0 \qquad \text{at } x = 0$$

If deflections are small relative to the length of the beam, then dy/dx will be small relative to other terms in the above model and might be neglected under some circumstances. The temptation to do this is especially strong when we realize later that to do so will give us a linear model. Such a linear model is precisely the one referred to as the "beam equation" in many textbooks.

Other coordinates and loadings are more useful in many engineering settings. One that becomes particularly useful is based on the intrinsic coordinate s traced along the neutral axis of the beam. The deflection coordinate v is traced normal to the unloaded neutral axis. The assumption of very small slopes, so the model can be made linear, as discussed in Example 2.12, is incorporated immediately. Development of this model is left to Example 2.13.

EXAMPLE 2.13

A cantilever beam with its neutral axis taken as the s coordinate is subjected to several generalized loads as indicated in the following. Let $M =$ moment, $w_t =$ transverse distributed load, $T =$ axial tension load, $r =$ local radius of curvature, $v =$ vertical displacement at s, $\theta =$ slope at s, $E =$ elastic modulus, $I =$ area moment of inertia of the cut surface, $w_a =$ axial distributed load, and $Q =$ local shear force. Also, from Example 2.12, let $ds = r\,d\theta$, and $d\theta/ds = 1/r = M/(EI)$. Assume linear behavior (i.e., small slopes) and set up a model for simulating v versus x. Furnish a sketch of the cutout (Figure 2.14) and identify all useful intermediate equations.

▶ Summing forces along the s-axis tangent at the midpoint of the cutout ($+$ to the right), and setting them to zero gives the following:

$$w_a\,ds + (T+dT)\cos\left(\frac{d\theta}{2}\right) - T\cos\left(\frac{d\theta}{2}\right) - Q\sin\left(\frac{d\theta}{2}\right) - (Q+dQ)\sin\left(\frac{d\theta}{2}\right) = 0$$

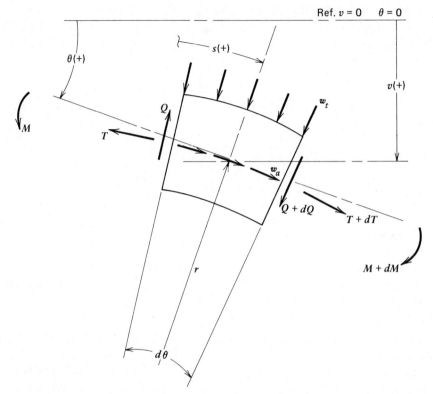

Figure 2.14 Beam cutout, Example 2.13.

Assuming $\cos(d\theta/2) \simeq 1$, $\sin(d\theta/2) \simeq d\theta/2$, and $dQ\,d\theta/2 \simeq 0$, gives Equation 2.22.

$$\frac{dT}{ds} - Q\frac{d\theta}{ds} = -w_a \qquad (2.22)$$

Summing forces along the s-axis normal at the cutout midpoint ($+$down) gives another equation.

$$w_t\,ds + (Q+dQ)\cos\left(\frac{d\theta}{2}\right) - Q\cos\left(\frac{d\theta}{2}\right) + (T+dT)\sin\left(\frac{d\theta}{2}\right) + T\sin\left(\frac{d\theta}{2}\right) = 0$$

Keeping the assumptions from above, and assuming further that $dT\,d\theta/2 \simeq 0$, gives Equation 2.23.

$$\frac{dQ}{ds} + T\frac{d\theta}{ds} = -w_t \qquad (2.23)$$

Summing moments around the neutral axis line at the left cut ($+$clockwise), and setting to zero gives another equation.

$$(M+dM) - M + (Q+dQ)\,ds + w_t\,ds\left(\frac{ds}{2}\right) = 0$$

Making the further assumptions that $dQ\,ds \simeq 0$, and $(ds)^2 \simeq 0$, gives Equation 2.24.

$$\frac{dM}{ds} = -Q \tag{2.24}$$

Combining Equations 2.23 and 2.24 gives

$$\frac{d^2M}{ds^2} - T\frac{d\theta}{ds} = w_t$$

Assuming that $d\theta/ds \simeq d^2v/dx^2$ for small slopes, and using $M = EI\,d\theta/ds$, substitution into the preceding gives one of the desired equations.

$$\frac{d^2(EI\,d^2v/dx^2)}{dx^2} - T\frac{d^2v}{dx^2} = w_t \tag{2.25}$$

Assuming $d(\;)/ds \simeq d(\;)/dx$, and using Equations 2.23 and 2.24 completes the model.

$$EI\frac{d^2v}{dx^2} = M$$

$$EI\frac{d^3v}{dx^3} = -Q$$

$$\frac{dv}{dx} = \theta$$

Equations 2.22, 2.23, and 2.24 are the differential equations of equilibrium for a beam that has bending, shear, and axial tension. Similar models can be set up for bending, and so on, in the other coordinate direction. For example, if we consider bending in and out of the page and call the displacement u, the counterpart of the v displacement equation above becomes

$$\frac{d^2(EI_z\,d^2u/dx^2)}{dx^2} - T\frac{d^2u}{dx^2} = g$$

where I_z = area moment around the neutral axis line perpendicular to
the line used for the v displacements
g = distributed transverse load in direction normal to the page

REFERENCES

1. J. R. Welty, C. E. Wicks, and R. E. Wilson, *Fundamentals of Momentum, Heat, and Mass Transfer*, 2nd ed., John Wiley & Sons, Inc., New York, 1976.

2. G. J. Van Wylen and R. E. Sonntag, *Fundamentals of Classical Thermodynamics*, 2nd ed., John Wiley & Sons, Inc., New York, 1973.

3. F. P. Incropera and D. P. De Witt, *Fundamentals of Heat Transfer*, John Wiley & Sons, Inc., New York, 1981.

4. J. P. Holman, *Heat Transfer, 4th ed.*, McGraw-Hill Book Company, New York, 1976.

5. F. S. Tse, I. E. Morse, and R. T. Hinkle, *Mechanical Vibrations, Theory and Applications, 2nd ed.*, Allyn & Bacon, Inc., Boston, Mass., 1978.

6. R. D. Kersten, *Engineering Differential Systems*, McGraw-Hill Book Company, New York, 1969.

7. J. P. Den Hartog, *Advanced Strength of Materials*, McGraw-Hill Book Company, New York, 1952.

8. S. Timoshenko and J. N. Goodier, *Theory of Elasticity*, McGraw-Hill Book Company, New York, 1951.

PROBLEMS

2.1 Re-pose each of the following basic flow formulas into the "universal flow formula" form and identify the flow resistances.

(a) $q = \dfrac{kA(T_h - T_c)}{L}$ (Fourier's algebraic law of heat conduction)

where
q = heat flow rate
k = thermal conductivity, a material property
T_h = high temperature
T_c = low temperature
A = area normal to heat flow
L = path length of heat flow

(b) $q = \dfrac{2\pi kl(T_h - T_c)}{\log_e(D_o/D_i)}$ (Fourier's conduction law in axisymmetric cylindrical coordinates)

where
q = radial heat flow rate
l = length of hollow cylinder
D_o = outer diameter
D_i = inner diameter
Others as given in (a)

(c) $q = hA(T_h - T_c)$ (Newton's law of convective cooling)

where
h = heat transfer coefficient of convection
Others as given in (a)

2.2 Re-pose the following well-known fluid flow laws into universal flow form and identify the flow resistances.

(a) The Hagen–Poiseuille law for laminar (volumetric) liquid flow through a circular cross-sectional tube.

$$VV = \frac{\pi(P_i - P_e)R^4}{8\mu L}$$

where $\quad\quad\quad\quad VV=$ volumetric flow rate

$P_i, P_e=$ pressures at the inlet and exit, respectively

$R=$ tube radius

$\mu=$ dynamic (Newtonian) viscosity

$L=$ flow-path length

(b) The volumetric flow rate of liquid flowing through a packed bed.

$$VV=\frac{(P_i-P_e)(A\varepsilon^3 D_p^2)}{[72\mu L(1-\varepsilon)^2]}$$

where $\quad\quad\quad A=$ cross-sectional area (gross) normal to flow

$\varepsilon=$ void fraction

$D_p=$ particle average diameter

Others as given in (a)

(c) The Buckingham–Reiner equation for volumetric flow of a non-Newtonian liquid through a circular cross-sectional tube.

$$VV=\left[\frac{(P_i-P_e)R^4}{8\mu_0 L}\right]\left[1-\frac{4(\tau_0/\tau_r)}{3}+\frac{(\tau_0/\tau_r)^4}{3}\right]$$

where $\quad\quad\tau_0=(P_i-P_e)r_0/(2L)$

$r_0=$ radius of plug-flow region

$\tau_r=$ shear stress in the fluid next to the wall

$\mu_0=$ Newtonian equivalent viscosity next to the wall

Others as given in (a)

2.3 Re-pose the following basic equations from radiant heat transfer into universal flow form and identify the flow resistances.

(a) The radiant heat exchange between surface 1 and surface 2.

$$q_{12}=A_1 F_{12}(J_1-J_2)$$

where $\quad\quad q_{12}=$ heat exchange rate between surfaces 1 and 2

$A_1=$ area of surface 1

$F_{12}=$ geometric shape (configuration) factor, surface 1 to 2

$J_1, J_2=$ radiosity of surfaces 1 and 2, respectively. Radiosity is defined as the rate at which radiant energy leaves a surface due to both emission and reflection.

(b) The net heat transfer for a surface in a radiant transfer enclosure

$$q_1=A_1\varepsilon_1\frac{E_{b1}-J_1}{1-\varepsilon_1}$$

where $\quad\quad q_1=$ net radiant heat transfer rate for surface 1

$\varepsilon_1=$ emissivity for surface 1, a property of the surface

$E_{b1}=$ black body emissive power of surface 1 $=\sigma T_1^4$

$T_1=$ absolute temperature of surface 1

$\sigma=$ Stefan-Boltzmann constant (a universal constant)

Others as in (a)

2.4 The flow of thermal energy (heat) from a hot vertical surface to the cool air in contact with it is often computed from the following equation. Re-pose this equation into universal flow form and identify the flow resistance.

$$q = A\left(\frac{0.21k}{L}\right)\left[\frac{\beta g\rho^2 c_p L^3}{\mu k}\right]^{0.4}(T_s - T_a)^{1.4}$$

where
- q = heat transfer rate
- A = area of the surface
- L = vertical height of the surface
- β = bulk thermal expansivity modulus of the air, approximately $1/T_a$
- g = gravitational acceleration
- T_a = absolute temperature of the air
- T_s = absolute temperature of the surface
- k = thermal conductivity of the air
- c_p = specific heat (at constant pressure) of the air
- ρ = density of the air
- μ = dynamic viscosity of the air

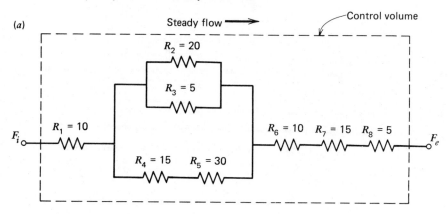

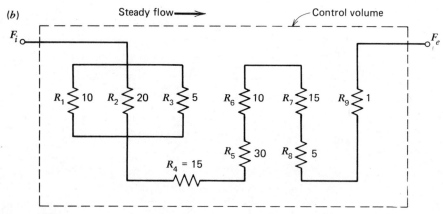

Figure P2.5 Control volume schematics, Problem 2.5

2.5 Determine the equivalent overall resistances to flow for the Figure P2.5 control volumes.

2.6 A sampler system in a process liquid flow line is shown in Figure P2.6.

(a) Sketch the resistance analog network assuming steady flow. Label appropriately.

(b) If the bypass line resistance is $237.61(\Delta P)^{0.5}$, and the sampler side resistance is $11.65(\Delta P)^{0.5}$, what is the system resistance?

(c) Assume consistent units and compute the volumetric flow rate, VV, if $P_1 = 30$ and $P_2 = 0$. Will the flow rate double if we increase P_1 to 60?

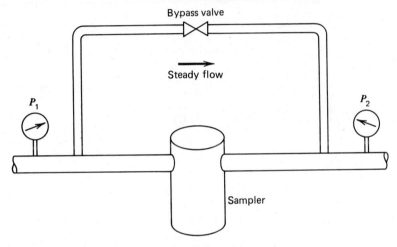

Figure P2.6 Sampler system, Problem 2.6.

2.7 A plumbing network for water flow through some unit fixtures is shown in Figure P2.7.

(a) Assume that the only significant flow resistances are the ones labeled and provide a sketch of the analogous steady flow network.

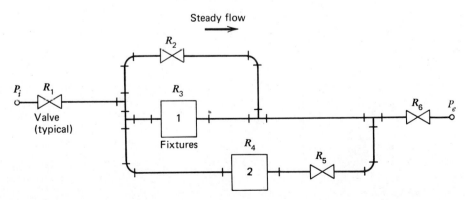

Figure P2.7 Flow system, Problem 2.7.

(b) Given the following values:

$$R_1 = 2.0 \qquad R_2 = 1.5$$
$$R_3 = 5.0 \qquad R_4 = 12.0$$
$$R_5 = 1.0 \qquad R_6 = 2.0$$
$$P_i = 50.0 \qquad P_e = 10.0$$

compute the overall flow resistance of the system.
(c) Assume consistent units and compute the flow rate through the system.
(d) Compute the flow rate through R_4.
(e) Compute the pressure between R_4 and R_5.

2.8 A steady heat flow through a three-layer wall with air films on either side is shown in Figure P2.8. Notice that the right-hand layer is a parallel path composite.

(a) Give a sketch of the analogous steady flow resistance network. Label appropriately.
(b) Given the following resistance values:

$$R_A = 0.1 \qquad R_B = 0.5$$
$$R_C = 5.0 \qquad R_D = 3.0$$
$$R_E = 6.0 \qquad R_F = 2.0$$
$$R_G = 0.07$$

compute the overall equivalent flow resistance of the wall.

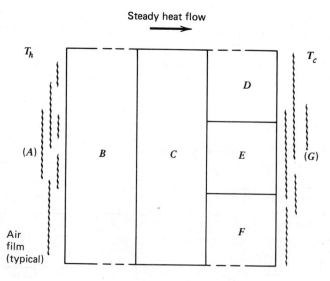

Figure P2.8 Steady heat flow system, Problem 2.8.

(c) Assume consistent units and compute the heat transfer rate through the wall if $T_h = 80$ and $T_c = -20$.

(d) Compute the temperature between layers B and C using the foregoing results and a single formula.

2.9 The resistance network for a liquid fuel supply system is shown in Figure P2.9. Set up the set of equations necessary to determine the pressures P_A and P_B. Can an equivalent overall system flow resistance be determined in this case?

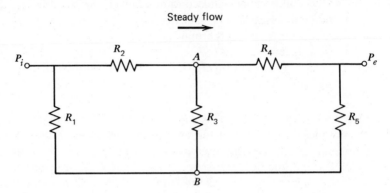

Figure P2.9   Fuel supply system, Problem 2.9.

2.10 A vertical cylindrical tank is being filled with water, while at the same time water is being drained as shown in Figure P2.10. Give a sketch of the analogous flow network using a capacitor symbol to indicate liquid volume storage.

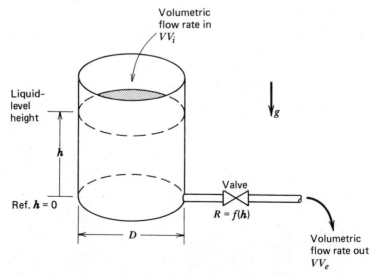

Figure P2.10   Cylindrical tank system, Problem 2.10.

2.11 A cover consisting of a solid slab of metal, initially at room temperature, T_r, is suddenly placed over an opening where the air on the outside is at $T_o(T_o < T_r)$, and the air on the inside is at $T_i(T_i > T_r)$. Assume that you can lump the thermal energy storage (mass × specific heat) of the cover, and call it MC. Call the thermal resistances presented by the air films R_i and R_o, and give a sketch of the analogous heat flow network. Use a capacitor symbol to indicate the thermal energy storage.

2.12 Use the building wall and conditions given in Example 2.3, except let the outside temperature vary with time, $T_c = T_h + A \sin 0.262t + Bt$, where A and B are constants and let thermal energy storage occur equally at the inner and outer surface of layer 7 (call each storage capacity C). Initially, all temperatures are at a uniform T_h. Use capacitor symbols to represent thermal storage and give a sketch of the analogous heat flow network. (*Hint:* Example 2.5.)

2.13 A tank with a volume V is supplied with compressed air at a rate that is a complicated function of the tank air pressure, $\dot{m}_i = f(P)$. At the same time pressurized air is withdrawn from the tank at a mass flow rate that depends on the exit line flow resistance, $R = a(P - P_e)^b$, where a and b are constants, P is the instantaneous tank pressure, and P_e is the instantaneous use pressure. Let C represent the mass storage capacity of the tank; use a capacitor symbol to represent it and give a sketch of the analogous flow network.

2.14 Consider the four-layer wall shown in Figure P2.14. Initially, the temperature of all layers is a uniform T_0; but at time $t = 0$, the left-hand surface is exposed to a high temperature T_h, and at its right-hand surface to a low temperature T_c, where $T_c < T_0 < T_h$. Assume that thermal energy storage can be lumped at the center of each layer and use the symbols $(MC)_A (MC)_B$, and so on, to represent the storage. Use the symbols R_A, R_B, and so on, to represent the unit area thermal energy transfer resistances and show a sketch of the analogous flow network.

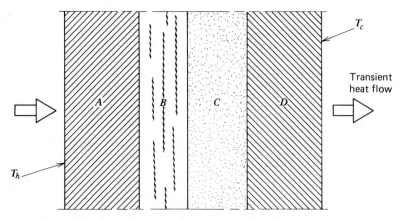

Figure P2.14   Transient heat transfer system, Problem 2.14.

2.15 A multiple-tank liquid lubricant supply system is shown in Figure P2.15. The volumetric flow rate into the system, VV_i, is a complicated transient function, and the outlet flow, VV_e, is to a low pressure environment (P_e). The significant flow resistances are labeled R_1, R_2, and so on; and the liquid-level heights in each tank are labeled h_A, h_B, and h_C. Use C_A, C_B, and C_C to represent storage capacity of each of the tanks and show a sketch of the analogous flow network.

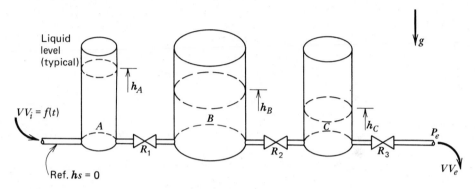

Figure P2.15 Transient liquid flow system, Problem 2.15.

2.16 Refer to Problem 2.10. Let $h=$ liquid level height; $t=$ time; $R=988.1(h)^{0.5}$; $VV_i=0.5+0.5\cos 0.05t$, inlet flow; $D=2.5$, tank diameter; $\gamma=60$, liquid specific weight; and $h_0=10$, initial h. Assume that the units are consistent and set up an appropriate model for simulating h versus t.

2.17 Repeat Problem 2.16, except change the tank to conical form with a diameter of 1.0 at the top and 3.0 at the bottom, with an overall height of 5.0

2.18 A small metal gear blank is being heat-treated by immersing it suddenly in a cool reservoir of oil. Use "lumped" analysis, the following variables and parameters, and set up a differential model for simulating T versus t assuming consistent units. Let $T=$ temperature, $t=$ time, $M=$ mass, $C=$ specific heat, $R=$ resistance between oil and blank, $R=a+b(T+T_0)^e$, $T_0=$ temperature of oil, $T_i=$ initial temperature of blank, and a and $b=$ constants.

2.19 A trough tank as shown in Figure P2.19 is being filled with liquid paint. If the volumetric inlet flow rate is $VV_i=A(1+\cos Bt)$, where A and B are constants and the initial liquid level height is $h(0)=h_0$, set up a differential model for simulating liquid level height, h, versus time, t.

2.20 Refer to Problem 2.11 and set up a differential model for simulating cover temperature, T, versus time, t.

2.21 (a) Refer to Problem 2.12 and set up a differential model for simulating the temperature of the left-hand surface of slab 7, T_{7L}, versus time, t.
(b) Refer to Problem 2.13 and set up a model for simulating the tank pressure, P, versus time, t. Assume that the perfect gas law, $PV=mGT$, where G is the universal gas constant, holds.

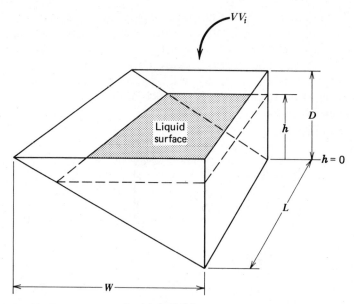

Figure P2.19 Tank system, Problem 2.19.

2.22 A can of beverage (mass $=0.6$), initially at a temperature of $T_0 = 40$, is brought into a convective surroundings at a temperature of $T_a = 90$. Some engineering estimates give a thermal resistance between the can and the surroundings as $R = 0.5$. The specific heat of the combined beverage and can is $C = 0.98$. Assume that lumped analysis is appropriate and that all units given are consistent, and set up a differential model for simulating the beverage temperature, T, versus time, t. Show a sketch of the analogous network.

2.23 A small metal pin is fitted into an undersized hole during a manufacturing process that assembles a pulley on a shaft. In order to get the desired shrink fit on the pin, the pin is cooled before insertion by placing it into a cryogenic fluid. The thermal resistance between the fluid and pin is $R = 2.8$. The mass and specific heat of the pin are m and c, respectively. If the initial temperature of the pin is T_r, and the fluid temperature is T_c, set up the model necessary to simulate the lumped temperature of the pin, T, versus the cool-down time, t. Show a sketch of the lumped analogous flow network. Assume consistent units.

2.24 In part (a) of the sketch of the electrical system shown in Figure P2.24, E_g represents an absolute driving voltage for charging the capacitor, C, when the switch, S, is closed. Let Q represent the charge on the capacitor. The instantaneous voltage across the capacitor is $E_c = Q/C$. Set up a differential model for simulating the voltage E_c versus time, t. In the system (b) let the effective current through the capacitor be $I_c = dQ/dt$ and set up the differential model for simulating the circuit current, I, versus time, t, when $E_g = 2 \sin 20t$. Assume that all units are consistent.

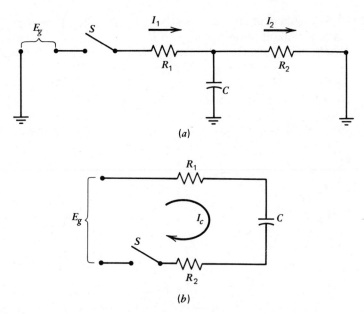

(a)

(b)

Figure P2.24 Electrical system, Problem 2.24

2.25 In the sketch of the electrical system in Figure P2.25, E_g represents the driving voltage difference for the inductive circuit. Let the instantaneous voltage drop across the inductor ($L=$ inductance value) be $E_L = dI/dt$, where $I =$ current through the inductor and $t =$ time. If $E_g = 2 \sin 20t$, set up a differential model for simulating I versus t.

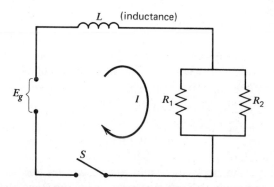

Figure P2.25 Transient inductive electrical system, Problem 2.25.

2.26 An underground cylindrical storage tank ($D=$ diameter; $L=$ length) with flat ends is oriented with its axis horizontal. It is being filled with fuel oil at the volumetric rate of VV_i. Initially, the liquid level height is h_0. Set up an appropriate model for simulating the liquid level height, h, versus time from the beginning of the filling process, t.

2.27 A large spherical tank is connected through a drain line at its bottom with a conical tank as shown in Figure P2.27. Initially, $h_1(0) = 0.8D_s$, and $h_2(0) = 0.01h_1(0)$. If the drainline resistance (with valve open) is R, set up an appropriate model for simulating liquid level heights in both tanks, h_1 and h_2, versus time after the valve was opened, t. Assume that $R = C(|h_1 - h_2|)^{0.5}$, $D_c = 12.0$, $D_s = 10.0$ and that all units are consistent.

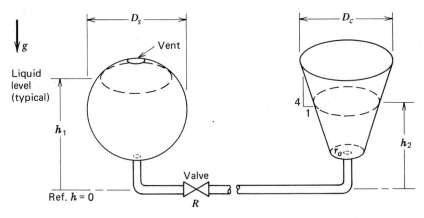

Figure P2.27 Transient liquid level system, Problem 2.27.

2.28 A cup of hot coffee, initially at T_0, is set on a cool table at a temperature of T_a. The composite heat transfer resistance between the hot coffee and the table and surroundings is R. The initial mass of the coffee is m_0, and the specific heat is C (we assume that the energy storage capacity of the lightweight insulating cup is negligible relative to the coffee). Shortly after the coffee is put on the table, some cream at a temperature T_a is poured into the cup at the rate $\dot{m}_c$ (the specific heat of the cream is almost the same as the coffee). Assume a well-stirred cup, consistent units, and lumped analysis, and set up a differential model for simulating the temperature of the coffee, T, versus time from placement of the cup on the table, t.

2.29 A biological observation is that a certain insect population increases at a rate directly proportional to the instantaneous number in the population. However, a natural control predator decreases the population as a complicated function of time. Set up a differential model that could be used to simulate the population number, N, versus time, t. Discuss other natural and man-made systems that might have the same model.

2.30 Your local banker agrees to give you interest on your savings account of r percent of the instantaneous amount in your account, D. Set up a differential model for representing this continuous compounding. Discuss the model in relationship to actual interest rate schemes and to flow systems in general.

2.31 The rate of disintegration of a radioactive substance is directly proportional to the instantaneous amount of the substance present, N. Set up a differential model for simulating the amount, N, versus time, t. Compare this model with other flow problem models.

2.32 A small control volume from within a porous wall is shown in Figure P2.32. The flow resistance is $R = (A + BP)^{-1}$, and the exit volumetric flow is $VV_e = VV_i + [d(VV)/dx] \, \Delta x$. If $\Delta P = P_i - (P_i - (dp/dx) \, \Delta x)$, set up a differential model for simulating local (lumped) pressure, P, versus position, x, within the wall.

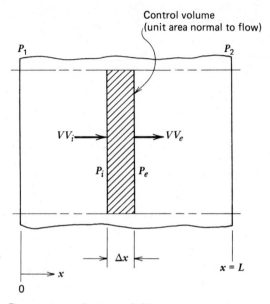

Figure P2.32 Flow system, Problem 2.32.

2.33 A metal bar is heated at both ends but loses heat to the cooler air blown across it, as shown in Figure P2.33. The thermal resistance between the bar and the air is R_c, and the thermal resistance through the small control volume within the bar is $R = (A + BT)^{-1}$, where A and B are constants. Let $q_e = q_i + (dq/dt) \, \Delta x$, and $\Delta T = T_i - (T_i - (dT/dx) \, \Delta x)$, and set up a differential model for simulating the locally lumped temperature of the control volume, T, versus position, x.

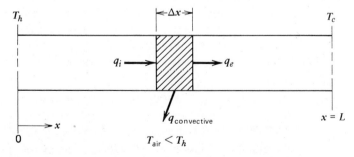

Figure P2.33 Heat transfer system, Problem 2.33.

2.34 A box (volume $= 8$, weight $= 300$) is inadvertently dropped off a work platform into water, hitting the water surface at a velocity of $V_i = 20$. The water drag on the container is $F_d = 3V^{1.2}$. Set up a differential model for simulating container vertical position (from the water surface), x, versus time, t. Assume a water specific weight of 62.4 and consistent units throughout.

2.35 A shipping container (weight $= W$) is suspended from a long hoisting cable as shown in Figure P2.35. A large temporary displacement of the crate causes pendulumlike movement. The air drag force on the crate is $F_d = CV^{1.5}$, where C is a constant and V is the tangential velocity. Assume consistent units and set up an appropriate differential model for use in simulating angular displacement, θ, versus time, t.

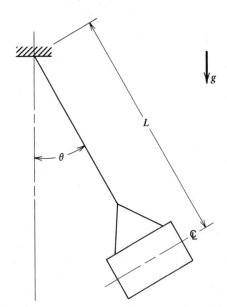

Figure P2.35   Suspended container system, Problem 2.35.

2.36 A hailstone falls straight down. Evaporation (sublimation) from its surface occurs at the mass rate of $\dot{m} = At$, where A is a constant and t is time. The air drag force on the falling ice sphere is given by $F_d = CD^2V^{1.5}$, where C is a constant, D is the sphere diameter, and V is the velocity. The initial weight of the hailstone is W_0. Assume consistent units and set up a differential model for simulating downward vertical position, x, and vertical downward velocity, V, with respect to time.

2.37 Neglect the evaporation in Problem 2.36 but include a horizontal wind velocity $V_w = f$ (elevation). Set up an appropriate two-degree-of-freedom model that can be used to simulate the $x - y$ movement of the hailstone.

2.38 Set up a differential model for simulating the movement (both displacement s and speed $V = ds/dt$) of a truck moving up a constant-slope (angle $= \theta$)

portion of a hill. The weight of the truck is W; its rolling resistance is $F_r = C_1 + C_2 V^a$; the air drag is $F_d = C_3 V^b$; and the propulsive traction from the engine is $F_p = C_4 + C_5 V^d + C_6 V^e$. The C's and a, b, d, and e are constants.

2.39 A special vehicle traveling along a level road, with a following wind, suddenly has its propulsion source disconnected. Using the following variables and parameters, set up an appropriate differential model for simulating the vehicle's speed variation with time (dV/dt) during the coastdown to a stop, where

V = instantaneous vehicle speed
t = time
V_w = wind speed, a statistical function of time
F_d = air drag force, CV_r^2
C = air drag constant
V_r = relative velocity between wind and vehicle
F_r = rolling resistance, a function of the road surface and tire pressure

2.40 A weighted cylindrical barrel used as a buoy marker floats upright in water. At equilibrium its bottom is submerged to a depth of 2. If its vertical position is temporarily displaced, and the vertical water-drag force on the barrel is $F_d = ACV^{1.5}$, where A is the submerged (vertical) area, C is a drag parameter, and V is the vertical velocity, dx/dt, set up an appropriate differential model to simulate up-and-down movement of the barrel. The barrel has a diameter of 2, an overall length of 3.5, and a weight of W. Assume a specific weight for water of γ and consistent units throughout.

2.41 A bicycle is ridden with steady propulsive effort (force $= F_p$) over several small hills, where the local slope of the road surface, θ, can be supplied as a tabulated function with respect to local position, S. If the air drag force is given by $F_d = CV^n$, where C is a drag constant, V is a velocity (dS/dt), and n is parameter, set up a differential model for simulating position, S, versus time, t. The rolling resistance is F_r, a week function of tire pressure and road surface roughness. Assume a weight for the bicycle and rider of W, and consistent units throughout. What happens if the vector dV/dt at the bottom of the hill is large enough that the rolling resistance changes significantly? Should this be a two-degree-of-freedom model?

2.42 A vehicle (weight $= W$) traveling at a high speed, V_0, suddenly has its brakes applied and locked because of an emergency. Assuming straight-line skidding with a bracking force of $F_b = C_b V^s$, where C_b and s are skid parameters, and an air drag force of $F_d = C_d V^2$, where C_d is a drag parameter, set up an appropriate differential model to simulate position, x, and speed, dx/dt, of the vehicle with respect to time, t. Assume consistent units. Is it reasonable to believe that a one-degree-of-freedom model is adequate in this case?

2.43 The printing head on a small computer graphics printer is positioned by a lightweight steel cable. The printing head weighs W and rides on three horizontal sliders that provide a combined sliding resistance of $F_s = CV^a$, where C and a are sliding parameters, and $V = dx/dt$ is the horizontal linear velocity of the head. The instantaneous horizontal forces exerted on the head by the cable are F_r and F_l in opposite directions. Assume consistent units throughout and set up an appropriate model to simulate head position, x, versus time, t.

2.44 A speedboat crosses a starting line at a speed of $V_0 = 20$. The air drag force on the boat is $F_d = 1.6V^{1.5}$, and the water drag is approximated by $F_w = 11.3V^{1.8}$ (but we observe that the actual drag is a very complicated function of the velocity, and generally requires detailed engineering effort to identify). The propulsive force is approximated by $F_p = 26 \times 10^3[1 - \exp(-8.2t)] + 5201$, where t represents time. The boat's weight is $W = 2000$. Assume consistent units and set up an appropriate differential model for simulating the boat's position, x, and speed, $V = dx/dt$, with respect to time, t.

2.45 A firehose stream of water is directed vertically upward with a nozzle velocity of $V = dy/dt = 60$. Consider the stream to be a series of lumped weights with an air drag force per lump of $F_d = CV^n$, where C and n are drag parameters. Assume consistent units and set up an appropriate differential equation model for simulating water particle vertical position, y, and velocity, $dy/dt = V$, versus time, t, from nozzle exit. What type of model would be required if the stream were directed upward at an angle of θ with respect to the horizontal?

2.46 The polar mass moment of inertia of an experimental lightweight rotor in an electric motor is J. The air drag torque on the rotor depends on speed and is approximated by $M_d = Cw^a$, where w is the angular velocity $d\theta/dt$, and C and a are drag parameters. The bearing drag moment on the rotor is $M_b = Bw^b$, where B and b are bearing drag parameters. During startup, the torque supplied to the rotor by the stator system is approximated by $M_s = C_1 + C_2 \cos(\alpha t)$, where C_1, C_2, and α are driving parameters, and t represents time. Assume consistent units and set up a differential model for simulating rotor speed, w, versus time from startup.

2.47 The pivoting behavior of a vehicle skidding on an ice-covered road surface is to be simulated. A simplified sketch of the situation is shown in Figure P2.47. The polar mass moment of inertia about the pivot point of the vehicle is J_0. Assume consistent units throughout and set up an appropriate differential model for simulating θ versus time, t. Both F_f and F_r are tabulated functions of a translational position x.

2.48 A large display board is mounted on a vertical column that has an angular spring (torque) rate of k_t. The polar mass moment of inertia of the board with respect to an axis that coincides with the twist axis of the support column is J_b. A turning moment, M_w, is exerted on the board by the wind. This moment is a statistical function of time, but typical tabulated values are available. Some rotational damping is provided by air drag, where the damping torque, M_d, is given by $M_d = C(d\theta/dt)^{1.8}$; and C is referred to as the rotational

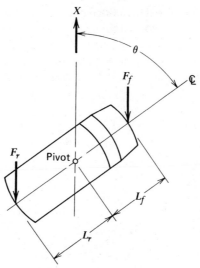

Figure P2.47 Pivoting system, Problem 2.47.

damping constant. The $d\theta/dt$ is the instantaneous angular velocity. Assume consistent units and set up an appropriate model for simulating angular position of the board, θ, with respect to time, t.

2.49 The dynamic insensitivity of pressure measurement manometers is well known. Consider the inclined tube manometer system illustrated in Figure P2.49. The damping force resisting liquid column movement is approximated by $F = C(dX/dt)^{0.7}$, where C is a constant. Set up an appropriate differential model for simulating the reading position, z, versus time, t, if the driving pressure, P, is a statistical function of time.

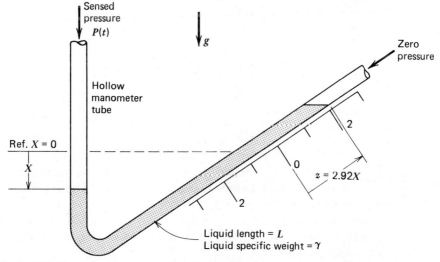

Figure P2.49 Manometer system, Problem 2.49.

2.50 A sketch of an unbalanced drive motor is shown in Figure P2.50. Set up a differential model for simulating up-and-down movement of the motor centerline with respect to time. Take the weight of the motor as W, the rotational speed as w, the rate of the linear spring as $k = 800$, and the damping force as $F_d = 1.2\,(dx/dt)^{1.2}$, where x is vertical position and t is time. The unbalance force is given by $F_u = 0.09w^2 \sin wt$. Assume consistent units.

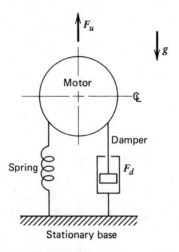

Figure P2.50 Unbalanced motor system, Problem 2.50.

2.51 The damage protection performance of a proposed shipping container has to be simulated. A simplified sketch of the system is shown in Figure P2.51. The packing material exhibits damping forces that are weak functions of displacement rates and springlike forces that are nonlinear.

$$F_{ds} = C_s[d(x_s - x_c)/dt]^{0.3}$$
$$F_{dc} = C_c[d(x_c - x_t)/dt]^{0.6}$$
$$F_{ss} = W_s + k_s(x_c - x_s)^{1.1}$$
$$F_{sc} = (W_c + W_s) + k_c(x_t - x_c)^{1.17}$$

C_s and C_c = damping parameters
k_s and k_c = equivalent spring rates
W_s and W_c = weight of object and container, respectively
x_s and x_c = displacements of object and container, respectively
x_t = external displacement imposed on outside of container = $f(t)$

Assume consistent units and set up an appropriate differential model for simulating object and container displacements with respect to time, t. Neglect any air drag forces.

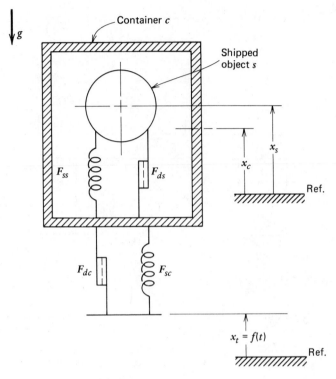

Figure P2.51 Container system, Problem 2.51.

2.52 A system consisting of some heavy skid-mounted compressors and tanks are mounted on a flexible base, which in turn is clamped to a floating construction platform, as shown in Figure P2.52. The following parameters are important.

$F_d = CA(dx_p/dt)^{1.9}$ = water drag on the floats
C = water drag constant
A = instantaneous submerged vertical area of the floats
γ = specific weight of the water
x_e and x_p = local vertical displacements of the equipment and platform
k = flexible base equivalent spring rate
W_e and W_p = weight of the equipment and platform
L = submerged depth (to bottom of floats) at equilibrium
D = diameter of floats

Assume consistent units and set up an appropriate differential model for simulating vertical movements of the platform and equipment with respect to time, t, after a temporary additional load is removed from the platform. Neglect the effects of any air drag forces. Are two degrees of freedom sufficient for simulations in this case?

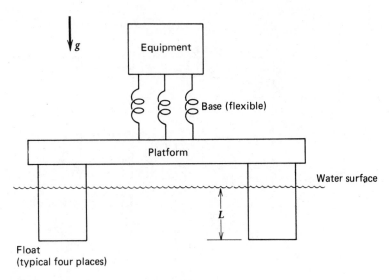

Figure P2.52 Floating system, Problem 2.52.

2.53 Set up a differential model for simulating the vertical movements of the two weights shown in Figure P2.53. One of the springs is nonlinear; $k_1 = 26.1 x_1^{0.3}$. All the other springs and dampers are linear, with the constants shown. Neglect any air drag forces. Assume consistent units.

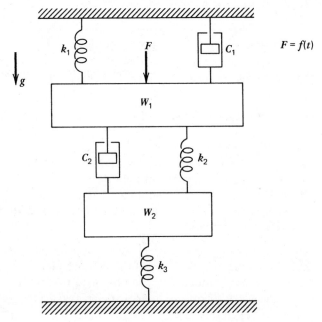

Figure P2.53 Two-Mass spring-damper system, Problem 2.53.

2.54 Set up the appropriate model for the simulations asked for in Problem 2.53, except assume nonlinear dampers, that is, $F_d = C_i |\dot{x}_i|^{0.78}(\dot{x}_i)$.

2.55 A shock absorber system (a nonlinear damper and spring working in parallel) is shown schematically in Figure P2.55. The damping force can be approximated by the following: $F_d = C(dy/dt)^{1.8}$, where y is the damper displacement and C is a constant. Set up an appropriate differential model for simulating vertical motion, x, of the weight, W, after it has been depressed and released. Ignore any rotational effects and air drag effects.

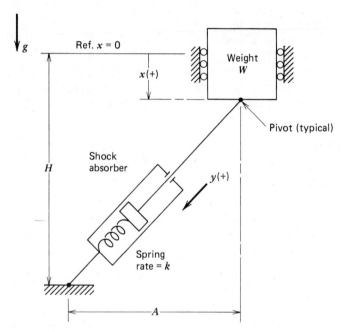

Figure P2.55   Shock absorber system, Problem 2.55.

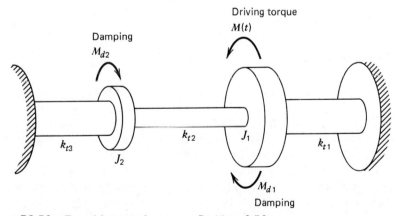

Figure P2.56   Two-Mass shaft system, Problem 2.56.

2.56 A power transmission shaft set carries two large pulleys as shown in Figure P2.56. The angular spring rates are k_{t1}, k_{t2}, k_{t3}, and the polar mass moments of inertia are J_1 and J_2. Damping torques are given by $M_{d1} = C_1(d\theta_1/dt)^{0.1}$ and $M_{d2} = C_2(d\theta_2/dt)$, where θ_1 and θ_2 are the relative angular displacements of the two pulleys. Assume consistent units and set up an appropriate differential model for simulating θ_1 and θ_2 versus time, t, if the instantaneous driving torque is $M(t)$.

2.57 A lumped spring dashpot system is sometimes used to study suspension systems. Consider the diagram shown in Figure P2.57. The damping force is approximated by $F_d = D\ d(x - x_0)/dt$, where x and x_0 are the vertical displacements of the suspended weight and road surface, respectively; t is time; and the damping coefficient is $D = 160[1 + 10(d(x-x_0)/dt)^{0.55}]$. The suspended weight is $W = 3500$; the linear spring rate is $k = 500$; $x_0 = 2[1 - \cos 2t]$, and all units are consistent. Set up an appropriate differential model for simulating the x displacement with respect to time.

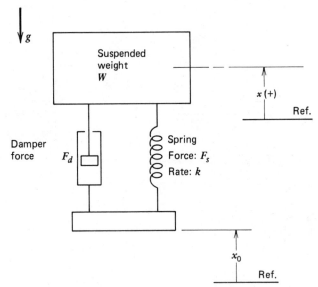

Figure P2.57   Suspension system, Problem 2.57.

2.58 A flowmeter system consists of a tapered cylindrical tube with a spherical indicator inside, as shown in Figure P2.58. The drag force that partially suspends the indicator sphere is given by $F_d = CV_r^{1.5}$, where V_r is the relative velocity between the falling sphere and the upflowing fluid moving with a velocity of V, and C is a drag constant. The tube taper (slope) is S; the indicator weight is W; the indicator diameter is D; the fluid specific weight is γ; and the fluid velocity around the sphere is a relatively complicated function of the volumetric flow rate; $V = f(VV)$. Assume consistent units and set up an appropriate differential model for simulating the indicator position, h, versus time, t, if the volumetric flow rate is varying $VV = f(t)$.

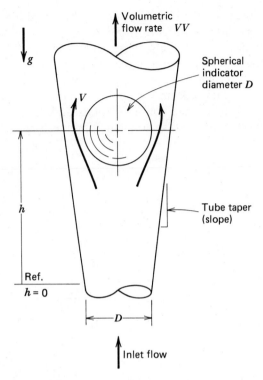

Figure P2.58 Flow indicator system, Problem 2.58.

2.59 The force required to give a uniform cross-sectional cantilever beam a small deflection is $F = 3EIy/L^3$, where EI is the product of the elastic modulus and moment of inertia; y is the deflection at the free end (i.e., where $x = L$), where the force is applied; and L is the length of the beam. A horizontal cantilever beam is used to hold a traffic marker; but the marker is subjected to horizontal wind drag forces, F_w, which are a complicated function of time, t. Damping forces can be approximated by $F_d = C(dy/dt)^{1.5}$, where C is a constant. The weight of the marker is W, which is large compared with the beam weight. Assume consistent units and set up an appropriate differential model for simulating the transverse horizontal movement of the marker, y, with respect to time, t.

2.60 An electrically driven blower is mounted on some hollow elastomer support blocks. The blocks serve as both nonlinear springs (part of the supporting force coming from entrapped compressed air) and damping devices. The spring force is given by $F_s = (a_0 - a_1 x)^{-1}$, where a_0 and a_1 are constants and x is displacement. The damping force is given by $F_d = a_2(dx/dt)^{1.1}$, where a_2 is a constant. The connecting cable and air resistance provides another small damping force given by $F_{da} = a_3(dx/dt)^{1.5}$, where a_3 is another constant. Set up a differential model for simulating vertical displacement, x, and velocity, dx/dt, versus time, t. Assume consistent units.

2.61 A floating platform is held in position by stretching two equal-length spring cables as shown in Figure P2.61. A relatively constant propulsive force F_p is provided to hold the platform against a rapidly varying stream velocity $V_s = f(t)$. The water drag force on the platform can be approximated by $F_s = C(V_r)^{1.9}$, where V_r is the relative velocity between the stream and the platform. The platform weight is W, and the tension preload force in the cables is T_0. Assume consistent units and set up a differential model for simulating platform displacement, X, versus time, t.

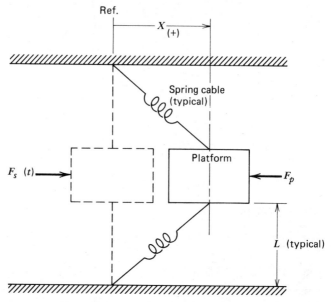

Figure P2.61   Floating platform system, Problem 2.61.

2.62 A cantilever beam is uniformly loaded, so heavily that large deflections are likely. Assume elastic behavior and consistent units, and set up a differential model that can be used to simulate deflections along the beam. Use the following variables and parameters. Neglect any shear deflections.

$x =$ position from fixed end along the beam axis
$y =$ vertical deflection of the beam's neutral axis at x
$L =$ length of the beam
$E =$ elastic (Young's) modulus of the beam material at x
$I =$ area moment of inertia of the beam's cross section at x
$w =$ load per unit length
$M =$ moment at $x = -w(L-x)^2/2$

2.63 The simply supported beam shown in Figure P2.63 exhibits large deflections, y, under a concentrated load W. The weight of the beam is insignificant in comparison with W. Assume consistent units, ignore any vertical shear deflections, and set up a differential model for simulating deflections at any x.

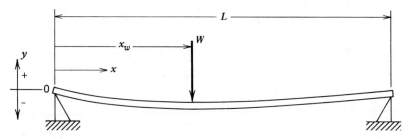

Figure P2.63 Simple beam system, Problem 2.63.

2.64 A circular cross-sectional shaft, with a pulley load W, is shown in Figure P2.64. Set up an appropriate differential model for simulating large elastic deflections, y at any position, x, from the left bearing support. Ignore any vertical shear deflections and assume consistent units.

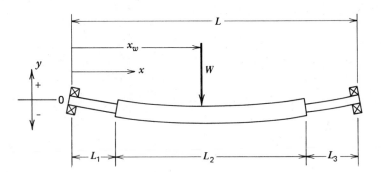

Figure P2.64 Shaft system beam, Problem 2.64.

2.65 Set up an appropriate differential model that can be used for simulating deflections, y, of a simply supported uniform beam 10 ft long weighing 150 lbf. It has a linear load distribution that varies from 60 lbf/ft at the left end to 15 lbf/ft at the right end. Assume large deflections and leave results in terms of EI units. How are axial stresses determined?

2.66 A simply supported beam is loaded transversely with a concentrated load F_t, located $L/3$ from its left end support, where L is the beam length. The beam is also loaded axially with a compressive load of F_a. The beam is nonuniform with EI a function of axial position, x. Assume large deflections and consistent units, and set up a differential model for simulating transverse displacements, y, versus position, x.

2.67 A simply supported beam is shown in Figure P2.67. Assume large deflections and a nonuniform cross section such that *EI* is a tabulated function of *X*, and set up an appropriate differential model for simulating deflections, *y*, versus position, *X*. Use singularity functions in the model setup if the instructor so designates. Assume consistent units.

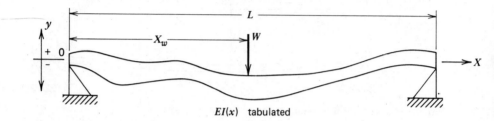

Figure P2.67   Beam system, Problem 2.67.

3
Setups Involving Partial Derivatives

FLOW SYSTEMS

In the model setups that eventually display partial-derivative symbols, multiple independent variables are always present, but these independent variables are not quite the same as those associated with multiple degrees of freedom in Chapter 2. Although some of the independent variables may now represent spatial coordinates in a manner similar to before, we are, in fact, dealing now with an infinite number of degrees of freedom and use the statement that we are working with "continuous systems" as opposed to "distributed systems." Actually, we have already slipped in a little work with continuous systems in Chapter 2 by working with an infinitesimal cutout in setting up the beam models and by never showing a deliberate "lumping" of time in any of the models. Basically, then, partial derivatives start showing up whenever we allow quantities to vary continuously through space and time, and the independent variables become the space and time coordinates. In this book we emphasize the simplest of the multiple independent variable problems—those involving no more than two spatial coordinates, and possibly time. Many of the concepts and procedures described previously for single independent variable (ordinary-derivative) models can be applied directly to the multiple independent variable model setups. These will be discussed within the contexts of the previous categories—flow systems, moving object systems, and structural systems—in the following sections. The reader interested in more background than is given here is directed to the references.[1-11]

In flow system problems, the concept of the control volume introduced in Chapter 2 is still the starting point for model setups. However, now the flow system will always be larger than a setup control volume, since the control volume will always be an infinitesimal region in space. But the formulations and equations given previously, describing conservation and universal flow principles, that is, Equations 2.1 through 2.11, are still good. The only difference in the equations will be partial derivative symbols appearing in place of the ordinary derivative symbols.

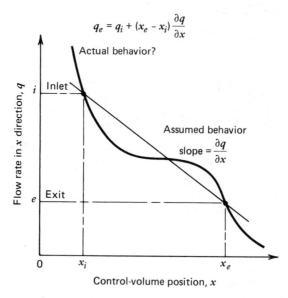

Figure 3.1   Inlet and exit flows for an infinitesimal control volume.

Consistent with the continuous system idea, we now introduce an extremely important stratagem, as shown in Figure 3.1, and symbolized in Equation 3.1. Using Equation 2.3 and the fact that we have an extremely small control volume, we assume that the flow rate change, while passing through the control volume in the x direction, will be linear, with a slope of $\partial q/\partial x$. Thus we can state the exit flow rate in the x direction without really knowing what it is! Of course, as the control volume is visualized smaller and smaller, the $x_e - x_i$ becomes the differential dx, and we have a true infinitesimal.

$$q_e = q_i + (x_e - x_i)\frac{\partial q}{\partial x} \tag{3.1}$$

Equation 2.6 still applies, but it now becomes convenient to acknowledge that the flow resistance always contains a flow path length of some kind and normalize with respect to this path length. Furthermore, since we will be taking derivatives of the resistance reciprocal, a single symbol is easier to deal with. So, we let $S = 1/R$, where R is presumed to be based on a unit length. Also it is apparent that the finite driving force difference in Equation 2.6 becomes a negative differential for the infinitesimal control volume, that is, ΔF becomes $-dF$. Thus the last term in Equation 3.1 can be re-posed as shown in Equation 3.2.

$$\frac{\partial q}{\partial x}\,dx = -\frac{\partial[S(\partial F/\partial x)]}{\partial x}\,dx \tag{3.2}$$

Equation 3.2 can in turn be re-posed by doing the differentiation. The result is shown in Equation 3.3.

$$\frac{\partial q}{\partial x}\,dx = -S\frac{\partial^2 F}{\partial x^2}\,dx - \frac{\partial S}{\partial x}\frac{\partial F}{\partial x}\,dx \tag{3.3}$$

But, if the flow resistance is assumed to be constant, then the last term in Equation 3.3 goes to zero. On the other hand, it may depend more on F than on x, and then we have to invoke the chain rule from calculus to give the version shown in Equation 3.4.

$$\frac{\partial q}{\partial x}\,dx = -S\,\frac{\partial^2 F}{\partial x^2}\,dx - \frac{\partial S}{\partial F}\left(\frac{\partial F}{\partial x}\right)^2\,dx \tag{3.4}$$

where $S = 1/R$ in Equations 3.2, 3.3, and 3.4; and R is based on unit length flow path.

The concepts discussed in the foregoing have had a significant impact on how engineering models have been set up in the past; the ideas extend beyond flow systems and will be used in other model setups still to be considered. However, for the flow systems, we see that now very localized flows within a large continuous flow system can be posed in a very general model form, where dependencies can be taken one independent variable at a time. The actual setup tasks can be described best by example from here on, and Examples 3.1 and 3.2 should be consulted.

EXAMPLE 3.1

A large vertical slab, L units thick, is at a uniform temperature of T_0. At time $t = 0$ the temperature on the left-hand face of the slab is raised to T_1, and the temperature on the right-hand surface is lowered to T_2, where $T_2 < T_0 < T_1$. The surface temperatures are then held constant. The thermal capacitance of the material in the slab is C_u, based on a unit volume. The flow resistance of the slab material is R_u, based on a unit length. Give a sketch of an infinitesimal control volume in the slab (use a unit area normal to the slab surfaces), label x direction heat flows normal to the slab surfaces, and assume that heat flows in the y and z directions are negligible. Set up a model that is appropriate for simulating temperatures at all points within the slab as they change with time, as in Figure 3.2.

▶ See Figure 3.2.

The storage capacity of the infinitesimal control volume is $C = C_u\,dx$. Recognizing that F in this case is T and letting $S = 1/R_u$, we apply the conservation principle through the partial-derivative versions of Equations 2.2, 2.6, and 2.11. However, we realize that we cannot get away from the "lumped analysis" concept, even though dealing with an infinitesimal control volume, and give the entire control volume a single temperature $T = T(x, t)$. This seems a little inconsistent in view of the three separate control volumes we used in Chapter 2 under similar circumstances. But here we are treating such a small volume that we argue it makes little difference where we locate the storage lump and simply pretend that it exists at either side of the dx. This is one reason we choose not to show a flow schematic here; but an alternate to using the dx wide control volume is discussed later, and there the schematic again is consistent and

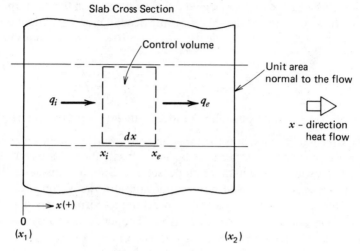

Figure 3.2 Thermal system infinitesimal control volume, Example 3.1.

useful. For now the results are

$$q_i - q_e = C \frac{\partial T}{\partial t}$$

Applying Equation 3.1 gives

$$-(x_e - x_i) \frac{\partial q}{\partial x} = C \frac{\partial T}{\partial t}$$

This can be simplified by using Equation 3.3, assuming that S is a constant, and doing a little algebra. Thus the desired model becomes

$$\frac{\partial^2 T}{\partial x^2} = \frac{C_u}{S} \frac{\partial T}{\partial t}$$

with

$$T(x, 0) = T_0$$

$$T(0, t) = T_1$$

$$T(L, t) = T_2$$

This will be recognized by many as the classical one-dimensional heat equation.

EXAMPLE 3.2

A long built-in rail on a hoisting system has the cross section illustrated in Figure 3.3. Surface temperatures TT, TR, TB, and TL vary around the rail cross section,

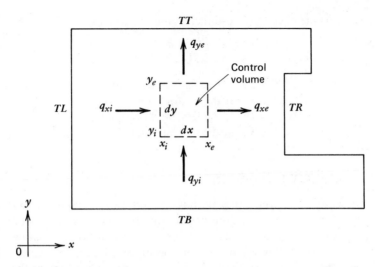

Figure 3.3 Thermal system infinitesimal control volume for steady flow, Example 3.2.

but not along its axis, nor with time. The thermal resistance of the rail material is R_u, based on a unit length. The R_u is constant and nondirectional; that is, the beam material is isotropic. Using the control volume shown, set up an appropriate model for simulating the steady state interior temperatures of the beam cross section.

▶ Although obviously a two-independent-variable problem, this may not be recognized immediately as a flow system problem, but we see that to get at any interior temperatures some type of flow equation will have to be used. The steady nature of the problem suggests that we use the Equation 2.3 form of the conservation principle, summing the inlet and exit thermal energy flow rates.

$$(q_{xi}+q_{yi})-(q_{xe}+q_{ye})=0$$

Substituting Equations 3.1 and 3.3, with F replaced by T, gives

$$(q_{xi}+q_{yi})-\left[q_{xi}-S\left(\frac{\partial^2 T}{\partial x^2}\right)dx+q_{yi}-S\left(\frac{\partial^2 T}{\partial y^2}\right)dy\right]=0$$

where
$$S=1/R_u$$
$$dx=x_e-x_i$$
$$dy=y_e-y_i$$

Simplification gives the desired model.

$$\left(\frac{\partial^2 T}{\partial x^2}\right)dx+\left(\frac{\partial^2 T}{\partial y^2}\right)dy=0$$

If we were to choose a square cross sectional control volume, we can simplify the foregoing to the classical two-dimensional Laplace equation.

$$\frac{\partial^2 T}{\partial x^2} + \frac{\partial^2 T}{\partial y^2} = 0$$

with $\qquad\qquad$ $T = TR$ at the right-hand boundaries
$\qquad\qquad\qquad$ $T = TB$ at the bottom boundaries
$\qquad\qquad\qquad$ $T = TL$ at the left-hand boundaries
$\qquad\qquad\qquad$ $T = TT$ at the top boundaries

MOVING-OBJECT SYSTEMS

The free-body subsystems that we have to use now, in contrast to those used in Chapter 2, are very small regions cut out (in imagination only, of course) from the overall body of the main system. There are so many of these infinitesimal cutouts that the number of degrees of freedom becomes infinite; and we say that we are treating a "continuous" system. Consequently, we cannot look at the relationship between degrees of freedom and the movement coordinates in the same manner as we did in Chapter 2. Instead we will use partial derivative symbols and rely on no more than four independent variables, one of which will always be time, with the others being classical orthogonal coordinates of some kind. To simplify the situation a bit, we will limit our emphasis to one or two space coordinates only, and time. The central idea here is that the partial-derivative expressions we can develop for a *single* infinitesimal free body will automatically take care of the coupling between subsystems that was of concern in Chapter 2.

It should be apparent that the setup task list (Table 2.2) will have to be modified slightly in regard to identification of degrees of freedom. Furthermore, the determination of the external forces and/or moments on a free body cutout will now be related to material properties through stress and strain expressions. Much of the effort in getting the appropriate partial derivatives put down is embodied in the stratagem discussed along with Equation 3.1 of the last section. Of course, now we replace the q's with forces or moments. And these forces and/or moments will be related to displacements through equations similar to the structural system Equations 2.18 and 2.19. The difficulty that many students have here is recognizing that shear stresses on the imaginary cut surfaces can be the dominant force contributors, and then if they are recognized, deciding in which direction they act. A comprehensive first course in engineering strength of materials is helpful, as are the references.[8-11] The implementation of the task list enumerated in Chapter 2 (Table 2.2), with the exceptions noted above for continuous systems, is presented in Examples 3.3 and 3.4.

EXAMPLE 3.3

A long horizontal steel beam, length $= L$, cross-sectional area $= A$, is fixed to a heavy foundation plate at its left end. The right end is clamped to a tension cable, which exerts a longitudinal force F_0 on the beam. At time $t=0$, the clamp fails, releasing the force on the beam. The beam will not recover its unloaded length immediately; local displacements will "travel" in effect through the beam until they become negligible. Assume the specific weight of the beam material to be γ and the modulus of elasticity to be E, and set up an appropriate partial-derivative model that can be used to simulate the local longitudinal displacements in the beam versus time after the clamp failure.

▶ We choose the x coordinate to coincide with the longitudinal axis of the beam. Because the displacements will be in the same coordinate, we superimpose another variable, namely u, which represents our local displacement and becomes our dependent variable $u(x, t)$, as shown in Figure 3.4.

Using Equations 2.12 and 2.13, and the concept discussed along with Equation 3.1, we get

$$F + \frac{\partial F}{\partial x} dx - F - \frac{A \, dx\gamma}{g} \frac{\partial^2 u}{\partial t^2} = 0$$

But from Equations 2.17 and 2.18 we have

$$S = \frac{F}{A} = E \left(\frac{\partial u}{\partial x} \right)$$

Substituting back gives

$$\frac{\partial}{\partial x} \left(EA \frac{\partial u}{\partial x} \right) dx - \frac{A\gamma \, dx}{g} \frac{\partial^2 u}{\partial t^2} = 0$$

Re-posing and simplifying gives the desired model.

$$\frac{\partial^2 u}{\partial x^2} = \gamma (gE)^{-1} \frac{\partial^2 u}{\partial t^2}$$

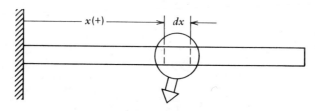

Cutout (free body)

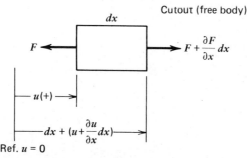

Figure 3.4 Infinitesimal free body, Example 3.3.

with
$$u(0, t)=0; \qquad \frac{\partial u(L, t)}{\partial x}=0$$

$$u(x, 0)=\frac{F_0 x}{AE}; \qquad \frac{\partial u(x, 0)}{\partial t}=0$$

EXAMPLE 3.4

A long horizontal wire is stretched (force $=F_0$) between two fixed points. At time $t<0$ the wire's vertical position y is given by $y=ax$ for $x\leqslant L/2$, and $y=a(L-x)$ for $x>L/2$, where $x=$ the horizontal distance from the left-hand anchor point, $a=$ constant, and $L=$ the distance between anchor points. Displacements and slopes are small enough that infinitesimal arc lengths, ds, along the wire's axis do not differ much from dx. The wire cross-sectional area is A, and the specific weight of the wire material is γ. At time zero the wire is released and allowed to vibrate transversely. Assume that the tension force remains constant, that the local radius of curvature is $r=(\partial^2 y/\partial x^2)^{-1}$, and set up an appropriate partial-derivative model for simulating local wire displacements, $y(x, t)$, versus time. Give a "cutout" free-body sketch showing forces, as in Figure 3.5.

▶ Using the idea discussed with Equation 3.1, the slope on the right side of the cutout is posed as a partial-derivative expression, $\theta+(\partial\theta/\partial x)\,dx$, even though we really do not know what it is. The slopes at the ends of the free body can then be used to resolve the tangential tension forces into the desired vertical components. Incorporating the results into the partial-derivative versions of

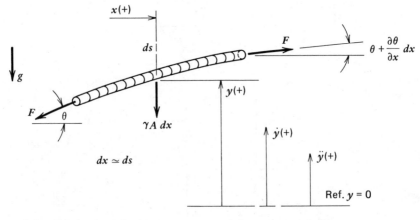

Figure 3.5 Infinitesimal free body, Example 3.4.

Equations 2.12 and 2.13 gives

$$-F \sin \theta + F\left[\sin \theta + \frac{\partial \theta}{\partial x}\,dx\right] - \gamma A\,dx - \frac{\gamma A\,dx}{g}\frac{\partial^2 y}{\partial t^2} = 0$$

But for small angles, which we presumably have, $\sin \theta \simeq \theta$, and $(\partial \theta / \partial x)\,dx \simeq d\theta$, and we get a much simpler equation.

$$F\,d\theta - \gamma A\,dx - \frac{\gamma A\,dx}{g}\frac{\partial^2 y}{\partial t^2} = 0$$

However, from the calculus we get $d\theta = ds/r$, and from the curvature information given in the problem statement, $ds/r = ds(\partial^2 y / \partial x^2)$. Substitution into the foregoing gives the desired model, where $F = F_0$, and $ds \simeq dx$.

$$\frac{\partial^2 y}{\partial x^2} = \frac{\gamma A}{F_0} + \gamma A(F_0 g)^{-1}\frac{\partial^2 y}{\partial t^2}$$

with

$$y(0, t) = 0; \qquad y(L, t) = 0$$

$$y(x, 0) = \text{given}; \qquad \frac{\partial y(x, 0)}{\partial t} = 0$$

STRUCTURAL SYSTEMS

The procedure for setting up these models is very similar to that just described for the moving-object systems, except that no inertia forces are involved.[8,9] Again, as in Chapter 2, we are excluding the treatment of multiple finite entities (since many fine structures books concentrate on this area), and instead concentrate on a single entity like a beam or plate. We isolate by visualization a very small region cut out from the overall system. This becomes our "cutout" (some analysts prefer to call this a "free body," just as in the moving-object setups); and upon sketches of this we attempt to show all the forces and moments acting on the external (cut) surfaces. However, this is usually easier said than done. The first problem is that, in general, forces in the interior of a large complicated domain will not be known. Consequently, we rely on the same scheme described for Equation 3.1. We simply represent the forces symbolically and then apply partial-derivative expressions (Equation 3.1 with the q's replaced) to show how these forces vary in particular directions through our infinitesimal cutout.

The second problem is identifying the significance and directions of the shear stress forces acting on the cut surfaces. Alternate visualizations of these forces are useful in many cases to give the analyst a feel for the problem, but usually a "best" picture emerges for certain types of problems. Unfortunately, the "best" is on occasion more connected with what can be solved with classical mathematics than with the true physical picture. Nevertheless, a good background in the engineering strength of materials will help us get past this problem. Once forces and/or moments have been identified, simple static balance equations (e.g., an algebraic summation of all forces in a particular coordinate direction must be zero), as illustrated by Equations 3.5 through 3.10, can be employed to get the model.

$$F_{1x} + F_{2x} + \cdots + F_{nx} = 0 \tag{3.5}$$

$$F_{1y} + F_{2y} + \cdots + F_{ny} = 0 \tag{3.6}$$

$$F_{1z} + F_{2z} + \cdots + F_{nz} = 0 \tag{3.7}$$

$$M_{1xy} + M_{2xy} + \cdots + M_{nxy} = 0 \tag{3.8}$$

$$M_{1yz} + M_{2yz} + \cdots + M_{nyz} = 0 \tag{3.9}$$

$$M_{1xz} + M_{2xz} + \cdots + M_{nxz} = 0 \tag{3.10}$$

where F = force

M = moment

x, y, or z subscript indicates force aligned with the like coordinate

xy, yz, or xz subscript indicates moment in like planes

As might be expected from the foregoing balance equations, the independent variables in these models will be the space coordinates. Since these are static systems, time does not enter. In most cases, both local stresses and displacements will be the objective of the simulations. But these dependent variables have intrinsic connections with each other through materials properties—namely, the elastic moduli and Poisson's ratio. So, typically, additional equations have to be given on a case by case basis to complete a model setup. This aspect is slightly beyond the scope of what we are willing to work with here, but the interested student can consult several classic books in strength of materials.[8,9, etc.] Examples 3.5 through 3.7 show the procedures that have been discussed here.

EXAMPLE 3.5

A thin plastic cover (thickness $= t$) is placed over a small opening into a clean room where a positive gage pressure of P is maintained. The plastic is so thin that it can support no bending stresses, and the only stresses occur in the plane of the membrane. The local tensile stresses (S_x and S_y) are taken to be the same and approximately uniform throughout the membrane, where the subscripts indicate the coordinate of action. Sometimes the x direction is referred to as the tangential direction; the y direction is referred to as the meridional direction. Assume that the weight of the membrane can be ignored and set up an appropriate partial derivative model for simulating transverse displacements, w, of the membrane. Show a sketch of a $dx\,dy$ cutout with the forces labeled.

▶ The tension and pressure forces are symbolically represented on the cutout shown in Figure 3.6, where the scheme discussed with Equation 3.1 is used to determine the slope variations.

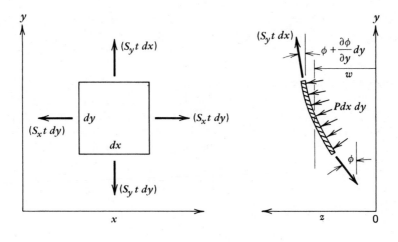

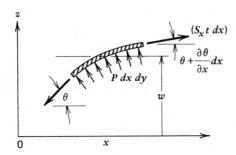

Figure 3.6 Membrane cutout, Example 3.5.

Assuming $dx = dy$ and summing forces in the z direction (Equation 3.7), we obtain

$$S_x t\, dy \left(-\theta + \theta + \frac{\partial \theta}{\partial x}\, dx \right) + S_y t\, dx \left(-\phi + \phi + \frac{\partial \phi}{\partial y}\, dy \right) + P\, dx\, dy = 0$$

where the approximation, $\sin \theta \simeq \theta$, for small angles has been used. Simplification using $\theta = \partial w / \partial x$ and $\phi = \partial w / \partial y$ gives

$$S_x \frac{\partial^2 w}{\partial x^2} + S_y \frac{\partial^2 w}{\partial y^2} = -\frac{P}{t}$$

If we let our uniform membrane stress be $S_m = S_x = S_y$, then our required model becomes

$$\frac{\partial^2 w}{\partial x^2} + \frac{\partial^2 w}{\partial y^2} = -\frac{P}{S_m t}$$

with four conditions.

EXAMPLE 3.6

A moderately thick rectangular plate, whose bending resistance is significant, has its edges at $x=0$ and L_x and at $y=0$ and L_y welded to a rigid opening containing a gage pressure of P on the underside. The unit-length shear stresses S_{1x} and S_{1y} on the edges of a $dx\,dy$ cutout of the plate (thickness $=t$) can be related to the vertical (z direction) deflection w by the following equations.[9]

$$S_{1x}=D\left[\frac{\partial^3 w}{\partial y^3}+\frac{\partial^3 w}{\partial x^2\,\partial y}\right]$$

$$S_{1y}=D\left[\frac{\partial^3 w}{\partial x^3}+\frac{\partial^3 w}{\partial y^2\,\partial x}\right]$$

where

$S_{1x}=S_{sx}t$, and likewise for S_{1y}
$D=$ plate stiffness $=Et^3[12(1-\mu^2)]^{-1}$
$E=$ modulus of elasticity
$\mu=$ Poisson's ratio

Set up an appropriate partial-derivative model for simulating the plate deflections $w(x, y)$. Give a sketch showing the zx and zy elevation views of the cutout, the shear forces (acting clockwise), and the uniform pressure force acting from underneath, as in Figure 3.7. Assume that the bending moments are already accounted for in the above equations.

▶ Using Equation 3.7 and summing the z direction forces ($+$ upward) gives

$$S_{1x}\,dx-\left(S_{1x}+\frac{\partial S_{1x}}{\partial y}\,dy\right)dx+S_{1y}\,dy-\left(S_{1y}+\frac{\partial S_{1y}}{\partial x}\,dx\right)dy+P\,dx\,dy=0$$

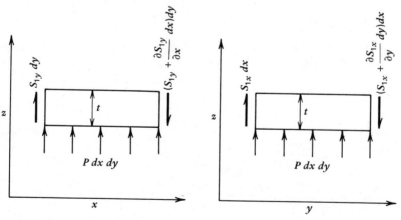

Figure 3.7 Plate cutout views, Example 3.6.

or

$$\frac{\partial S_{1y}}{\partial x} \, dx \, dy + \frac{\partial S_{1x}}{\partial y} \, dy \, dx = P \, dx \, dy$$

Substituting the problem statement equations gives the desired model.

$$\frac{\partial^4 w}{\partial x^4} + 2 \frac{\partial^4 w}{\partial x^2 \, \partial y^2} + \frac{\partial^4 w}{\partial y^4} = \frac{P}{D}$$

with eight conditions.

The foregoing will be recognized as the classical biharmonic partial-differential equation.[11]

EXAMPLE 3.7

The stresses in a steel column subjected to pure twisting are to be simulated. The column has a rectangular cross section, a modulus of rigidity of G, and a twist angle of θ per unit length. Assume that the axis of twist is the origin for $x-y$ coordinates on a cross section at z from the base, that w is the displacement in the z direction, and that all the cross-sectional stresses are zero except the local xz and yz plane shear stresses given by the following equations.

$$S_{xz} = G\left(\theta y + \frac{\partial w}{\partial x}\right)$$

$$S_{yz} = G\left(-\theta x + \frac{\partial w}{\partial y}\right)$$

Set up an appropriate partial-derivative model. Give a sketch showing $dx \, dz$ and $dy \, dz$ cutouts with all forces labeled as in Figure 3.8. Assume that $dx = dy = dz$, and conditions $\partial w/\partial y = -\theta x$ at $y = \pm y_b$ and $\partial w/\partial x = \theta y$ at $x = \pm x_b$.

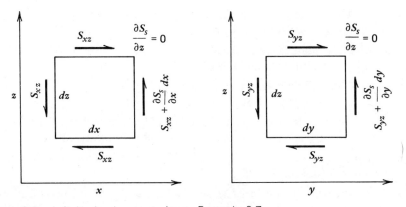

Figure 3.8 Infinitesimal cutout views, Example 3.7.

► All moment summations are seen to be zero if it is assumed that $(dx)^2$ and $(dy)^2$ are negligible relative to other quantities. Using Equation 3.7 and summing the z forces (+ upward on the sketch) gives

$$-S_{xz}\,dz\,dy+\left(S_{xz}+\frac{\partial S_{xz}}{\partial x}\,dx\right)dz\,dy-S_{yz}\,dz\,dx+\left(S_{yz}+\frac{\partial S_{yz}}{\partial y}\,dy\right)dz\,dx=0$$

Simplification gives

$$\frac{\partial S_{xz}}{\partial x}+\frac{\partial S_{yz}}{\partial y}=0$$

Differentiating the first equation given in the problem statement with respect to x, and the second one with respect to y, and substituting into the foregoing gives our desired partial-derivative model. Results have to be used with the given equations to get the desired stresses.

$$\frac{\partial^2 w}{\partial x^2}+\frac{\partial^2 w}{\partial y^2}=0$$

with four conditions.

The foregoing model will be recognized as the classical Laplace equation, which has been encountered in a previous example.

REFERENCES

1. J. P. Holman, *Heat Transfer, 4th ed.*, McGraw-Hill Book Company, New York, 1976.

2. R. B. Bird, W. E. Stewart, and E. N. Lightfoot, *Transport Phenomena*, John Wiley & Sons, Inc., New York, 1960.

3. R. M. Olson, *Essentials of Engineering Fluid Mechanics, 2nd ed.*, International Textbook Co., Scranton, Pa., 1966.

4. G. J. Van Wylen and R. E. Sonntag, *Fundamentals of Classical Thermodynamics, 2nd ed.*, John Wiley & Sons, Inc., New York, 1973.

5. C. M. Haberman, *Engineering Systems Analysis*, Charles E. Merrill Publishing Company, Columbus, Ohio, 1965.

6. R. V. Churchill, *Fourier Series and Boundary Value Problems*, McGraw-Hill Book Company, New York, 1941.

7. F. S. Tse, I. E. Morse, and R. T. Hinkle, *Mechanical Vibrations, Theory and Applications, 2nd ed.*, Allyn & Bacon, Inc., Boston, Mass., 1978.

8. S. Timoshenko and J. N. Goodier, *Theory of Elasticity*, McGraw-Hill Book Company, New York, 1951.

9. J. P. Den Hartog, *Advanced Strength of Materials*, McGraw-Hill Book Company, New York, 1952.

10. T. Baumeister, E. A. Avallone, and T. Baumeister, III, eds., *Mark's Standard Handbook for Mechanical Engineers, 8th ed.,* McGraw-Hill Book Company, New York, 1978.

11. R. D. Kersten, *Engineering Differential Systems,* McGraw-Hill Book Company, New York, 1969.

PROBLEMS

3.1 An engineering fire insurance analysis requires simulations of the temperature versus time patterns along an insulated metal support column. The lower end of the column is subjected to flame conditions that give a temperature increasing with time: $T(0, t) = a + bt$, where a and b are constants. Initial temperatures within the column are a uniform $T = T_\infty$. The cross-sectional area of the column, A, is uniform; density of the column material is ρ; specific heat of the column material is c_p; thermal conductivity of the column material is k; height of the column is L; and the periphery of the column is perfectly insulated. Assume that the top of the column is held at $T(L, t) = T_\infty$, that all units are consistent, and set up an appropriate differential model for simulating the interior temperatures, $T(y, t)$, with respect to time, t. Is the model elliptic or parabolic?

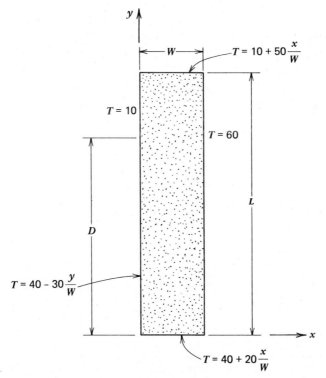

Figure P3.2 Wall cross section, Problem 3.2.

3.2 Engineering analysis of the thermal stresses in a long concrete wall require simulation of steady state temperatures within a cross section of the wall, as illustrated in Figure P3.2. Thermal conductivity of the wall material is k, a constant. Assume consistent units; use a finite control volume with $\Delta x = \Delta y$; and set up an appropriate differential model for simulating the interior temperatures, T, with respect to interior location, x and y. Is your model elliptic, parabolic, or hyperbolic?

3.3 Repeat Problem 3.2 except use a control volume with $\Delta x \neq \Delta y$. Is the model hyperbolic, parabolic, or elliptic?

3.4 Repeat Problem 3.1 except let the thermal conductivity be a function of temperature: $k = k_0 + aT$, where k_0 and a are constants.

3.5 Repeat Problem 3.2 except let the thermal conductivity be a tabulated function of the x position in the wall (i.e., a composite wall).

3.6 Repeat Problem 3.2 except let the thermal conductivity be a function of temperature: $k = k_0 + a_1 T + a_2 T^2$, where k_0, a_1, and a_2 are constants.

3.7 A basic empirical relationship governing incompressible steady fluid flow through porous media is Darcy's law, shown here in differential form:

$$-VV_x = \frac{KA_x}{\mu} \frac{dP}{dx}$$

where VV_x is the volumetric flow rate in the x direction through an area normal to x of A_x; K is the permeability of the flow system's solid medium (length squared units), μ is the dynamic viscosity of the fluid (force × time/ length squared units); and P is the fluid pressure (force/length squared units). Consider an aquifer of uniform thickness as shown in Figure P3.7. Assume consistent units and set up an appropriate differential model for simulating steady flows and pressures within the aquifer. Use a finite control volume with $\Delta x = \Delta y$.

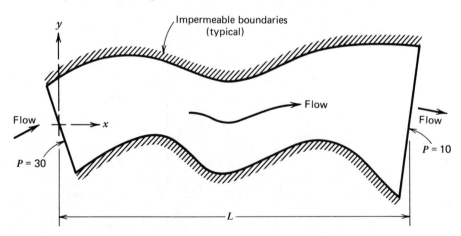

Figure P3.7 Aquifer system, Problem 3.7.

3.8 Repeat Problem 3.7 except let the permeability be a tabulated function of x.

3.9 Repeat Problem 3.7 except use a finite control volume with $\Delta x = 2\,\Delta y$.

3.10 Consider an infinitesimal two-dimensional control volume, with sides dx and dy and a thickness of one unit, taken from a large flow system. Let the mass flow rates entering the control volume from the left and bottom be $\rho U\,dy$ and $\rho V\,dx$, where U and V are the x and y velocity components, respectively, and ρ is density. Assume steady flow, and using Equation 2.3, show that the conservation principle in this case becomes the following continuity equation model:

$$\frac{\partial(\rho U)}{\partial x} + \frac{\partial(\rho V)}{\partial y} = 0$$

Show the continuity equation for incompressible flow.

3.11 Repeat Problem 3.10 except assume unsteady flow and use Equation 2.2 instead of Equation 2.3.

3.12 The thermal resistance for a ceramic material is $R = a_0 + a_1 T$ per unit flow path length, where a_0 and a_1 are constants and T is temperature. Consider the system shown in Example 3.2 except assume that it is constructed of our ceramic material. Assume consistent units and set up an appropriate differential model.

3.13 Repeat Problem 3.12 except use the system shown in Example 3.1 instead of the one shown in Example 3.2.

3.14 Consider the system in Example 3.3, except let the beam cross section be a function of x: $A = a_0 + a_1 x$, where a_0 and a_1 are constants. Set up an appropriate differential model for simulating the local longitudinal displacements, u, in the beam versus time, t.

3.15 A long metal pushrod is used in a power actuator mechanism. Assume that the large axial movements are slow enough that inertia effects for these can be ignored; however, local axial compressions, u's, are of interest as the driving force on the rod end varies with time: $F_0 = f(t)$. Assume a uniform rod cross section and set up an appropriate differential model for simulating the u's with respect to time, t.

3.16 Consider the system shown in Example 3.4 except include a damping effect owing to air friction of $F_d = C(\partial y/\partial x)^{1.4}\,dx$, where C is a damping constant. Set up an appropriate differential model for simulating local wire displacements with respect to time, t. What happens to the model if the approximation of $\tan\theta = \partial y/\partial s$ is used for small angles?

3.17 Show a free-body sketch, and demonstrate that the transverse vibrations (small z displacements only) of a thin membrane under uniform tension, F_t (force per unit length units), in all directions, can be simulated from the following model:

$$\frac{\partial^2 z}{\partial x^2} + \frac{\partial^2 z}{\partial y^2} = \frac{\gamma t}{F_t} + \left(\frac{\gamma t}{F_t}\right)\left(\frac{1}{g}\right)\frac{\partial^2 z}{\partial t^2}$$

where γ is the specific weight of the membrane material and t is the thickness. (*Hint:* See Example 3.5 and Problem 3.18.)

3.18 Consider Example 3.5. Assume that the opening is horizontal with the pressure on the lower side, and that the weight of the membrane (specific weight $=\gamma$) is significant. Assume consistent units and set up a differential model for simulating transverse, z, displacements.

3.19 If the variations of stresses in the membrane in Example 3.5 are accounted for, we get the following equation.

$$\frac{\partial^2 S_x}{\partial x^2} + \frac{\partial^2 S_y}{\partial y^2} = 0$$

The relationship between S_x and S_y can be shown to be the following.[9]

$$\frac{S_x}{R_x} + \frac{S_y}{R_y} = -\frac{P}{t}$$

where R_x and R_y represent the local tangential and meridional curvatures of the membrane. For some membrane configurations the R's are empirical constants; for example, a balloon may have $R_x = R_y = 2.0$. Set up an appropriate model for simulating the stresses in the latter case. What additional equation, or equations, would be needed to complete the model if we could not specify R_x and R_y beforehand?

3.20 Consider Example 3.6. Using the definitions of the unit stresses, that is, $S_{1x} = S_{sx}t$ and $S_{1y} = S_{sy}t$, write differential equation models for simulating the shear stresses, S_{sx} and S_{sy}, in terms of the displacement w (i.e., $t^2 w$).

3.21 Consider Example 3.6. Local moments in the xz and yz planes are given by the following equation:

$$M_x = \left(\frac{D}{t}\right)\left(\frac{\partial^2 w}{\partial x^2} + \mu \frac{\partial^2 w}{\partial y^2}\right)$$

$$M_y = \left(\frac{D}{t}\right)\left(\frac{\partial^2 w}{\partial y^2} + \mu \frac{\partial^2 w}{\partial x^2}\right)$$

Write equations for the maximum bending stresses in the plate material. [*Hint:* bending stress at the outer fibers is given by $S_{max} = 6M/t$.

3.22 A stress function, Θ, is often defined in certain structural problems, as follows:

$$S_{xz} = -\frac{\partial \Theta}{\partial y}$$

$$S_{yz} = \frac{\partial \Theta}{\partial x}$$

Using the information in Example 3.7, show that an alternate differential model can be set up, as follows, for simulating stresses and displacements on a twisted cross section, where $\theta =$ twist angle.

$$\frac{\partial^2 \Theta}{\partial x^2} + \frac{\partial^2 \Theta}{\partial y} = -2G\theta$$

with four conditions.

PART TWO
SIMULATION METHODS

4
Model Identifications and Choice of Simulation Method

The best method to be used in simulating the behavior of an engineering system is dictated largely by certain characteristics of the equations making up the model and by the computational resources (including software, hardware, and expertise) available. This is especially true when we are working with systems represented by differential models. In some cases classical mathematics will be the most cost-effective method (e.g., where certain linear models exist), and it would be foolish to employ involved computer procedures. In rare cases, building an actual physical model, or prototype, and testing it in the laboratory, might be the most effective (e.g., where large uncertainties exist in the mathematical model or where complete capability to set up a good model is missing). Currently, however, in most engineering settings, the most cost-effective simulation procedures involve the use of appropriate numerical methods applied through the digital computer. But even here wide differences in procedure accuracies and costs exists; and tradeoffs are inevitable. Thus the difficult task repeatedly faced by the engineer analyst is to make a choice of a "best-for-the-situation" simulation method.

Evidence abounds that "overkill," that is, using a method and equipment too sophisticated for the objectives, usually at too great an expense, is a continuing problem in many engineering organizations. Usually, the choice of a "best" simulation method is made much less difficult if the analyst engineer is able to pose his differential models in a particular form, which we call the "standard form," and from this form identify certain characteristics. Clarification of what is meant by "standard form," and characteristics, are discussed in the following paragraphs. For additional information on this and allied topics in categorizing models and choosing simulation methods, the reader is urged to consult the references.[1-14]

MODELS CONTAINING ORDINARY-DERIVATIVE EQUATIONS

A Standard Form

Almost without exception in engineering-analysis situations where models are being constructed, the initial differential model set up to represent behavior of a system will *not* be in any easily recognizable form. Consequently, it is common practice to re-pose (re-formulate) the model. Unfortunately, at present, the form to which it is re-posed typically depends more on the background of the individual engineer doing the re-posing than on usefulness for making simulation decisions. The re-posing technique recommended herein forces the analyst to make some critical identifications necessary for any later use of the computer. However, a process used at the professional level, but which is beyond the scope of this book, is also generally a part of this re-posing process. This is a "normalization" and "ordering" process, which leaves the model in dimensionless form. In this book we will assume that such a normalization has already been done if no units are furnished with the model. The student who desires more background in this area is encouraged to investigate some advanced courses in his or her specific discipline. The recommended procedure for re-posing a differential model into standard form follows.

1. Determine Dependent Variable(s)

Determine the primary dependent variable in each equation of the model by observing the upper variable in each derivative present. The other variable (the lower one in each derivative) is the independent variable. In a one-equation model there will be one dependent, and one independent, variable.

2. Arrange Derivatives by Order

In each equation, find the term containing the highest-order derivative. Place this term as the leftmost term of the equation. Find the next highest in order term (same dependent variable), and place it as the next most leftmost term. Repeat for lower-order terms until terms containing only the dependent variable (no derivatives), if they exist, are located as the rightmost terms before the equal sign. If more than one dependent variable exists, locate terms of equal order next to each other. However, remember that there can be only one primary dependent variable in each equation, and its highest-order derivative must be the leftmost term.

3. Place Nondependent Variable Terms

If any terms remain from the foregoing, they will not contain any dependent variables. All of these terms should be placed on the right-hand side of the equal sign in each equation. If there are no such terms, place a zero on the right-hand side of the equal sign.

4. Simplify Equation(s)

Use algebra rules to simplify the equation(s). For example divide through by the coefficient of the leftmost term in each equation. Occasionally, the term(s) on the right-hand side of the equal sign will have to be shifted back to the left-hand side of the equal sign because the divisor contained the dependent variable. If this happens, replace the right-hand side by a zero. If the leftmost term is raised to some power (do not confuse this with order), shift everything else temporarily to the right so that a root can be taken to bring the exponent on the leftmost term to one. Thus, finally, for each equation in the model, the leftmost term will be the highest-order derivative with no coefficients, and no exponents except one.

5. Assemble Conditions

Assemble conditions and constraints for each equation making up the model. Make sure that as many conditions exist on the dependent variable for each equation as the order of the derivative in the leftmost term. Most engineering models will have redundant conditions. Choose the best by judgment; usually conditions giving values of the dependent variable itself, rather than one of its derivatives, are preferred for systems that do not have moving objects, but others may allow easier choices of simulation methods later.

With the model in standard form we observe that two important identifications have already been made. We know the dependent variable(s) and the order of each equation making up the model. Although not strictly true, the order of the model is sometimes said to be the same as the highest-order derivative occurring in any equation making up the model. An example of re-posing a single-equation differential model into standard form is given as Example 4.1.

EXAMPLE 4.1

Following the procedures of Chapter 2, a model has been set up for a moving-object system, as given below. The dy/dx represents a velocity, and in this particular system can take on negative values. Using the procedures described above, re-pose the model into "standard form."

$$(A + B \sin qx) + W - C\left(\frac{dy}{dx}\right)^{1.9} - (ky^{1.1} + W) - \frac{W}{g}\frac{d^2y}{dx^2} = 0$$

with
$$y = 10 \quad \text{at } x = 0$$

$$\frac{dy}{dx} = 0 \quad \text{at } x = 0$$

$$y = 82.1 \quad \text{at } x = 13.1$$

▶ **1.** *Determination of Dependent Variable*

The upper variable in the derivatives is the dependent variable, y. The independent variable is x.

2. *Arrange Derivatives by Order*

The highest order of any of the derivative terms is 2. This term becomes our leftmost term as in the following equation. Other terms are placed in decreasing order of their derivative.

$$-\frac{W}{g}\frac{d^2y}{dx^2} - C\left(\frac{dy}{dx}\right)^{1.9} - ky^{1.1} = \underline{\quad}$$

3. *Place Nondependent Variable Terms*

We observe that two of the W's cancel. The other terms, none of which contain the dependent variable, are placed on the right-hand side of the equal sign.

$$-\frac{W}{g}\frac{d^2y}{dx^2} - C\left(\frac{dy}{dx}\right)^{1.9} - ky^{1.1} = -(A + B \sin qx)$$

4. *Simplify Equation(s)*

To get the leftmost term to the first power, and with a coefficient of unity, we divide through by $-(W/g)$. However, we also recognize from the setup context that we need to modify the second term to preserve the sign in all situations. The results are

$$\frac{d^2y}{dx^2} + \frac{gC}{W}\left|\frac{dy}{dx}\right|^{0.9}\frac{dy}{dx} + \frac{gk}{W}|y|^{0.1}y = \frac{g}{W}(A + B \sin qx)$$

5. *Add Conditions*

We check the conditions given with the setup and observe that there are more than enough (we need 2 to match the highest-order derivative). We choose the two conditions given at $x=0$ even though we might prefer two values of the dependent variable. The reason will become evident later.

with $\qquad\qquad y(0) = 10$

$$y'(0) = 0 \qquad \text{or} \qquad \frac{dy}{dx}\bigg|_{(0)} = 0$$

Features to Identity

Usually, the following specific characteristics of the standard-form differential model must be determined before proceeding with a choice of a simulation method. It should be recognized that identification of the dependent variable(s) and the order of the equation(s) making up the model have already been accomplished in the re-posing process.

1. Dependent Variable. This has already been discussed as item number (1) under the re-posing procedure.

2. Order of the Model. This has also been discussed in the re-posing procedure. Generally, what is meant by the "order of the model" is the order of the

highest-order derivative belonging to any equation in the model. However, this is not strictly true because simultaneous equations making up a model can often be combined algebraically to give a smaller set with higher-order equations than the original. But since we are interested primarily in choosing simulation procedures from this information, the former designation is adequate.

3. Linearity. A differential-model equation is linear if none of the dependent-variable terms exhibit products of the dependent variable and/or any of its derivatives, or powers other than unity on the dependent variable and/or any of its derivatives. Furthermore, the equation is linear if the dependent variable, and/or any of its derivatives, does not occur any place as an argument for a transcendental function. Thus, generally speaking, a model equation is *nonlinear* if any equation making up the model has any dependent-variable (or any of its derivative) terms raised to any power other than one; if any dependent variable, or any of its derivatives, are used as arguments in any transcendents, or if any product-type terms containing the dependent variable, or any of its derivatives, occur. A single nonlinear equation in a model makes the complete model nonlinear. The foregoing also holds for all condition equations that are part of the model. More rigorous definitions of linearity exist; the interested reader is urged to consult the mathematics literature.

4. Homogeneity. Generally, if a zero exists on the right-hand side of the equal sign in a standard-form model equation, the equation is said to be "homogeneous." If something other than zero exists on the right-hand side, this term is called a "driving term," or "forcing function," and the equation is said to be "nonhomogeneous." A model is considered to be homogeneous if all equations making up the model are homogeneous. In rare nonlinear equation cases, the final simplification for the standard form brings what would otherwise be a driving term as part of another term, to some power other than one, over to the left-hand side of the equation (see item 4 under the re-posing procedure). This is sometimes referred to as pseudohomogeneity and should be viewed more as a highly nonlinear nonhomogeneous situation for any classical manipulations.

5. Condition Type. If all conditions for a model equation are specified at a constant value of the independent variable, the conditions are said to be "initial conditions," or "initial values." If the specifications are at two, and no more, values of the independent variable, the conditions are said to be "boundary conditions," or "boundary values." In the latter case, the span between the two values of the independent variable is called the intrinsic simulation domain. If more than two values of the independent variable show up, a mixed condition model exists. In this book we avoid the complications of the latter, but this does not imply that they never occur.

6. Coefficient Variability. If any of the coefficients on the terms on the left-hand side of a model equation contain the independent variable, then we have a "variable coefficient" equation. Notice that this specifically excludes any coeffi-

cients that may be variable from having a dependent variable because this simply causes nonlinearity.

7. Driving Term Type. If the nonhomogeneous term on the right-hand side of the equal sign in the standard-form equation can be evaluated as a continuous, classical mathematical function, then it is said to be an "analytic driving term." This includes, of course, the trivial cases of a constant and the homogeneous zero. On the other hand, simply a functional notation, showing dependency on the independent variable, may be given along with tabular information from the field or laboratory. This is said to be a "tabular function driving term."

8. Set or Single. If in the process of re-posing and identifying characteristics of the model, unaccounted-for symbols show up, it is likely that they are additional dependent variables. This means that additional equations may be needed to complete the model setup. Thus a set of simultaneous equations is a strong possibility; and the model is said to be an "equation-set model," as compared with a "single-equation model." In this book we concentrate on single-equation models to avoid diversions from the main principles, but again this does not imply that equation-set models are rare. In fact, most professional-setting modeling situations involve equation-set models.

The process of identifying the foregoing characteristics is illustrated for several cases in Example 4.2. Typically, these identifications become part of the jargon used to choose and discuss various simulation schemes that will be described in the next few chapters. However, a few comments in regard to choices can be made at this point. If a model is nonlinear, has complicated variable coefficients, has tabular driving terms, or consists of a relatively large set, a classical mathematics approach to the simulation may not be very productive. If nonlinear, the model may not be as amenable to matrix-method, or finite-element, numerical procedures as first appears.

EXAMPLE 4.2

The following single-equation differential models have been setup and re-posed into standard form. Perform the identification procedure as described in this chapter, using the eight items listed.

Part a

$$\frac{d^3q}{dz^3} + 2.3z^2 \frac{dq}{dz} - \sin\frac{q}{2} = 3e^{0.1z}$$

with
$$q(0) = 0$$
$$q'(0) = 2.4$$
$$q''(0) = 0.59$$

Part b

$$\frac{d^4y}{dx^4} + 5.9y = G(x)$$

x	$G(x)$
-1.6	23.1
-0.1	16.4
2.7	6.8
...	...
...	...
89.7	-3.74

with

$$y(0) = 1.2$$
$$y'(10) = 0$$
$$y''(0) = -0.2$$
$$y'''(10) = 0.9$$

Part c

$$\frac{d^2u}{dy^2} + 3u\frac{du}{dy} + 5.2u = 8.2\cos 10.2 + 0.53y$$

with

$$u(0) = 0$$
$$u(20) = 0.1$$

▶ Identifications are made as follows:

Part a	Part b	Part c
(1) Dependent variable: q	(1) Dependent variable: y	(1) Dependent variable: u
(2) Order: 3	(2) Order: 4	(2) Order: 2
(3) Linearity: *NL*	(3) Linearity: *L*	(3) Linearity: *NL*
(4) Homogeneity: *NH*	(4) Homogeneity: *NH*	(4) Homogeneity: *NH*
(5) Conditions: *IC*	(5) Conditions: *BC*	(5) Conditions: *BC*
(6) Coefficient: *VC*	(6) Coefficient: *CC*	(6) Coefficient: *CC*
(7) D. Term: Analytic	(7) D. Term: Data set	(7) D. Term: Analytic
(8) Model: single equation	(8) Model: Single equation	(8) Model: Single equation

The following typical abbreviations have been used in the above: *NL* for nonlinear, *NH* for nonhomogeneous, *IC* for initial condition, *BC* for boundary condition, *VC* for variable coefficient, and *CC* for constant coefficient.

MODELS CONTAINING PARTIAL-DERIVATIVE EQUATIONS

A Standard Form

As with the models containing ordinary derivatives, the partial-derivative models that are initially set up by an engineer analyst are seldom in any easily recognized form. Moreover, the task of putting them into some useful standard form is complicated by multiple independent variables. Fortunately, most of these models have as independent variables time and/or geometric space coordinates; in most cases a very useful arrangement for simulations separates the space coordinate terms to the left-hand side of the equation and the time terms to the right. Traditionally, the space-variable terms are arranged from left to right consistent with the classical pattern of stating the coordinates, for example, x, y, and z for the rectangular Cartesian system. From here, basically the same procedure as described for the ordinary-derivative models can be used. Consequently, it is quite common to see the highest-order partial-derivative term in x in the leftmost position of an equation. It is also useful to arrange, on the right-hand side, the time-variable terms in decreasing order of the time derivatives. Obviously in this process, the dependent variable(s) will be identified as described before.

Features to Identify

Referring to the list of characteristics given before for the ordinary-derivative models, we see that some items have to be interpreted a bit differently. For example, the order of the model may be stated for each of the independent variables. Linearity, an extremely important characteristic for any simulation work, will be determined by the same rules. Homogeneity is handled in the same way, but the type of coefficient may be difficult to determine because of the number of different symbols representing the independent variables. As before, the driving term can be either an analytic term or a tabular function. Also, any equation may be only one from a set making up the model. A re-posing and identification of a typical one-equation model containing partial derivatives is illustrated in Example 4.3.

EXAMPLE 4.3

In an engineering analysis of a flow system, the following equation resulted from the model setup process. Re-pose the model into standard form and identify characteristics consistent with the list of items discussed above. The symbols a, b, and c represent constants in the model, and $S(x)$ is a parameter.

$$9.8 \times 10^6 (1 + cy) \frac{\partial y}{\partial t} - 1.52(1 + by^2)(1 + ax) \frac{\partial^2 y}{\partial x^2}$$

$$= 1.52a(1 + by^2) \frac{\partial y}{\partial x} - 3.04(1 + ax)by \left(\frac{\partial y}{\partial x}\right)^2 + S(x)$$

with

$$y = 300 \quad \text{at } t = 0, \text{ all } x\text{'s}$$
$$y = 500 \quad \text{at } t > 0, x = 0$$
$$y = 26.1 \quad \text{at } t > 0, x = 28$$

▶ Re-posing the model into standard form gives

$$\frac{\partial^2 y}{\partial x^2} + a(1+ax)^{-1}\frac{\partial y}{\partial x} + 2by(1+by^2)^{-1}\left(\frac{\partial y}{\partial x}\right)^2 + \frac{0.65789S(x)}{1+by^2+ax+abxy^2}$$

$$= 6.4474 \times 10^6(1+cy)(1+by^2)^{-1}(1+ax)^{-1}\frac{\partial y}{\partial t}$$

with
$$y(0, x) = 300$$
$$y(t, 0) = 500$$
$$y(t, 28) = 26.1$$

Identifications follow.

1. Dependent variable: y.
2. Order: 2.
3. Linearity: nonlinear.
4. Homogeneity: homogeneous.
5. Conditions: initial and boundary conditions.
6. Coefficients: variable.
7. Driving term: analytic.
8. Model type: single equation.

Classifications of Second-Order Models

Another identification can be made in regard to second-order-in-space models found in many setups for simulating physical systems. Strictly speaking, this classification is for linear equations only, but many useful features associated with it can be extended to nonlinear models as well. We can classify second-order models as "parabolic," "elliptic," or "hyperbolic," where the general form for each of the foregoing is given by Equations 4.1, 4.2, and 4.3, respectively. The f on the right-hand side in each case represents a general function that may contain x, y, u, partial first-order derivatives of u, and constants. Sometimes the cross derivatives may also show up. The a in the following represents a constant.

Parabolic:
$$\frac{\partial^2 u}{\partial x^2} = f \qquad (4.1)$$

Elliptic:
$$\frac{\partial^2 u}{\partial x^2} + a\frac{\partial^2 u}{\partial y^2} = f \qquad (4.2)$$

Hyperbolic:
$$\frac{\partial^2 u}{\partial x^2} - a\frac{\partial^2 u}{\partial y^2} = f \qquad (4.3)$$

Some specific classically known equations are also given for reference purposes. Four of the best-known ones are shown as Equations 4.4 through 4.7. It should be noted that Equation 4.4 is parabolic; Equations 4.5 and 4.6 are elliptic; and

Equation 4.7 is hyperbolic. The symbols a, b, and c represent constants.

Diffusion (or heat) equation:
$$\frac{\partial^2 u}{\partial x^2} = a\,\frac{\partial u}{\partial t} \qquad (4.4)$$

Poisson equation:
$$\frac{\partial^2 u}{\partial x^2} + \frac{\partial^2 u}{\partial y^2} = b \qquad (4.5)$$

Laplace equation:
$$\frac{\partial^2 u}{\partial x^2} + \frac{\partial^2 u}{\partial y^2} = 0 \qquad (4.6)$$

Wave equation:
$$\frac{\partial^2 u}{\partial x^2} = c\,\frac{\partial^2 u}{\partial t^2} \qquad (4.7)$$

Biharmonic equation:
$$\frac{\partial^4 u}{\partial x^4} + 2\,\frac{\partial^4 u}{\partial x^2 \partial y^2} + \frac{\partial^4 u}{\partial y^4} = 0 \qquad (4.8)$$

Equation 4.8 has been included in this second-order group because it has so many characteristics in common with the Laplace equation. Typically, it evolves from structural model setups, where u is used to represent an artificial quantity known as a stress function. Although the classifications presented in the foregoing give only an introduction to a large and diverse area, the information is sufficient for most smaller-model setup problems. For readers who find themselves more involved than this, additional information can be found in the literature.[1,4,5,6,7,8,11]

PICKING THE SIMULATION METHOD

Ordinary-Derivative Models

If the model is linear, a classical solution should be investigated first. Some of the classical methods are reviewed in Appendix A1. Laplace transform methods, although not reviewed, are also recommended. If a large system is involved, a finite-element method may be the best choice (see Chapter 7). If, on the other hand, the model is nonlinear, one of the numerical methods described in the next few chapters will probably prove to be the best choice.[7-10] The Euler method, one of the Runge–Kutta methods, or one of the Adams predictor–corrector methods, can be applied efficiently to an initial value model. However, the Euler method is a first-estimate method and is generally used to give a rough simulation quickly. For more accuracy, one of the higher-order Runge–Kutta or predictor–corrector methods is preferred. If the model is a boundary value type, either a matrix method, or one of the above methods can be used. A nonlinear model will give a nonlinear matrix to cope with, however, and only one of the iterative solution schemes can be applied directly. Difficulties in getting solutions are always present, and we recommend as an alternate a shooting method. In the "shooting method," one of the initial value methods mentioned earlier is used in a root-seeking fashion. Convergence to a good simulation is usually quite fast. The matrix methods, including finite-element schemes, should not be ruled out, however, for linear models that exhibit the complications of tabular driving terms and/or involved coefficients; they can be more efficient than the shooting methods.

Partial-Derivative Models

Finite-difference substitutions will always lead to some matrixlike behaviors.[7-14] However, for both the parabolic and hyperbolic models, forward-differencing the time derivative (or equivalent) will lead to so-called explicit, or "marching," methods. Within the constraints that must be used, the marching schemes are fast and accurate, and do not require a lot of computer memory to manipulate matrices. Therefore, they should always be considered first. Unfortunately, outside the constraints, which usually limit the size of the time (or equivalent) increment, the explicit method can be unstable. To avoid the latter problem, one of the implicit methods should be used. For elliptic and biharmonic models, straight matrix methods, or finite-element methods, are the best choice. For Laplace equation models the differencing leads to dominant diagonal sparse matrices, and an iterative solution technique like the Gauss–Seidel method can be a best choice, particularly for large sets. Alternately, a finite-element method may be easier to implement (see Chapter 7). Nonlinear models will lead to nonlinear matrices, and an iterative method is about the only choice. Much of the nonlinear treatment is still exploratory at this time and is clearly beyond the scope of this book.

REFERENCES

1. M. R. Spiegel, *Applied Differential Equations*, 2nd ed., Prentice-Hall, Inc., Englewood Cliffs, N.J., 1967.

2. J. J. Tuma, *Engineering Mathematics Handbook*, McGraw-Hill Book Company, New York, 1970.

3. T. Baumeister, E. A. Avallone, and T. Baumeister, III, eds., *Marks' Standard Handbook for Mechanical Engineers*, 8th ed., McGraw-Hill Book Company, New York, 1978.

4. E. Kreyszig, *Advanced Engineering Mathematics, 2nd ed.*, John Wiley & Sons, Inc., New York, 1967.

5. R. V. Churchill, *Fourier Series and Boundary Value Problems*, McGraw-Hill Book Company, New York, 1941.

6. R. D. Kersten, *Engineering Differential Systems*, McGraw-Hill Book Company, New York, 1969.

7. R. W. Hornbeck, *Numerical Methods*, Quantum Publishers, Inc., New York, 1975.

8. J. H. Ferziger, *Numerical Methods for Engineering Application*, John Wiley & Sons, Inc., New York, 1981.

9. K. E. Atkinson, *An Introduction to Numerical Analysis*, John Wiley & Sons, Inc., New York, 1978.

10. J. M. Ortega and W. G. Poole, Jr., *An Introduction to Numerical Methods for Differential Equations*, Pitman Publishing Corporation, Belmont, Calif., 1981.

11. V. Vemuri and W. J. Karplus, *Digital Computer Treatment of Partial Differential Equations*, Prentice-Hall, Inc., Englewood Cliffs, N.J., 1981.

12. W. F. Ames, *Numerical Methods for Partial Differential Equations*, 2nd ed., Academic Press, Inc., New York, 1977.

13. T. E. Shoup, *A Practical Guide to Computer Methods for Engineers*, Prentice-Hall, Inc., Englewood Cliffs, N.J., 1979.

14. S. L. S. Jacoby and J. S. Kowalik, *Mathematical Modeling with Computers*, Prentice-Hall, Inc., Englewood Cliffs, N.J., 1980.

PROBLEMS

4.1 Re-pose the following model into standard form and identify.

$$\frac{32}{y} - 1 = \frac{4}{y}\frac{d^2y}{dx^2}; \qquad \begin{aligned}y(0)&=0\\y(4)&=66\end{aligned}$$

4.2 Re-pose the following model into standard form and identify.

$$t^{-1} - \frac{d[\ln(t)]}{dy} = t; \qquad t(0)=0$$

4.3 Re-pose the following model into standard form and identify.

$$\frac{1}{z}\frac{dx}{dt} - \sqrt{z} = \frac{x}{z}; \qquad x(0)=1$$

$$z^2 + \frac{dz}{dt} - y = x; \qquad z(0)=2$$

$$\frac{dy^2}{dt} - 2yx^2 = 2; \qquad y(0)=1$$

4.4 Re-pose the following model into standard form and identify.

$$y^{0.1} = \frac{1}{y}\frac{d^2y}{dx^2}; \qquad \begin{aligned}y(0)&=0\\y(1.5)&=5\end{aligned}$$

4.5 Re-pose the following model into standard form and identify.

$$xy - \frac{xd^2y}{dx^2} = \frac{dy}{dx}; \qquad \begin{aligned}y(0.1)&=1.0\\y'(0.9)&=0.0\end{aligned}$$

4.6 Re-pose the following model into standard form and identify.

$$23.1x^{1.09} - y^{0.08}\left(\frac{dx}{dy}\right)^{2.1} = f(y) - 1.1\frac{d^2x}{dy^2}$$

y	$f(y)$	
0.0	0.1	$x(0) = 0.0$
16.0	-11.9	$x'(0) = 0.2$
83.0	-3.1	

4.7 Re-pose the following model into standard form and identify.

$$20 - 3y\left(\frac{dx}{dy}\right)^2 = 1.5x^2; \qquad \text{with several conditions}$$

4.8 Re-pose the following model into standard form and identify.

$$\ln\left(\frac{1}{x^{1.2}+y}\right) - y^2\frac{d^3x}{dy^3} = x^2 + 0.18x + 10.5; \qquad \begin{aligned} x(0) &= 0.0 \\ x'(0) &= 0.2 \\ x(25) &= 0.0 \end{aligned}$$

4.9 Re-pose the following model into standard form and identify.

$$7\frac{dy}{dt} + 3\frac{d^2y}{dt^2} + \frac{d^3y}{dt^3} = -5y; \qquad \begin{aligned} y(0) &= 0.0 \\ y'(0) &= 1.0 \\ y''(0) &= 2.0 \end{aligned}$$

4.10 Re-pose the following model into standard form and identify.

$$70.2\sin 2x - \frac{d^4y}{dx^4} - 6.3y = 0; \qquad \begin{aligned} y(0) &= 0.0 \\ y(10) &= 2.0 \\ y'(0) &= 0.1 \\ y'(10) &= 0.1 \end{aligned}$$

4.11 Re-pose the following model into standard form and identify.

$$45.2x - 0.32\left(\frac{dx}{dy}\right)^{1.8} + \frac{d^2x}{dy^2} = 10\cos 5y; \qquad \begin{aligned} x(0) &= 0.0 \\ x'(0) &= 5.0 \end{aligned}$$

4.12 Re-pose the following model into standard form and identify.

$$15\frac{d^3x}{dy^3} + 30x\frac{dx}{dy} - 15e^{3.1y} = 0; \qquad \begin{aligned} x(0) &= 0.0 \\ x'(0) &= 2.0 \\ x''(0) &= 1.0 \end{aligned}$$

4.13 Re-pose the following model into standard form and identify.

$$\sin\theta = \cos 3t - 93\frac{d^2\theta}{dt^2}; \qquad \begin{aligned} \theta(0) &= 0.0 \\ \theta'(0) &= 2.3 \end{aligned}$$

4.14 Re-pose the following model into standard form and identify.

$$\cos 3y \, \frac{d^3x}{dy^3} - x \frac{dx}{dy} + 30 = 0; \qquad \text{three conditions at } y = 10$$

4.15 Re-pose the following model into standard form and identify.

$$\left(\frac{dy}{dx}\right)^3 - 14.8 \tan x = g(x)$$

x	0.0	10.0	80.0	
$g(x)$	2.1	67.3	1158.0	$y(0) = 6$ $y'(0) = 2$

4.16 Re-pose the following model into standard form and identify.

$$10.6 \, \frac{\partial^2 a}{\partial b^2} - \frac{\partial a}{\partial c} = 0; \qquad \begin{aligned} a(b, 0) &= 6.0 \\ a(b_0, c) &= 2 + \sin 5c \\ a(b_\infty, c) &= 6.0 \end{aligned}$$

4.17 Re-pose the following model into standard form and identify.

$$\frac{\partial^2 x}{\partial y^2} = -\frac{\partial^2 x}{\partial w^2}; \qquad \text{with sketch showing conditions}$$

4.18 Re-pose the following model into standard form and identify.

$$1.3 \, \frac{\partial^2 y}{\partial x^2} = \frac{\partial^2 y}{\partial z^2}; \qquad \text{with sketch showing conditions}$$

4.19 Re-pose the following model into standard form and identify.

$$3P - \frac{\partial^2 z}{\partial x^2} = \frac{\partial^2 z}{\partial y^2}; \qquad P = \text{parameter}$$

with sketch showing conditions

4.20 Re-pose the following model into standard form and identify.

$$6 \, \frac{\partial A}{\partial z} - 30 - \frac{\partial^2 A}{\partial x^2} + 1.2 \, \frac{\partial^2 A}{\partial y^2} = 2.4$$

with conditions shown on sketch

4.21 Re-pose the following model into standard form and identify.

$$1.9 - 6.3b \, \frac{\partial c}{\partial a} + \frac{\partial^2 c}{\partial b^2} = 4.2b; \qquad \text{with conditions described}$$

4.22 Re-pose the following model into standard form and identify.

$$4.3 \, \frac{\partial^4 B}{\partial A^2 \partial C^2} + 2.15 \, \frac{\partial^4 B}{\partial A^4} = -2.15 \, \frac{\partial^4 B}{\partial C^4}; \qquad \text{with conditions described}$$

4.23 Re-pose the following model into standard form and identify.

$$2.6 \sin 3x + \frac{\partial^2 C}{\partial x^2} + \frac{\partial^2 C}{\partial y^2} = 0; \qquad \text{with conditions described}$$

5

Numerical Simulations from Ordinary-Derivative Models[1]

FIRST-ORDER MODELS

An Introduction—The Euler Method

The Euler method formula is simply an approximate equation for the first derivative, in which a forward finite difference has been substituted for the derivative, and which has been algebraically re-formed. If the derivative $dy/dx = f(x, y)$ is approximated by $(y_{i+1} - y_i)/\Delta x = f(x_i, y_i)$, then the Euler formula, Equation 5.1, results. This equation can also be viewed as a severely truncated, forward expanded Taylor series for dy/dx around point x_i.

$$y_{i+1} = y_i + \Delta x\, f(x_i, y_i) \tag{5.1}$$

Now let us consider *any* first-order differential model equation that has had all of its terms, except the derivative, transferred to the right-hand side of the standard-form equation. Obviously, the right-hand side is just exactly the function $f(x, y)$ needed to make the Euler formula work. The condition that comes with the model equation, $y(0) = y_0$, gives us the information we need to substitute into the $f(x, y)$. The whole idea can be illustrated geometrically as shown in Figure 5.1. A calculator can be used effectively for a few steps of the method if a tabular algorithm is employed. Such a short simulation is illustrated in Example 5.1. A flow chart for implementing the method with FORTRAN on a computer is illustrated in Figure 5.2. Programs in BASIC and FORTRAN are illustrated in Example 5.2.

The accuracy of the Euler method is low, even with relatively small increments of the independent variable. Consequently, in engineering simulations, it is used mainly for "quick rough" estimates of behaviors over small spans of the independ-

---

[1] For additional information, including error analysis topics, the reader should consult the references.[1-10]

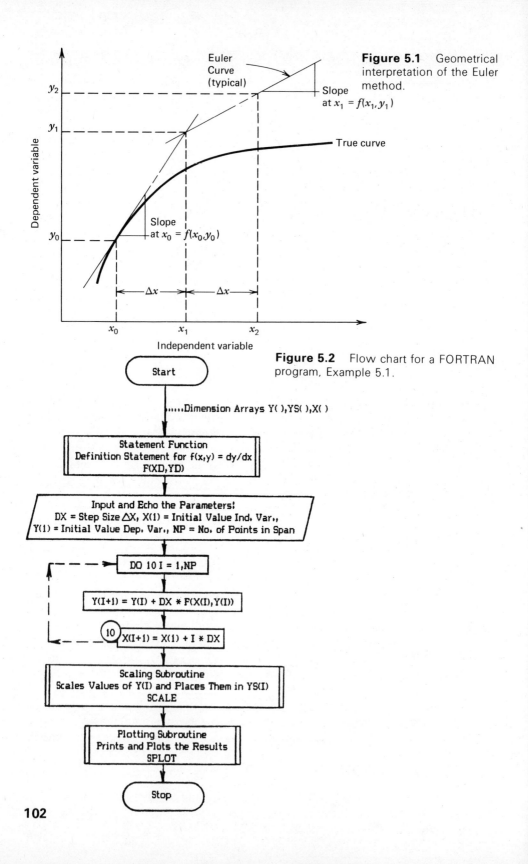

Figure 5.1 Geometrical interpretation of the Euler method.

Euler Curve (typical)

Slope at $x_1 = f(x_1, y_1)$

True curve

Slope at $x_0 = f(x_0, y_0)$

y_2

y_1

y_0

x_0 x_1 x_2

Δx Δx

Dependent variable

Independent variable

Figure 5.2 Flow chart for a FORTRAN program, Example 5.1.

Start

......Dimension Arrays Y( ),YS( ),X( )

Statement Function
Definition Statement for f(x,y) = dy/dx
F(XD,YD)

Input and Echo the Parameters:
DX = Step Size ΔX, X(1) = Initial Value Ind. Var.,
Y(1) = Initial Value Dep. Var., NP = No. of Points in Span

DO 10 I = 1,NP

Y(I+1) = Y(I) + DX * F(X(I),Y(I))

10 X(I+1) = X(1) + I * DX

Scaling Subroutine
Scales Values of Y(I) and Places Them in YS(I)
SCALE

Plotting Subroutine
Prints and Plots the Results
SPLOT

Stop

ent variable. It is evident from Figure 5.1 that the error associated with each extrapolation step accumulates; and the overall (global) error at the end of a lengthy simulation may be very significant. Each step contributes an error of the order of $(\Delta x)^2$ and leads to global error of the order of Δx. The Euler method is called a "single step," or "self-starting" method, since it is only necessary to know y_i to compute y_{i+1}.

EXAMPLE 5.1

The setup of a model for a physical system gives the following single-equation model. Demonstrate how the Euler method works by generating a simulation of y versus t over the span $t=0$ to $t=0.5$. Use a t increment of 0.1 and a tabular algorithm.

$$\frac{dy}{dt} - 3y^{1.2} = t \qquad y(0) = 1.0$$

▶ The required formulation is

$$y_{i+1} = y_i + \Delta t\, f(y_i, t_i)$$

For this specific model, the formula becomes

$$y_{i+1} = y_i + 0.1(t_i + 3y_i^{1.2})$$

Our computed values go into a table, Table 5.1. Sample calculations are given from the entry at $i=2$ in Table 5.1.

$$t_2 = t_1 + \Delta t = 0.0 + 0.1 = 0.1$$

$$y_2 = y_{1+1} = 1.3 \qquad \text{(from the preceding step)}$$

$$f(y_2, t_2) = t_2 + 3y_2^{1.2} = 0.1 + 3(1.3)^{1.2} = 4.21$$

$$y_3 = y_2 + \Delta t\, f(y_2, t_2) = 1.3 + 0.1(4.12) = 1.721$$

Table 5.1 Computational Results,
Fxample 5.1

i	t_i	y_i	$f(t_i, y_i)$	y_{i+1}
1	0.0	1.0	3.0	1.3
2	0.1	1.3	4.21	1.721
3	0.2	1.721	5.955	2.317
4	0.3	2.317	8.523	3.169
5	0.4	3.169	12.374	4.406
6	0.5	4.406	—	—

```
100     REM EXAMPLE 5.2 BASIC PROGRAM - EULER METHOD
110     CLS: CLEAR20: DEFINT I-N
120     DIM Y(201),YS(201),T(201)
130     DEF FNF(TD,YD) = -11.141/(20.* SQR(YD) - YD^1.5) + 1./(1.9 + .6 * TD)
140     READ DT,T(1),Y(1),NP : N = NP - 1
150       DATA 0.2, 0.0, 19.5, 201
160     REM COMPUTATION LOOP
170       FOR I = 1 TO N
180         Y(I+1) = Y(I) + DT * FNF(T(I),Y(I))
190         T(I+1) = T(1) + DT * I
200       NEXT I
210     REM SCALE Y(I) VALUES TO BETWEEN 0.0 AND 1.0 : STORE IN YS(I)
220     GOSUB 1000
230     REM PRINT AND TREND-PLOT SIMULATION RESULTS
240     GOSUB 2000
250     END
```

```
1000    REM SCALES Y(I) TO: 0.0 < YS(I) < 1.0
1010    REM GIVES MINIMUM Y(I) AS U1, MAXIMUM Y(I) AS U2
1020    U1 = Y(1): U2 = Y(1)
1030    REM SCREENING LOOP FOR HIGH AND LOW VALUES
1040      FOR I = 1 TO NP
1050        IF Y(I) > U2 THEN U2 = Y(I): GOTO 1070
1060        IF Y(I) < U1 THEN U1 = Y(I)
1070      NEXT I
1080    REM SCALING LOOP
1090      FOR I = 1 TO NP
1100        YS(I) = (Y(I) - U1)/(U2 - U1)
1110      NEXT I
1120    RETURN
1130    END
```

```
2000    REM PRINT HEADINGS
2010    LP = 4
2020    LPRINT TAB(20)"SIMULATION RESULTS"
2030    LPRINT TAB(7)"T";TAB(16)"Y";TAB(27)"MINIMUM=";U1;"MAXIMUM=";U2
2040    REM PRINT-PLOT LOOP
2050      FOR I = 1 TO NP
2060        K = YS(I) * 45
2070        LPRINT TAB(5)T(I);TAB(15)Y(I);TAB(27)"!";
2080        IF K = 0 THEN LPRINT "*": GOTO 2100
2090          FOR IP = 1 TO K: LPRINT "-"; : NEXT IP: LPRINT "*"
2100        LP = LP + 1: IF LP > 54 THEN LP = 1: FOR IP = 1 TO 12: LPRINT: NEXT IP
2110      NEXT I: LPRINT: LPRINT
2120    RETURN
2130    END
```

Figure 5.3   BASIC program, Example 5.2.

```
                7
C EXAMPLE 5.2 FORTRAN PROGRAM - EULER METHOD
        DIMENSION Y(201),YS(201),T(201)
        F(TD,YD) = -11.141/(20. * SQRT(YD) - YD ** 1.5) + 1./(1.9 + .6 * TD)
        READ*, DT, T(1), Y(1), NP
        PRINT*, 'DT=',DT,' T(1)=',T(1),' Y(1)=',Y(1),' NP=',NP
        N = NP - 1
C COMPUTATION LOOP
        DO 10 I=1,N
            Y(I+1) = Y(I) + DT * F(T(I),Y(I))
            T(I+1) = T(1) + I * DT
  10        CONTINUE
C SCALE Y(I) VALUES TO BETWEEN 0.0 AND 1.0; STORE IN YS(I)
        CALL SCALE(Y,YS,NP,U1,U2)
C PRINT AND PLOT SIMULATION RESULTS
        CALL SPLOT(T,Y,YS,NP,U1,U2)
        STOP
        END
```

```
        SUBROUTINE SCALE(Y,YS,N)
        DIMENSION Y(N),YS(N)
        U1 = Y(1)
        U2 = Y(1)
C SCREENING LOOP FOR HIGH AND LOW VALUES
        DO 100 I = 1,N
          IF(Y(I) .GT. U2) THEN
            U2 = Y(I)
            GO TO 100
          ENDIF
          IF(Y(I) .LT. U1) U1 = Y(I)
  100     CONTINUE
C SCALING LOOP
        DO 200 I = 1,N
            YS(I) = (Y(I) - U1)/(U2 - U1)
  200     CONTINUE
        RETURN
        END
```

```
        SUBROUTINE SPLOT(T,Y,YS,N,US,UL)
        DIMENSION T(N),Y(N),YS(N)
        CHARACTER*80 PLOT
        DATA M/80/
C 'M' IS FIELD WIDTH FOR PLOT; 'A' FORMAT SPEC MUST BE CONSISTENT
        WRITE(6,50) US,UL
  50    FORMAT(1H1,50X,'SIMULATION RESULTS'//
     110X,'T',20X,'Y',10X,'MINIMUM=',E16.7,'  MAXIMUM=',E16.7)
C PRINT-PLOT LOOP
        DO 20 J = 1,N
          DO 10 I = 1,M
  10        PLOT(I:I) = ' '
          K = YS(J) * (M - 1) + 1.5
          IF(K .LT. 0 .OR. K .GT. M) THEN
            WRITE(6,60)
  60        FORMAT(5X,'ERROR ENCOUNTERED IN A SCALED VALUE OF Y')
            GO TO 30
          ENDIF
  15      PLOT(K:K) = '*'
          WRITE(6,100) T(J),Y(J),PLOT
  100     FORMAT(2E20.7,5X,A80)
  20      CONTINUE
  30    RETURN
        END
```

```
0.2, 0.0, 19.5, 201
```

Figure 5.4 FORTRAN program, Example 5.2.

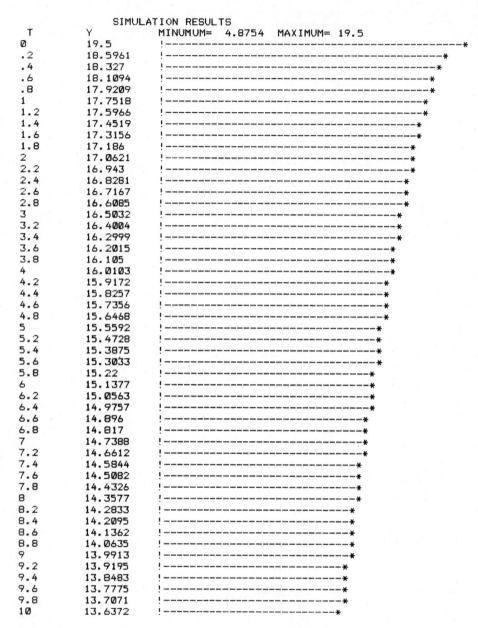

```
                     SIMULATION RESULTS
     T          Y          MINUMUM=  4.8754   MAXIMUM= 19.5
     0         19.5        !---------------------------------------------------*
     .2        18.5961     !--------------------------------------------------*
     .4        18.327      !-------------------------------------------------*
     .6        18.1094     !------------------------------------------------*
     .8        17.9209     !------------------------------------------------*
     1         17.7518     !-----------------------------------------------*
     1.2       17.5966     !-----------------------------------------------*
     1.4       17.4519     !----------------------------------------------*
     1.6       17.3156     !---------------------------------------------*
     1.8       17.186      !--------------------------------------------*
     2         17.0621     !--------------------------------------------*
     2.2       16.943      !-------------------------------------------*
     2.4       16.8281     !------------------------------------------*
     2.6       16.7167     !------------------------------------------*
     2.8       16.6085     !-----------------------------------------*
     3         16.5032     !----------------------------------------*
     3.2       16.4004     !----------------------------------------*
     3.4       16.2999     !---------------------------------------*
     3.6       16.2015     !--------------------------------------*
     3.8       16.105      !-------------------------------------*
     4         16.0103     !-------------------------------------*
     4.2       15.9172     !------------------------------------*
     4.4       15.8257     !-----------------------------------*
     4.6       15.7356     !-----------------------------------*
     4.8       15.6468     !----------------------------------*
     5         15.5592     !---------------------------------*
     5.2       15.4728     !--------------------------------*
     5.4       15.3875     !--------------------------------*
     5.6       15.3033     !-------------------------------*
     5.8       15.22       !------------------------------*
     6         15.1377     !------------------------------*
     6.2       15.0563     !-----------------------------*
     6.4       14.9757     !-----------------------------*
     6.6       14.896      !----------------------------*
     6.8       14.817      !---------------------------*
     7         14.7388     !---------------------------*
     7.2       14.6612     !--------------------------*
     7.4       14.5844     !-------------------------*
     7.6       14.5082     !-------------------------*
     7.8       14.4326     !------------------------*
     8         14.3577     !------------------------*
     8.2       14.2833     !-----------------------*
     8.4       14.2095     !-----------------------*
     8.6       14.1362     !----------------------*
     8.8       14.0635     !----------------------*
     9         13.9913     !---------------------*
     9.2       13.9195     !---------------------*
     9.4       13.8483     !--------------------*
     9.6       13.7775     !-------------------*
     9.8       13.7071     !-------------------*
     10        13.6372     !------------------*
```

Figure 5.5 Typical simulation results (BASIC Program), Example 5.2.

EXAMPLE 5.2

The water level in a storage tank is to be simulated for an emergency draindown process. The single-equation model has been set up and re-posed into the standard form given below. Simulation of y versus t is to be accomplished with the Euler method over the span $t=0$ to $t=40$, using an increment of $\Delta t=0.2$. Develop two computer programs, one in BASIC and one in FORTRAN, to do the simulation.

$$\frac{dy}{dt}+(11.141)(20y^{0.5}-y^{1.5})^{-1}=(1.9+0.6t)^{-1}; \qquad y(0)=19.5$$

► The BASIC program is given as Figure 5.3; the FORTRAN program is given as Figure 5.4. Typical simulation results are given as Figure 5.5.

The Runge-Kutta Methods

The Runge–Kutta methods are used where accuracy better than can be obtained from the Euler method is desired. All Runge–Kutta methods are one-step, self-starting methods and have recursive formulations. They quickly become too complicated for any significant manual use. As a consequence we will discuss tabular simulation examples with only one or two independent variable steps. The simplest Runge–Kutta method is called the "second-order Runge–Kutta method," and for the first-order equation $dy/dx=f(x, y)$, can be given a geometric interpretation as shown in Figure 5.6. More rigorous justification can be developed based on Taylor series expansions but is not given here. The required formula for applying the second-order Runge–Kutta method to the first-order model equation $dy/dx=f(x, y)$ is given by Equation 5.2. The a_1 and a_2 equations are called the recursion equations and always have to be computed before the main equation can be evaluated.

$$y_{i+1}=y_i+\frac{(a_1+a_2)}{2} \tag{5.2}$$

where
$$a_1=\Delta x\, f(x_i, y_i)$$

$$a_2=\Delta x\, f(x_i+a_1, x_i+\Delta x)$$

Higher-order Runge–Kutta methods can be developed to give even better simulation accuracies. Our purpose here is not to demonstrate this development (the interested reader is directed to the references[1–10] for sources showing the developments) but to summarize application details of the most widely used methods. Although third-, and higher-, order Runge–Kutta methods are available, the one most used in engineering simulation applications to date is the fourth-order method. The fourth-order Runge–Kutta method seems to be the

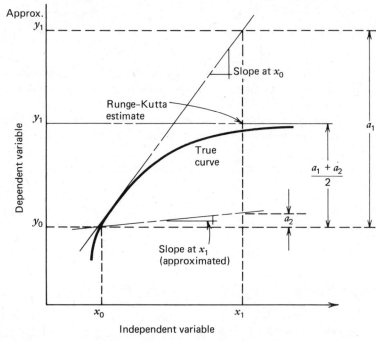

Figure 5.6 Geometrical interpretation of the second-order Runge–Kutta method.

best compromise between complexity during application and maximization of the types of accuracy desired in most engineering problems. The required formulation for applying the method to a simulation is given by Equation 5.3.

$$y_{i+1} = y_i + \frac{a_1 + 2a_2 + 2a_3 + a_4}{6} \tag{5.3}$$

where

$$a_1 = \Delta x \, f(x_i, y_i)$$

$$a_2 = \Delta x \, f(x_i + \Delta x/2, \, y_i + a_1/2)$$

$$a_3 = \Delta x \, f(x_i + \Delta x/2, \, y_i + a_2/2)$$

$$a_4 = \Delta x \, f(x_i + \Delta x, \, y_i + a_3)$$

An alternate, but equivalent, formulation is sometimes more efficient for setting up programs and is given by Equation 5.4.[1]

$$y_{i+1} = y_i + \Delta x \frac{f_1 + 2f_2 + 2f_3 + f_4}{6} \tag{5.4}$$

where

$$f_1 = f(x_i, y_i)$$

$$f_2 = f(x_i + \Delta x/2, \, y_i^*)$$

$$y_i^* = y_i + 0.5 \, \Delta x \, f(x_i, y_i)$$

$$f_3 = f(x_i + \Delta x/2, y_i^{**})$$

$$y_i^{**} = y_i + 0.5 \Delta x\, f(x_i + \Delta x/2, y_i^*)$$

$$f_4 = f(x_i + \Delta x, y_{i+1}^*)$$

$$y_{i+1}^* = y_i + \Delta x\, f(x_i + \Delta x/2, y_i^{**})$$

An example of how the fourth-order method works using Equation 5.3 and a tabular algorithm with a calculator, for several steps in simulating y versus t, is shown in Example 5.3. A flow chart for computer execution in FORTRAN is illustrated in Figure 5.7. Computer programs replicating Example 5.2, except replacing the Euler method with the fourth-order Runge–Kutta method, are given as Figures 5.8 and 5.9. It should be noted that the per-step error for this method of the order of $(\Delta x)^5$ leads to global error of the order of $(\Delta x)^4$.

EXAMPLE 5.3

The following single-equation model from a system setup is to be used for simulating y versus t over the span $t=0$ to $t=0.5$. Demonstrate how the fourth-order Runge–Kutta method works by using a calculator and tabular algorithm with $\Delta t = 0.1$.

$$\frac{dy}{dt} - 3y^{1.2} = t; \qquad y(0) = 1.0$$

▶ We choose Equation 5.3 as the easiest formulation to use with the calculator, noting that our independent variable symbol has to be changed to t, and that the required function is $f(t, y) = t + 3y^{1.2}$. In this case keeping about four digits to the right of the decimal in all intermediate computations (we could use more, but it becomes a chore to write more down), and saving all the a_i recursive values, we enter the calculation results into Table 5.2.
Sample calculations are given for the entry at $i=2$ in Table 5.2.

$$a_1 = 0.1[0.1 + 3(1.3683)^{1.2}] = 0.4471$$

Table 5.2 Computational Results, Example 5.3

i	t_i	y_i	a_1	a_2	a_3	a_4	y_{i+1}
1	0.0	1.0	0.3	0.3598	0.3707	0.4481	1.3683
2	0.1	1.3683	0.4471	0.5391	0.5573	0.6786	1.9214
3	0.2	1.9214	0.6768	0.8230	0.8541	1.0512	2.7684
4	0.3	2.7684	1.0481	1.2886	1.3437	1.6768	4.1000
5	0.4	4.1000	1.6710	2.0826	2.1850	2.7733	6.2633
6	0.5	6.2633					

$$a_2 = 0.1[(0.1 + 0.1/2) + 3(1.3683 + 0.4471/2)^{1.2}] = 0.5391$$

$$a_3 = 0.1[(0.1 + 0.1/2) + 3(1.3683 + 0.5391/2)^{1.2}] = 0.5573$$

$$a_4 = 0.1[(0.1 + 0.1) + 3(1.3683 + 0.5573)^{1.2}] = 0.6786$$

$$y_3 = 1.3683 + (0.4471 + 2(0.5391 + 0.5573) + 0.6786)/6 = 1.9214$$

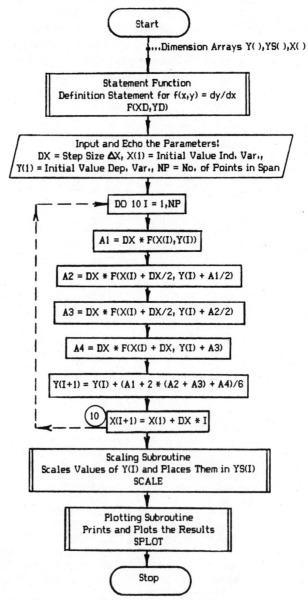

Figure 5.7 Flow Chart for FORTRAN program based on fourth-order Runge-Kutta method, Example 5.3.

Main Program

```
100        REM EXAMPLE 5.2 BASIC PROGRAM
110        REM 4TH-ORDER RUNGE-KUTTA METHOD
120        DIM Y(201),YS(201),T(201)
130        DEF FNF(TD,YD) = -11.141/(20*SQR(YD) - YD↑1.5) +1/(1.9 + .6*TD)
140        READ DT,T(1),Y(1),NP: N = NP-1
150          DATA 0.2, 0.0, 19.5, 201
160        REM COMPUTATION LOOP
170          FOR I = 1 TO N
180            A1 = DT * FNF(T(I),Y(I))
190            A2 = DT * FNF(T(I)+DT/2, Y(I)+A1/2)
200            A3 = DT * FNF(T(I)+DT/2, Y(I)+A2/2)
210            A4 = DT * FNF(T(I)+DT, Y(I)+A3)
220            Y(I+1) = Y(I) + (A1 + 2*(A2 + A3) + A4)/6
230            T(I+1) = T(1) + I*DT
240          NEXT I
250        REM CALL SCALING AND PLOTTING SUBROUTINES TO DISPLAY RESULTS
260        REM SEE EULER-METHOD PROGRAM FOR SUBROUTINE SOURCE LISTINGS
270        GOSUB 1000
280        GOSUB 2000
290        END
```

Figure 5.8 BASIC program based on fourth-order Runge—Kutta method, Example 5.2 problem.

Main Program

```
        7
C EXAMPLE 5.2 FORTRAN PROGRAM
C 4TH-ORDER RUNGE-KUTTA METHOD
        DIMENSION Y(201),YS(201),T(201)
        F(XD,YD) = -11.141/(20*SQRT(YD) - YD**1.5) + 1/(1.9 + .6*XD)
        READ*, DT,T(1),Y(1),NP
        PRINT*,'DT=',DT,' T(1)=',T(1),' Y(1)=',Y(1),' NP=',NP
        N = NP-1
C COMPUTATION LOOP
        DO 10 I=1,N
          A1 = DT * F(T(I), Y(I))
          A2 = DT * F(T(I)+DT/2, Y(I)+A1/2)
          A3 = DT * F(T(I)+DT/2, Y(I)+A2/2)
          A4 = DT * F(T(I)+DT, Y(I)+A3)
          Y(I+1) = Y(I) + (A1 + 2*(A2 + A3) + A4)/6
          T(I+1) = T(1) + I*DT
10        CONTINUE
C CALL SCALING AND PLOTTING SUBROUTINES TO DISPLAY SIMULATION RESULTS
C SEE EULER-METHOD PROGRAM FOR SUBROUTINE SOURCE LISTINGS
        CALL SCALE(Y,YS,NP,U1,U2)
        CALL SPLOT(T,Y,YS,NP,U1,U2)
        STOP
        END
                    Place Subroutine SCALE Here
                    Place Subroutine SPLOT Here
                            Data
    0.2, 0.0, 19.5, 201
```

Figure 5.9 FORTRAN program based on fourth-order Runge—Kutta method, Example 5.2 problem.

The Adams Predictor–Corrector Methods

The Adams predictor–corrector methods are multiple-step methods requiring some other self-starting method to get a simulation started. Typically, in engineering a Runge–Kutta method of equivalent or better accuracy is used for starting. Once started the predictor–corrector (PC) method uses one equation at the beginning of each step to predict a value of the dependent variable. This equation is called the "open formula," or "predictor equation." Then another equation is used to successively correct the dependent-variable value to better accuracy. The latter equation is known as the "closed formula," or "corrector equation." The number of corrections applied at each step can vary, but large numbers are seldom used because the truncation accuracy of the formulas is reached rather quickly in most cases. One to three corrections per step are common. Of course, the multiple corrections allow convergence to a value to be monitored, and this can be used as an "error estimator," or "uncertainty indicator." Thus an executive scheme of some kind can be devised to use this indicator to change the step size to minimize computation time while still maintaining a prespecified uncertainty level in the simulation results. The foregoing feature, along with an efficient computation pattern, has made the predictor–corrector methods increasingly popular for engineering simulations. The computational efficiency of the methods is associated with the fact that only one evaluation of the derivative, $f(x, y)$, need be made for each correction because the required preceding $f(x, y)$'s can be saved from previous steps.

The required formulations can be developed by manipulating forward and backward Taylor series expansions. The interested reader is directed to the references[1-10] for several sources showing this development. The results are always given as equation pairs of stated truncation accuracy relative to the Taylor series. For example, assuming the standard one-equation model $dy/dx = f(x, y)$, the third-order accurate Adams predictor–corrector formulation is given as Equation 5.5:

$$y_{i+1}^p = y_i + \Delta x \frac{23f_i - 16f_{i-1} + 5f_{i-2}}{12} \tag{5.5a}$$

$$y_{i+1}^c = y_i + \Delta x \frac{5f_{i+1}^p + 8f_i - f_{i-1}}{12} \tag{5.5b}$$

We observe that the superscript p represents a prediction; and the superscript c represents a correction. The shortened notation for $f(x_i, y_i)$, that is, f_i, is used.

The popular fourth-order Adams predictor–corrector formulation is given as Equation 5.6. It is common to find the fourth-order Runge–Kutta method used to start both the third- and fourth-order PC methods.

$$y_{i+1}^p = y_i + \Delta x \frac{55f_i - 59f_{i-1} + 37f_{i-2} - 9f_{i-3}}{24} \tag{5.6a}$$

$$y_{i+1}^c = y_i + \Delta x \frac{9f_{i+1}^p + 19f_i - 5f_{i-1} + f_{i-2}}{24} \tag{5.6b}$$

An estimate of the truncation error for the fourth-order PC method can be computed from Equation 5.7.

$$\text{error}_i = \left(\frac{19}{270}\right)(y_i^c - y_i^p) \tag{5.7}$$

Per-step errors of the order of $(\Delta x)^5$ can be anticipated with one application of Equation 5.6.

The PC method is not suited for simulations using a calculator. However, for tracing the computation process for no more than a few steps, a tabular algorithm can be used along with calculator values. An example of such a demonstration of how the method behaves is given in Example 5.4. A flow chart for implementing the third-order PC method using Fortran is given as Figure 5.10. It should be noted that no attempt is made to monitor accuracy or to change step size. Computer programs replicating Example 5.2, except replacing the Euler method with a third-order Adams predictor–corrector method, are given as Figures 5.11 and 5.12.

EXAMPLE 5.4

The following single-equation model from a system setup is to be used to simulate y versus t over the span $t=0$ to $t=0.5$. Demonstrate how the third-order Adams Predictor–Corrector method (with fourth-order Runge–Kutta starter) works by using a tabular algorithm and a calculator. Use $\Delta t = 0.1$.

$$\frac{dy}{dt} - 3y^{1.2} = t; \qquad y(0) = 1.0$$

▶ Equation 5.5 becomes our required formulation pair. We note that our independent variable must be changed to t, and that our function is: $f_i = f(t_i, y_i) = t_i + 3y^{1.2}$. We cheat a little by getting our Runge–Kutta starting values from Example 5.3. The results of our computations are placed into Table 5.3.

Table 5.3 Computational Results, Example 5.4

i	t_i	y_i	f_i	y_{i+1}^p	y_{i+1}^c	y_{i+1}^c	y_{i+1}^c
1	0.0	1.0	3.0^a	—	—	—	—
2	0.1	1.3683	4.4706^a	—	—	—	—
3	0.2	1.9214	6.7684^a	2.7476	2.7683	2.7721	2.7728
4	0.3	2.7728	10.5005	4.0692	4.1066	4.1140	4.1155
5	0.4	4.1155	16.7843	6.2144	6.2872^b	6.3029	6.3063
6	0.5	6.3063^c	—	—	—	—	—

[a]Starting values obtained from the fourth-order Runge–Kutta method.
[b]Note that this value is close to fourth-order Runge–Kutta simulation results.
[c]$y_6 = 6.2633$ by the fourth-order Runge–Kutta method.

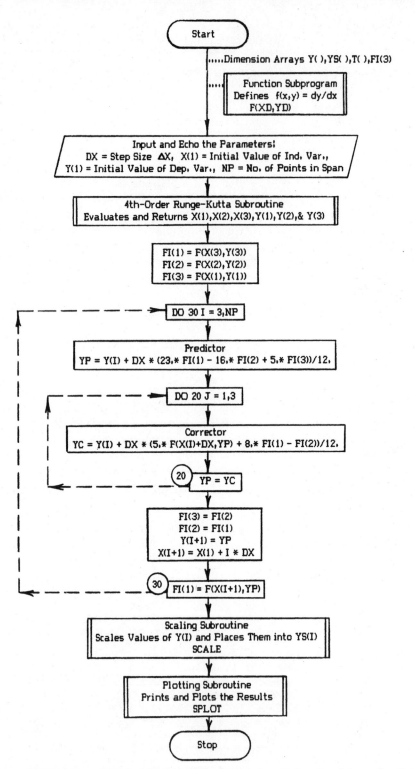

Figure 5.10 Flow chart for FORTRAN version of a third-order Adams Predictor–Corrector method, Example 5.4.

Sample calculations are given for the entry at $i=4$ in Table 5.3.

$$t_4 = 0.2 + 0.1 = 0.3; \ t_5 = 0.4$$
$$f_4 = 0.3 + 3(2.7728)^{1.2} = 10.5005$$
$$y_5^p = 2.7728 + 0.1(23(10.5005) - 16(6.7684) + 5(4.4706))/12 = 4.0692$$
$$f_5^p = 0.4 + 3(4.0692)^{1.2} = 16.5634$$

correct 1: $y_5^c = 2.7728 + 0.1(5(16.5634) + 8(10.5005) - 6.7684)/12 = 4.1066$
$$f_5^p = 0.4 + 3(4.1066)^{1.2} = 16.7418$$

correct 2: $y_5^c = 2.7728 + 0.1(5(16.5634) + 8(10.5005) - 6.7684)/12 = 4.1140$
$$f_5^p = 0.4 + 3(4.1140)^{1.2} = 16.7772$$

correct 3: $y_5^c = 2.7728 + 0.1(5(16.7772) + 8(10.5005) - 6.7684)/12 = 4.1155$

```
                        Main Program
100     REM EXAMPLE 5.2 BASIC PROGRAM
110     REM 3RD-ORDER ADAMS PREDICTOR-CORRECTOR METHOD
120     DIM Y(201),YS(201),T(201),FI(3)
130     DEF FNF(TD,YD) = -11.141/(20*SQR(YD) - YD↑1.5) + 1/(1.9 + .6*TD)
140     READ DT,T(1),Y(1),NP: N = NP - 1
150         DATA 0.2, 0.0, 19.5, 201
160     REM CALL 4TH-ORDER RUNGE-KUTTA SUBROUTINE TO GET METHOD STARTED
170     GOSUB 500
180     FI(1) = FNF(T(3),Y(3))
190     FI(2) = FNF(T(2),Y(2))
200     FI(1) = FNF(T(1),Y(1))
210     REM COMPUTATION LOOPS
220         FOR I = 3 TO N
230     REM PREDICTOR
240             YP = Y(I) + DT*(23*FI(1) - 16*FI(2) + 5*FI(3))/12
250     REM CORRECTOR LOOP
260                 FOR J = 1 TO 3
270                     YC = Y(I) + DT*(5*FNF(T(I)+DT,YP) + 8*FI(1) - FI(2))/12
280                     YP=YC
290                 NEXT J
300             FI(3) = FI(2)
310             FI(2) = FI(1)
320             Y(I+1) = YP
330             T(I+1) = T(1) + I*DT
340             FI(1) = FNF(T(I+1),YP)
350         NEXT I
360     REM CALL SCALING AND PLOTTING SUBROUTINES TO DISPLAY RESULTS
370     REM SEE EULER-METHOD PROGRAM FOR SUBROUTINE SOURCE LISTINGS
380     GOSUB 1000
390     GOSUB 2000
400     END
                        Runge-Kutta Subroutine
500     REM 4TH-ORDER RUNGE-KUTTA SUB FOR TWO STEPS
510         FOR I = 1 TO 2
520             A1 = DT*FNF(T(I),Y(I))
530             A2 = DT*FNF(T(I)+DT/2, Y(I)+A1/2)
540             A3 = DT*FNF(T(I)+DT/2, Y(I)+A2/2)
550             A4 = DT*FNF(T(I)+DT, Y(I)+A3)
560             Y(I+1) = Y(I) + (A1 + 2*(A2 + A3) +A4)/6
570             T(I+1) = T(1) + I*DT
580         NEXT I
590     RETURN
600     END
```

Figure 5.11 BASIC program based on third-order Adams Predictor–Corrector method, Example 5.2 problem.

```
                        7
C EXAMPLE 5.2 FORTRAN PROGRAM
C 3RD-ORDER ADAMS PREDICTOR-CORRECTOR METHOD
            DIMENSION Y(201),YS(201),T(201),FI(3)
C REMEMBER TO FURNISH A FUNCTION SUBPROGRAM FOR  F(T,Y) = DY/DT
            READ*, DT,T(1),Y(1),NP
            PRINT*,DT,T(1),Y(1),NP
            N = NP - 1
C CALL 4TH-ORDER RUNGE-KUTTA SUBROUTINE TO GET METHOD STARTED
            CALL RK4(T(1),T(2),T(3),Y(1),Y(2),Y(3),DT)
            FI(1) = F(T(3),Y(3))
            FI(2) = F(T(2),Y(2))
            FI(1) = F(T(1),Y(1))
C COMPUTATION LOOPS
            DO 30 I = 3,N
C PREDICTOR
               YP = Y(I) + DT*(23*FI(1) - 16*FI(2) + 5*FI(3))/12.
C CORRECTOR LOOP
               DO 20 J = 1,3
                  YC = Y(I) + DT*(5*F(T(I)+DT,YP) + 8*FI(1) - FI(2))/12.
                  YP=YC
20                CONTINUE
               FI(3) = FI(2)
               FI(2) = FI(1)
               Y(I+1) = YP
               T(I+1) = T(1) + I*DT
               FI(1) = F(T(I+1),YP)
30             CONTINUE
C CALL SCALING AND PLOTTING SUBROUTINES TO DISPLAY SIMULATION RESULTS
C SEE EULER-METHOD PROGRAM FOR SOURCE LISTINGS OF SUBROUTINES
            CALL SCALE(Y,YS,NP,U1,U2)
            CALL SPLOT(T,T,YS,NP,U1,U2)
            STOP
            END
```

F(T,Y) Function Subprogram
```
            FUNCTION F(XD,YD)
            F = -11.141/(20.*SQRT(YD) -YD**1.5) + 1./(1.9 + .6*XD)
            RETURN
            END
```

Runge-Kutta Subroutine
```
            SUBROUTINE RK4(X(1),X(2),X(3),Y(1),Y(2),Y(3),DX)
            DIMENSION X(3),Y(3)
            DO 10 I = 1,2
              A1 = DX*F(X(I),Y(I))
              A2 = DX*F(X(I)+DX/2, Y(I)+A1/2)
              A3 = DX*F(X(I)+DX/2, Y(I)+A2/2)
              A4 = DX*F(X(I)+DX, Y(I)+A3)
              Y(I+1) = Y(I) + (A1 + 2*(A2 + A3) + A4)/6.
              X(I+1) = X(1) + I*DX
10            CONTINUE
            RETURN
            END
```

Data
```
0.2,  0.0,  19.5,  201
```

Figure 5.12 FORTRAN program based on third-order Adams Predictor–Corrector method, Example 5.2 problem.

HIGHER-ORDER MODELS

Initial Condition Problems

Any higher-order differential equation with initial conditions can be broken down into a set of two or more coupled first-order differential equations, each with a condition. The additional dependent variables introduced in such a reduction are simply the derivatives of the original dependent variable up through one order less than the highest order in the original model. The resulting set will have as many equations as the order of the original equation. For example, in the following equation, we substitute z for dy/dx, and w for d^2y/dx^2.

$$\frac{d^3y}{dx^3} + A\frac{dy}{dx} + By^2 = G(x)$$

with

$$y(0) = y_0$$

$$y'(0) = y_0'$$

$$y''(0) = y_0''$$

After the substitution and breakdown, we get the following set of coupled first-order equations.

$$\frac{dw}{dx} + Az + By^2 = G(x); \qquad w(0) = y_0''$$

$$\frac{dz}{dx} = w; \qquad z(0) = y_0'$$

$$\frac{dy}{dx} = z; \qquad y(0) = y_0$$

This breakdown into a relatively simple set is highly significant for any numerical simulation efforts. *Any of the previously discussed methods for simulating with first-order models can be applied to each equation in the foregoing set.* Since the application has to be done for all the equations at each step of the independent variable, simultaneous solutions, which include all the original dependent variable derivatives, are obtained. The fact that all slopes are simultaneously simulated has significant appeal when working with engineering systems.

Manual computations are not practical for higher-order simulations in general. However, the operations involved in a few steps of any of the methods can be tracked using a tabular algorithm, as demonstrated for the first-order cases before. We will not attempt to illustrate the manual technique for all the methods in the following material, but we will give examples using either the Euler or fourth-order Runge–Kutta method. It is a relatively simple task for most students to extend the first-order examples for the methods not illustrated. Some of this is left for exercises. Example 5.5 illustrates the Euler method applied for a few steps with a tabular algorithm. A FORTRAN flow chart of the same problem, but for simulation over a greater span, is given as Figure 5.13. Use of the fourth-order Runge–Kutta method is illustrated in Example 5.6, with a FORTRAN flow chart given as Figure 5.14 and a typical BASIC program given as Figure 5.15.

EXAMPLE 5.5

The following single-equation model comes from the engineering analysis of a moving-object system. A quick estimate of the y versus x behavior is desired in the range $x=0$ to $x=0.4$. Demonstrate how the Euler method works in this situation by using a tabular algorithm and a step size of $\Delta x=0.1$.

$$\frac{d^2y}{dx^2} - 3y^{1.2} = x; \qquad y(0)=1.0, \quad y'(0)=0.0$$

▶ The second-order equation is broken down into a set of two first-order equations by substituting z for dy/dx. Putting the equations into the form required for application of the Euler method, we get the following:

$$\frac{dz}{dx} = x + 3y^{1.2}; \qquad z(0)=0.0$$

$$\frac{dy}{dx} = z; \qquad y(0)=1.0$$

Letting $f_i = x_i + 3y_i^{1.2}$, we apply Equation 5.1 to both of the equations in the set.

$$z_{i+1} = z_i + \Delta x\, f_i$$

$$y_{i+1} = y_i + \Delta x\, z_i$$

We then compute z and y at each step, keeping track of our results in a table, Table 5.4.

Table 5.4 Computational Results, Example 5.5

i	x_i	y_i	z_i	f_i	y_{i+1}	z_{i+1}
1	0.0	1.0	0.0	3.0	1.0	0.3
2	0.1	1.0	0.3	3.1	1.03	0.61
3	0.2	1.03	0.61	3.3083	1.091	0.94083
4	0.3	1.091	0.94083	3.6305	1.1851	1.3041
5	0.4	1.1851	1.3041			

Sample calculations are given for the entry at $i=2$ in Table 5.4.

$$x_2 = x_1 + \Delta x = 0.0 + 0.1 = 0.1$$

$y_2 = 1.0$, from the preceding row

$z_2 = 0.3$, from the preceding row

$f_2 = 0.1 + 3(1.0)^{1.2} = 3.1$

$y_3 = 1.0 + 0.1(0.3) = 1.03$

$z_3 = 0.3 + 0.1(3.1) = 0.61$

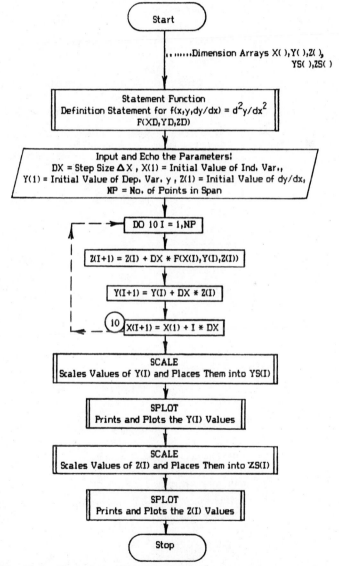

Figure 5.13 Flow Chart for a FORTRAN version of the Euler method applied to a second-order model, Example 5.5.

EXAMPLE 5.6

The following single-equation model has been set up during the engineering analysis of a heat transfer system. A quick estimate of how the fourth-order Runge–Kutta method might handle the simulation of y versus x is desired. Using a calculator and tabular algorithm, demonstrate the method for one step of the independent variable (i.e., $\Delta x = 0.1$).

$$\frac{d^2 y}{dx^2} - 3y^{1.2} = x; \qquad y(0) = 1.0, \quad y'(0) = 0.0$$

▶ The second-order equation is broken down into a two-equation set by substitution of z for dy/dx. These equations are immediately put into the form required for application of the Runge–Kutta method.

$$\frac{dz}{dx} = x + 3y^{1.2}; \qquad z(0) = 0.0$$

$$\frac{dy}{dx} = z; \qquad y(0) = 1.0$$

Table 5.5 Computational Results, Example 5.6

i	x_i	y_i	z_i	a's	b's	y_{i+1}	z_{i+1}
1	0.0	1.0	0.0	$a_1 = 0.3$ $a_2 = 0.305$ $a_3 = 0.3077$ $a_4 = 0.3155$	$b_1 = 0.0$ $b_2 = 0.015$ $b_3 = 0.01525$ $b_4 = 0.03077$	1.0152	0.3068
2	0.1	1.0152	0.3068	$a_1 = 0.31548$ $a_2 = 0.32603$ $a_3 = 0.32889$ $a_4 = 0.34252$	$b_1 = 0.03068$ $b_2 = 0.04645$ $b_3 = 0.04698$ $b_4 = 0.06357$	1.0621	0.6348

Sample calculations are given for the entry at $i = 2$ in Table 5.5.

$$x_2 = 0.0 + 0.1 = 0.1$$
$$y_2 = 1.0152 \text{ and } z_2 = 0.3068 \qquad \text{from the previous step}$$
$$a_1 = 0.1(0.1 + 3(1.0152)^{1.2}) = 0.31548$$
$$b_1 = 0.1(0.3068) = 0.03068$$
$$a_2 = 0.1[(0.1 + 0.1/2.) + 3(1.0152 + 0.03068/2.)^{1.2}] = 0.32603$$
$$b_2 = 0.1(0.3068 + 0.31548/2.) = 0.04645$$
$$a_3 = 0.1[(0.1 + 0.1/2.) + 3(1.0152 + 0.04645.)^{1.2}] = 0.32889$$
$$b_3 = 0.1(0.3068 + 0.32603/2.) = 0.046981$$
$$a_4 = 0.1[(0.1 + 0.1) + 3(1.0152 + 0.046981)^{1.2}] = 0.34252$$
$$b_4 = 0.1(0.3068 + 0.32899) = 0.06357$$
$$z_3 = 0.3068 + (0.31548 + 2(0.32603) + 2(0.32889) + 0.34252)/6. = 0.6348$$
$$y_3 = 1.0152 + (0.03068 + 2(0.04645) + 2(0.04698) + 0.06357)/6. = 1.0621$$

Letting $f(x_i, y_i) = x_i + 3y_i^{1.2}$ and referring to Equation 5.3, we write the specific Runge–Kutta equations required for this problem.

$$z_{i+1} = z_i + \frac{a_1 + 2a_2 + 2a_3 + a_4}{6} \qquad y_{i+1} = y_i + \frac{b_1 + 2b_2 + 2b_3 + b_4}{6}$$

$$a_1 = \Delta x \, f(x_i, y_i) \qquad\qquad b_1 = \Delta x \, z_i$$

$$a_2 = \Delta x \, f\left(x_i + \frac{\Delta x}{2}, y_i + \frac{b_1}{2}\right) \qquad b_2 = \Delta x \left(z_i + \frac{a_1}{2}\right)$$

$$a_3 = \Delta x \, f\left(x_i + \frac{\Delta x}{2}, y_i + \frac{b_2}{2}\right) \qquad b_3 = \Delta x \left(z_i + \frac{a_2}{2}\right)$$

$$a_4 = \Delta x \, f(x_i + \Delta x, y_i + b_3) \qquad b_4 = \Delta x(z_i + a_3)$$

The results of the calculations are shown in Table 5.5.

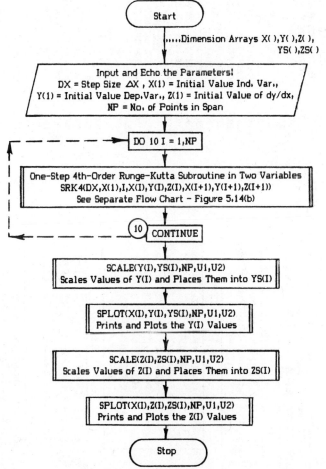

Figure 5.14a Flow Chart for FORTRAN version of a fourth-order Runge–Kutta method applied to a second-order model, Example 5.6.

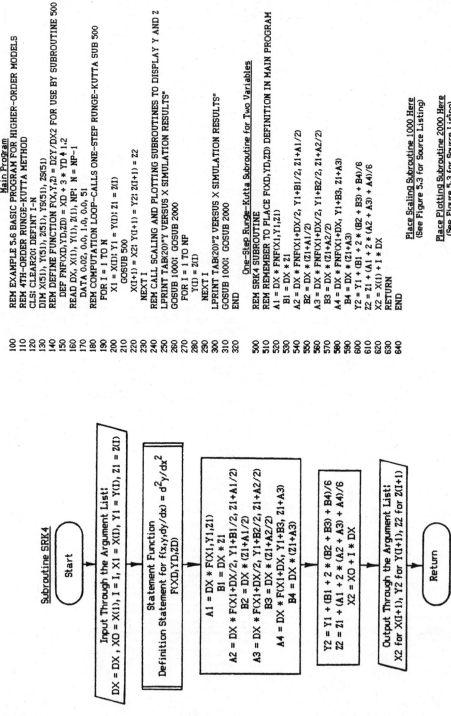

Main Program

```
100  REM EXAMPLE 5.6 BASIC PROGRAM FOR HIGHER-ORDER MODELS
110  REM 4TH-ORDER RUNGE-KUTTA METHOD
120  CLS: CLEAR20: DEFINT I-N
130  DIM X(51), Y(51), Z(51), YS(51), ZS(51)
140  REM DEFINE FUNCTION F(X,Y,Z) = D2Y/DX2 FOR USE BY SUBROUTINE 500
150  DEF FNF(X,YD,ZD) = XD + 3 * YD ↑ 1.2
160  READ DX, X(1), Y(1), Z(1), NP: N = NP-1
170  DATA 0.02, 0.0, 1.0, 0.0, 51
180  REM COMPUTATION LOOP-CALLS ONE-STEP RUNGE-KUTTA SUB 500
190  FOR I = 1 TO N
200  X1 = X(I): Y1 = Y(I): Z1 = Z(I)
210     GOSUB 500
220  X(I+1) = X2: Y(I+1) = Y2: Z(I+1) = Z2
230  NEXT I
240  REM CALL SCALING AND PLOTTING SUBROUTINES TO DISPLAY Y AND Z
250  LPRINT TAB(20)"Y VERSUS X SIMULATION RESULTS"
260  GOSUB 1000: GOSUB 2000
270  FOR I = 1 TO NP
280     Y(I) = Z(I)
290  NEXT I
300  LPRINT TAB(20)"Z VERSUS X SIMULATION RESULTS"
310  GOSUB 1000: GOSUB 2000
320  END

                        One-Step Runge-Kutta Subroutine for Two Variables
500  REM SRK4 SUBROUTINE
510  REM REMEMBER TO PLACE F(XD,YD,ZD) DEFINITION IN MAIN PROGRAM
520  A1 = DX * FNF(X1,Y1,Z1)
530  B1 = DX * Z1
540  A2 = DX * FNF(X1+DX/2, Y1+B1/2, Z1+A1/2)
550  B2 = DX * (Z1+A1/2)
560  A3 = DX * FNF(X1+DX/2, Y1+B2/2, Z1+A2/2)
570  B3 = DX * (Z1+A2/2)
580  A4 = DX * FNF(X1+DX, Y1+B3, Z1+A3)
590  B4 = DX * (Z1+A3)
600  Y2 = Y1 + (B1 + 2 * (B2 + B3) + B4)/6
610  Z2 = Z1 + (A1 + 2 * (A2 + A3) + A4)/6
620  X2 = X(I) + I * DX
630  RETURN
640  END
```

Place Scaling Subroutine 1000 Here
(See Figure 5.3 for Source Listing)

Place Plotting Subroutine 2000 Here
(See Figure 5.3 for Source Listing)

Figure 5.15 BASIC program based on a fourth-order Runge–Kutta method applied to a second-order model, Example 5.6.

Figure 5.14b FORTRAN flow chart for a fourth-order Runge–Kutta subroutine Example 5.6.

Boundary Condition Problems

The Shooting Method for Nonlinear Models

In this method, the desired boundary condition simulation is replaced with an initial condition simulation so that the powerful extrapolation methods just discussed can be used. As you will recall from our discussion of model identifications, boundary condition models are always of order two or higher and have conditions specified at both extremes of a simulation domain. This means that not enough natural conditions will be available at the beginning of a domain to start one of our now-familiar methods. Consequently, some tricks have to be used to guess, or otherwise determine, the missing initial condition(s).

One approach that has been surprisingly effective for engineers using an interactive computing system is to simply guess values for the missing condition, in an iterative manner, until the computed boundary value at the terminal boundary agrees, within some acceptably small amount, with the given boundary condition. The name "shooting" is most apt for this approach because of the similarities to adjusting the sights when shooting a gun at a target. Usually, a fairly good simulation can be accomplished within fewer than 10 trials, but obviously success depends upon how familiar the analyst is with his or her model. Another approach is to view the simulation as a root-search problem with an extremely complicated function evaluation (namely, a simulation across a complete domain) required. Although familiarity with the model is still a necessity to get good starting guesses, some of the root-finding algorithms can converge on an appropriate match of a boundary condition extremely fast. In particular, the secant method has been found to be very efficient. Examples of both approaches will be presented in the material that follows. However, another aspect of the shooting method has to be mentioned first.

On the surface, the shooting method does not seem to be a very good method because it requires much more participation from the analyst than seems reasonable. Although this assessment may be correct for linear problems (where use of the shooting method is certainly questionable on efficiency grounds), the typical nonlinear problem creates an iterative matrix situation if any of the alternative methods are chosen. At the time this is written, it seems that almost all nonlinear matrix solution schemes leave much to be desired for the analyst who likes to track what is happening. Furthermore, for many undergraduate engineering students, background in handling nonlinear matrix solutions is not required. Thus the shooting method, in spite of occasional unpredictable and time-consuming behavior in the hands of the novice, is the best choice for nonlinear boundary value problems.

One of the self-starting methods is probably the best choice for incorporating into a shooting-method algorithm. Even the Euler method does better than might be expected when used in shooting. This probably results from the fact that the cumulative extrapolation errors are transformed, through the iterative matching at the boundaries, into less noticeable, internally distributed errors. Many times a very fast, Euler-based shooting method can give an approximate simulation, good enough to make engineering decisions. In any event, if the decision is to make a refined simulation, with say a fourth-order Runge–Kutta method, the starting guesses for the missing condition(s) are easily extracted. As might be expected, manual implementation of the shooting method, even Euler-based, is not practical for any reasonably long simulation. However, a few shoots, with large independent variable increments, can be performed with a calculator and a tabular algorithm to demonstrate the behavior of the method. Such a demonstration, using the approach of just guessing the missing condition, is given as Example 5.7. A FORTRAN flow chart for treating an expanded version of the same problem is given as Figure 5.16. An interactive BASIC program is given as Figure 5.17.

EXAMPLE 5.7

The following single-equation model has been set up during the analysis of a flow system. A quick indication of the y versus x behavior is desired first; then a more refined simulation using a computer will be done. Using an Euler shooting method, with $\Delta x = 0.2$, demonstrate a rough simulation with a calculator and a tabular algorithm. Furnish a FORTRAN flow chart and a BASIC program, using a $\Delta x = 0.012$, for more refined simulations. Preliminary knowledge of the model leads us to believe that the initial missing slope should have a value between 0.0 and 1.5.

$$\frac{d^2y}{dx^2} + y^2 = x; \qquad y(0) = 0.0, \quad y(0.6) = 0.3$$

▶ Since some knowledge exists about how the model might behave, we choose the iterative guessing (trial and error) approach here. Performing the usual breakdown of the higher-order model equation by substituting z for dy/dx, and using the Euler formula on the resulting two-equation set, gives us the following working equations. We will assume that our subscripts began at $x = 0.0$.

$$z_{i+1} = z_i + \Delta x \, f_i; \qquad z_1 = ? \text{ (the missing condition)}$$

$$y_{i+1} = y_i + \Delta x \, z_i; \qquad y_1 = 0.0$$

where $f_i = x_i - y_i^2$.

For our first shoot, we will guess $z_1 = 0.1$; for the second, $z_1 = 1.0$. Calculated values are placed in Table 5.6. It is observed that we compare our computed y_4 at each shoot to our specified boundary value, $y(0.6) = 0.3$, to make the decision about what our z_1 should be on the next shoot.

Sample calculations are given for shoot no. 2, at $i = 2$:

$$x_2 = 0.0 + 0.2 = 0.2$$

$$y_2 = 0.2 \qquad \text{from the previous entry at } i = 1, \text{ shoot no. 2}$$

Table 5.6 Computational Results, Example 5.7

Shoot	i	x_i	y_i	z_i	f_i	y_{i+1}	z_{i+1}
1	1	0.0	0.0	0.1^a	0.0	0.02	0.1
	2	0.2	0.02	0.1	0.1996	0.04	0.13992
	3	0.4	0.04	0.13992	0.3984	0.06798	0.2196
	4	0.6	0.06798^b	0.2136			
2	1	0.0	0.0	1.0^a	0.0	0.2	1.0
	2	0.2	0.2	1.0	0.16	0.4	1.032
	3	0.4	0.4	1.032	0.24	0.6064	1.08
	4	0.6	0.6064^b	1.08			
3	1	0.0	0.0	0.5^a	0.0	0.1	0.5
	2	0.2	0.1	0.5	0.19	0.2	0.538
	3	0.4	0.2	0.538	0.36	0.3076	0.61
	4	0.6	0.3076^b	0.61			

[a] Guessed values.
[b] Computed boundary values. These values are compared to the given boundary condition $y(0.6) = 0.3$ to decide what the value of z_1 should be on the next shoot.

$$z_2 = 1.0 \quad \text{from the previous entry at } i = 1, \text{ shoot no. 2}$$
$$f_2 = 0.2 - (0.2)^2 = 0.16$$
$$y_3 = 0.2 + 0.2(1.0) = 0.4$$
$$z_3 = 1.0 + 0.2(0.16) = 1.032$$

The application of a computer through interactive FORTRAN is illustrated by the flow chart in Figure 5.16. An interactive application with a BASIC program is shown in Figure 5.17.

One way of refining the simulation in Example 5.7 further is to use the fourth-order Runge–Kutta method. However, at this point the other approach may have more to offer. Rather than make trial-and-error choices of the missing z_1, we can invoke a variation on the secant root-finding algorithm to give us "best" choices for z_1 on succeeding shoots. This scheme is encompassed in Equation 5.8.

$$(z_1)_{\text{new}} = \frac{A - B}{CB_b - CB_a} \tag{5.8}$$

where
$$A = (z_1)_a (CB_b - G)$$
$$B = (z_1)_b (CB_a - G)$$
CB_a = computed boundary value when using $(z_1)_a$
CB_b = computed boundary value when using $(z_1)_b$
G = the given boundary value

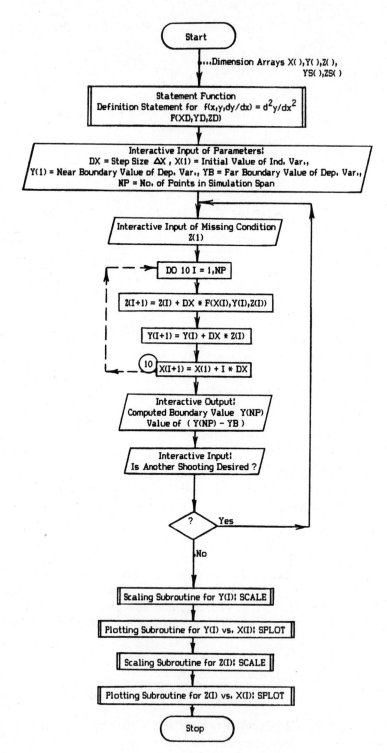

Figure 5.16 Flow chart for interactive FORTRAN program, Example 5.7.

```
                                    Main Program
100        REM EXAMPLE 5.7 BASIC PROGRAM FOR BOUNDARY VALUE PROBLEM
110        REM EULER SHOOTING METHOD WITH INTERACTIVE INPUT
120        CLS: CLEAR20: DEFINT I-N
130        REM SEE STATEMENT 170 FOR ARRAY DIMENSIONING
140        REM DEFINITION OF FUNCTION: F(X,Y,Z) = D2Y/DX2
150        DEF FNF(XD,YD,ZD) = XD - YD*2
160        INPUT"GIVE STEP SIZE 'DX' AND NO. OF POINTS 'NP'"; DX,NP
170        DIM X(NP+1),Y(NP+1),Z(NP+1),YS(NP+1),ZS(NP+1)
180        INPUT"GIVE BOUNDARY CONDITIONS: X(1),Y(1), AND YB"; X(1),Y(1),YB
190        INPUT"GIVE A GUESS FOR THE MISSING CONDITION  Z(1)";Z(1)
200        REM COMPUTATION LOOP
210          FOR I = 1 TO NP
220            Z(I+1) = Z(I) + DX * FNF(X(I),Y(I),Z(I))
230            Y(I+1) = Y(I) + DX * Z(I)
240            X(I+1) = X(1) + I * DX
250          NEXT I
260        CLS
270        PRINT"****** COMPUTED BOUNDARY VALUE =";Y(NP)
280        PRINT"********* DIFFERENCE: (Y(NP) - YB) ="; Y(NP) - YB
290        PRINT"*********** LAST GUESS OF MISSING CONDITION =";Z(1)
300        INPUT"DO YOU WISH TO SHOOT AGAIN ? -- Y/N"; C$
310          IF C$ = "Y" GOTO 190
320        REM CALL SUBROUTINES TO SCALE AND PLOT Y(I) VALUES
330        GOSUB 1000: GOSUB 2000
330          FOR I = 1 TO NP
330            Y(I) = Z(I)
340          NEXT I
350        REM CALL SUBROUTINES TO SCALE AND PLOT Z(I) VALUES
360        GOSUB 1000: GOSUB 2000
370        END
```

Place Scaling Subroutine 1000 Here
(See Figure 5.3 for Source Listing)

Place Plotting Subroutine 2000 Here
(See Figure 5.3 for Source Listing)

Figure 5.17 Interactive BASIC program, Example 5.7.

The shooting is stopped when the absolute value of $(CB_b - G)$ becomes less than some epsilon (a convergence parameter). It should be noted that the a and b subscripts shift cyclically on successive shoots. For example, we have to have two reasonably good guesses for z_1 to get started. We call these $(z_1)_a$ and $(z_1)_b$; the computed boundary values are CB_a and CB_b, respectively. Now, on the fourth shoot, $(z_1)_{new}$ becomes $(z_1)_b$, and the former $(z_1)_b$ becomes $(z_1)_a$. The same subscript shift is used on the CB's. Another important point is that the boundary condition given (G) does not have to be the dependent variable; it may be the slope of the dependent variable. Also, this method has been successfully used on so-called inverse problems. In these problems, the boundary and initial conditions are known, but a parameter within the model (for example, a coefficient on one of the terms) may not be known. This parameter can be treated as the missing condition in Equation 5.8, and the simulation carried out to determine a "good" parameter value. The latter aspects are left as exercises. An application of the secant approach is illustrated in Example 5.8.

EXAMPLE 5.8

Repeat Example 5.7, except incorporate the secant scheme with epsilon of 0.001 after the first two shoots.

▶ The computational setup is the same as in Example 5.7 and is not repeated. We apply Equation 5.8 to get an estimate for the third shoot z_1.

$$(z_1)_{\text{new}} = \frac{(A - B)}{0.6064 - 0.06798} = 0.48783$$

where
$$A = (0.1)(0.6064 - 0.3)$$
$$B = (1.0)(0.06798 - 0.3)$$

Applying the new z_1 for shoot no. 3 gives us $y_2 = 0.9757$, $z_2 = 0.48783$, $y_3 = 0.19514$, $z_3 = 0.52593$, $y_4 = 0.30033$, and $z_4 = 0.59831$. Since the absolute value of $(0.30033 - 0.3)$ is less than 0.001, we have our simulation in one application of the secant algorithm. Incorporation of the secant algorithm into computer programs is illustrated with a flow chart for FORTRAN (Figure 5.18), and a program for BASIC (Figure 5.19).

Matrix Methods for Linear Models

Many variations on these methods exist. However, the central theme is that a simulation of the dependent variable, at discrete regions or points over the complete domain of interest, can be accomplished in one step if an appropriate matrix equation can be generated and solved. The methods are restricted in most cases to systems describable by linear models. Where it might be desired to use one of these methods with a nonlinear system, the alternative to generating a nonlinear matrix equation is to break the domain down into subdomains for which the models are approximated by linear models. Then, after appropriate interpolations and substitutions have been made, the subdomain matrices can be combined, usually by some kind of local boundary matching, into a global matrix equation for the complete simulation, However, many complications can arise in the interpolations of the nonlinear effects and the combining of the matrices. For this reason, we do not treat the nonlinear cases by this method and recommend instead that the shooting method be used.

The first task in applying a matrix method is to discretize the domain or span. Of course, the span of the independent variable will always be given within the boundary condition statement furnished as part of the model. To accomplish this, one of two procedures is commonly used. In the first procedure, the span is divided up by superimposing a uniformly spaced grid, where the grid-line intersections with the independent variable coordinate give a set of "nodes," or "nodal points,"

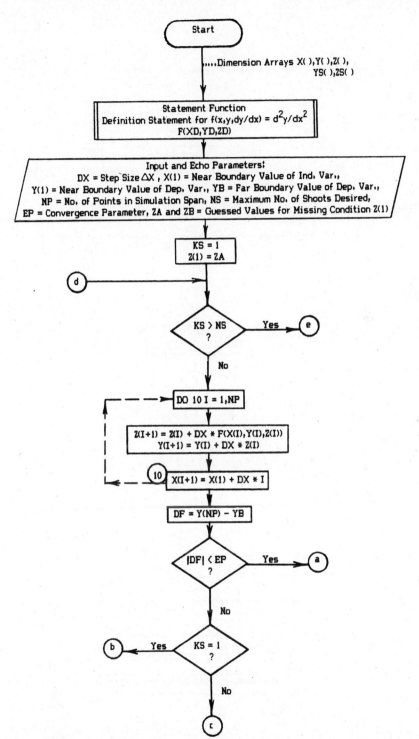

Figure 5.18*a* Flow Chart for FORTRAN version of secant-accelerated shooting method, Example 5.8.

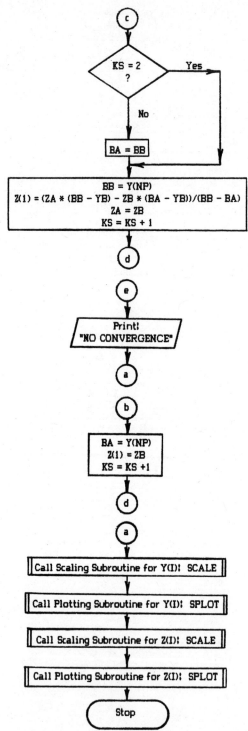

Figure 5.18*b* Flow chart for FORTRAN version of secant-accelerated shooting method, Example 5.8*(continued)*.

Main Program
```
100    REM EXAMPLE 5.8 BASIC PROGRAM FOR BOUNDARY VALUE PROBLEM
110    REM EULER SHOOTING METHOD WITH SECANT CONVERGENCE ACCELERATOR
120    CLS: CLEAR20: DEFINT I-N
130    DIM X(52), Y(52), Z(52), YS(52), ZS(52)
140    REM DEFINITION OF FUNCTION  F(X,Y,DY/DX) = D2Y/DX2
150    DEF FNF(XD,YD,ZD) = XD - YD↑2
160    READ DX,X(1),Y(1),YB,NP,NS,EP,ZA,ZB
170      DATA 0.012, 0.0, 0.0, 0.3, 51, 10, .0001, 0.1, 1.0
180    KS = 1: Z(1) = ZA
190    IF KS > NS THEN LPRINT"NO CONVERGENCE": GOTO 370
200    REM COMPUTATION LOOP
210      FOR I = 1 TO NP
220        Z(I+1) = Z(I) + DX * FNF(X(I),Y(I),Z(I))
230        Y(I+1) = Y(I) + DX * Z(I)
240        X(I+1) = X(1) + DX * I
250      NEXT I
260    PRINT"SHOOT NO.=";KS;" COMPLETED !"
270    DF = Y(NP) - YB:  PRINT"BOUNDARY DIFFERENCE =";DF
280    REM SECANT CONVERGENCE ACCELERATOR: ESTIMATES BEST NEW Z(1)
290      IF ABS(DF) <= EP GOTO 370
300      IF KS = 1 THEN BA=Y(NP): Z(1) = ZB: KS = KS + 1: GOTO 190
310      IF KS = 2 GOTO 330
320      BA = BB
330      BB = Y(NP)
340      Z(1) = (ZA * (BB - YB) - ZB * (BA - YB))/(BB - BA)
350      ZA = ZB: ZB = Z(1): KS = KS + 1
360      GOTO 190
370    REM CALL SUBROUTINES TO SCALE AND PLOT Y(I) VALUES
380    GOSUB 1000: GOSUB 2000
390      FOR I = 1 TO NP
400        Y(I) = Z(I)
410      NEXT I
420    REM CALL SUBROUTINES TO SCALE AND PLOT Z(I) VALUES
430    GOSUB 1000: GOSUB 2000
440    END
```
Place Scaling Subroutine 1000 Here
(See Figure 5.3 for Source Listing)

Place Plotting Subroutine 2000 Here
(See Figure 5.3 for Source Listing)

Figure 5.19   BASIC program for secant-accelerated shooting method, Example 5.8.

which are then numbered and become the simulation points. In the other procedure, the span is divided up into "cells," or "elements," usually of different sizes, which are then labeled and become the simulation "finite elements." We will concentrate on the former procedure herein; the latter is discussed in Chapter 7.

After the discretizing is done, a generic "finite difference" (FD) equation can be written to take the place of the model equation. This is done by substituting general finite-difference approximations for each of the derivatives in the original model equation, using centered differences wherever possible, and centering on the arbitrary node i. The result is a generalized algebraic equation for the node i.

Using the generic algebraic equation from the foregoing, we can write a specific algebraic equation for each node. If a slope-type boundary condition (Neumann condition) is part of the model, then another finite difference equation will have to be written for this boundary node. Either a forward or backward difference approximation, depending on whether it is for a near, or far, boundary will have to be used.

Finally, the equations are simplified and re-formed into a set that can be written as a matrix equation, and then the set is solved. Some solution methods do not require the set to be posed as a matrix equation; and if such a method is being used, then the re-forming as a matrix can be deleted. Many different methods exist for solving a matrix equation, and we assume the reader has some familiarity with at least one exact method and one iterative method. Our examples and problems will use the two most commonly known techniques: the basic, unmodified Gauss elimination method and the elementary Gauss–Seidel method. For the reader unfamiliar with these methods, a brief explanation is included in Appendix A. Additional details are given in Example 5.9.

The accuracy of the matrix approach is dependent on the approximation formulas substituted for the derivatives and on the size of spacing (increment) between the nodes. We can improve accuracy by decreasing increment size, but this means that the number of equations, and the size of our matrix, increases. Most solution methods, with maybe the exception of some of the iterative methods, run into roundoff problems on the computer as the matrix gets large. Furthermore, the iterative methods can run into convergence-time problems with large sets. Consequently, accuracy may be more of a tradeoff situation with this method than with the extrapolation methods discussed previously.

EXAMPLE 5.9

In the engineering analysis of a flow system, a linear, one-equation, boundary condition model has been set up, as shown below. Using a finite-difference matrix method, with first central difference approximations, and an increment size of $\Delta x = 0.05$, set up an appropriate matrix equation that could be used for simulating y versus x. Using tabular algorithms and a calculator, demonstrate a solution by the Gauss elimination method, and another solution by the Gauss–Seidel iterative method. Using $\Delta x = 0.005$, furnish a flow chart for a FORTRAN-language computer simulation, and a source listing for a BASIC-language computer simulation, of y versus x.

$$\frac{d^2 y}{dx^2} - \frac{y}{e^{2x}} = x \qquad y(0) = 1.0, \quad y(0.25) = 2.8$$

▶ We discretize the span into six nodes, numbered from left to right as shown below. An integer i index coordinate is superimposed on the physical x domain.

Physical domain: 0.0   0.05   0.10   0.15   0.20   0.25 $\rightarrow x$
Node number: 1   2   3   4   5   6 $\rightarrow i$

Table 5.7 Gauss Elimination Calculator Results, Example 5.9

Stage	Reduced Matrices					Auxiliary Equations			
1	-2.0023	1.0	0.0	0.0	-0.99988	1.0 -0.49943	0.0	0.0	0.49937
	1.0	-2.0020	1.0	0.0	0.00025	1.0 -0.49943	0.0	0.0	0.49937
	0.0	1.0	-2.0019	1.0	0.00038	0.0 0.0	0.0	0.0	0.0
	0.0	0.0	1.0	-2.0017	-2.7995	0.0 0.0	0.0	0.0	0.0
2	-2.0023	1.0	0.0	0.0	-0.99988				
	0.0	-1.5026	1.0	0.0	-0.49912	1.0	-0.66551	0.0	0.33217
	0.0	1.0	-2.0019	1.0	0.00038	1.0	-0.66551	0.0	0.33217
	0.0	0.0	1.0	-2.0017	-2.7995	0.0	0.0	0.0	0.0
3	-2.0023	1.0	0.0	0.0	-0.99988				
	0.0	-1.5026	1.0	0.0	-0.49912				
	0.0	0.0	-1.3364	1.0	-0.32837		1.0	-0.74828	0.24571
	0.0	0.0	1.0	-2.0017	-2.7995		1.0	-0.74828	0.24571
4	-2.0023	1.0	0.0	0.0	-0.99988				
	0.0	-1.5026	1.0	0.0	-0.49912				
	0.0	0.0	-1.3364	1.0	-0.32837				
	0.0	0.0	0.0	-1.2534	-3.0452				

Back Substitution

$$y_5 = -3.0452/(-1.2534) = 2.4296$$

$$y_4 = (-0.32837 - y_5)/(-1.3364) = 2.0637$$

$$y_3 = (-0.49912 - y_4)/(-1.5026) = 1.7056$$

$$y_2 = (-0.99988 - y_3)/(-2.0023) = 1.3512$$

Table 5.8 Gauss Seidel Calculator Results, Example 5.9

Iteration	y_2	Δy_2	y_3	Δy_3	y_4	Δy_4	y_5	Δy_5	F.Pts.
—[a]	2.0	—[b]	2.0	—	2.0	—	2.0	—	4[c]
1	1.4982	-0.5018	1.7472	-0.2528	1.8717	-0.1283	2.3336	0.3336	4
2	1.3720	-0.1262	1.6201	-0.1272	1.9748	0.1031	2.3851	0.0515	4
3	1.3085	-0.0635	1.6399	0.0198	2.0104	0.0356	2.4029	0.0178	4
4	1.3184	0.0099	1.6626	0.0227	2.0306	0.0202	2.4130	0.0101	3
5	1.3297	0.0113	1.6784	0.0158	2.0436	0.0129	2.4195	0.0065	3
6	1.3376	0.0079	1.6888	0.0104	2.0520	0.0084	2.4237	0.0042	1
7	1.3428	0.0052	1.6956	0.0868	2.0575	0.0055	2.4264	0.0027	0
	(1.34)		(1.70)		(2.06)		(2.43)		

[a]Initial values for the unknowns are simply good-judgment guesses.
[b]Difference between current and preceding iteration value of y_i (typical).
[c]Number of points (unknowns) failing the convergence test $|\Delta y_i| \leqslant \varepsilon$, where the convergence parameter has the value $\varepsilon = 0.01$.

Sample calculations are given for node 4 from iteration no. 3: The matrix is re-posed into a Liebmann-form equation set for computing the unknowns:

$$y_2 = (0.99988 + y_3)/(2.0023)$$

$$y_3 = (0.00025 - y_2 - y_4)/(-2.0020)$$

$$y_4 = (0.00038 - y_3 - y_5)/(-2.0019)$$

$$y_5 = (2.7995 + y_4)/(2.0017)$$

Using the third equation, we get

$$y_4 = (0.00038 - 1.6399 - 2.3851)/(-2.0019) = 2.0104$$

$$\Delta y_4 = 2.0104 - 1.9748 = 0.0356$$

Introduction of the finite-difference approximations for the derivatives gives

$$\frac{y_{i+1} - 2y_i + y_{i-1}}{(\Delta x)^2} - \exp(-2x_i)y_i = x_i$$

This can be re-posed for convenience, as a generic equation:

$$y_{i-1} - [2 + (\Delta x)^2 \exp(-2x_i)]y_i + y_{i+1} = (\Delta x)^2 x_i$$

This can be used to write a specific equation for each node ($i = 2$ through 5). The result can be shown as a matrix equation.

$$\begin{bmatrix} -2.0023 & 1.0 & 0.0 & 0.0 \\ 1.0 & -2.0020 & 1.0 & 0.0 \\ 0.0 & 1.0 & -2.0019 & 1.0 \\ 0.0 & 0.0 & 1.0 & -2.0017 \end{bmatrix} \begin{Bmatrix} y_2 \\ y_3 \\ y_4 \\ y_5 \end{Bmatrix} = \begin{Bmatrix} -0.99988 \\ 0.00025 \\ 0.00038 \\ -2.7995 \end{Bmatrix}$$

The calculator-based simulations are shown as Tables 5.7 and 5.8. The flow chart is given as Figure 5.20; and the BASIC program is given in Figure 5.21.

```
                            Main Program
100     REM EXAMPLE 5.9 BASIC PROGRAM FOR LINEAR BOUNDARY-VALUE PROBLEM
110     REM SIMULATION USING MATRIX METHOD WITH GAUSS SEIDEL SOLVER
120     CLS: CLEAR20: DEFINT I-N
130     DIM Y(51),YS(51),X(51)
140     READ NP,MX,EP,X(1),Y(1),Y(NP): N = NP - 1
150       DATA 51, 100, 0.001, 0.0, 0.0, 1.0, 2.8
160     REM SET STARTING ESTIMATES AND X VALUES
170       FOR I = 2 TO N: Y(I) = 2.0: X(I) = X(1) + DX * (I-1): NEXT I
180     REM ITERATION LOOP
190       FOR IT = 1 TO MX: NI = 0
200     REM COMPUTATION LOOP- USES MATRIX ARRANGED INTO LIEBMANN FORM
210         FOR I = 2 TO N
220           SA = Y(I)
230           Y(I) = - ( DX ↑ 2 * X(I) - Y(I-1) -Y(I+1))/(-2. - DX ↑ 2 * EXP(-2*X(I)))
240           DF = SA - Y(I)
250           IF ABS(DF) <= EP GOTO 270
260           NI = NI + 1
270         NEXT I
280         IF NI <= 0 GOTO 320
290           PRINT"ITERATION NO.";IT;" NO. OF POINTS FAILING=";NI
300       NEXT IT
310     LPRINT "**** NO CONVERGENCE !!!! ****"
320     LPRINT TAB(20)"Y VERSUS X SIMULATION RESULTS"
330     GOSUB 1000
340     GOSUB 2000
350     END
                        Place Scaling Subroutine 1000 Here
                        (See Figure 5.3 for Source Listing)

                        Place Plotting Subroutine 2000 Here
                        (See Figure 5.3 for Source Listing)
```

Figure 5.21 BASIC Program based on Gauss–Seidel method, Example 5.9.

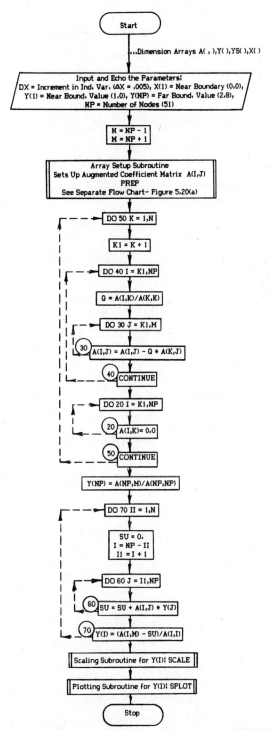

Figure 5.20*a* Flow chart for FORTRAN version of Gauss Elimination method, Example 5.9.

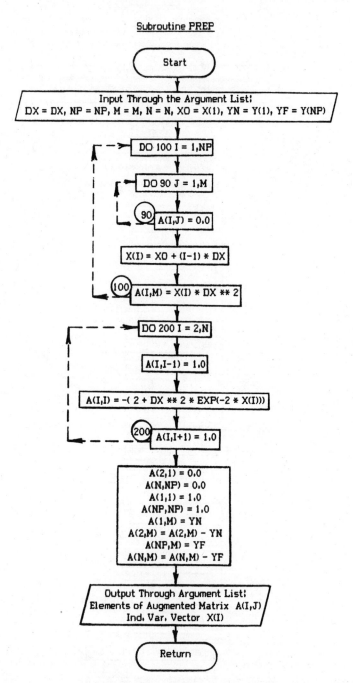

Figure 5.20*b*　Flow chart of PREP subroutine for FORTRAN program, Example 5.9.

THE USE OF A PROBLEM-ORIENTED LANGUAGE (POL)

Problem-oriented languages are usually referred to by the acronym POL. As its name implies, a POL is a special-purpose computer language directed at a specific type of simulation problem and/or a specific discipline. In the hierarchy of computer languages, it is generally considered to be a "higher-level" language, at a level higher than so-called general languages like FORTRAN or PL-1. Typically, a POL assumes that the user is familiar with the type of model, or problem, the POL is designed to handle. The language is constructed in a manner to make the actual programming task as short and convenient as possible. Consequently, many programming tradeoffs and unique procedures in model setups are found in POL packages. The price paid for these conveniences is more in-core storage requirements and longer CPU execution times, compared with equivalent simulations done with general-language programs.

Other penalties often include poor diagnostic support and equipment limitations. Nevertheless, the convenience feature has been so sought after that almost every major engineering-related organization (including universities) in the United States has developed, at one time or another, its own POL for handling simulations of the type we have discussed in this book. Among these is one developed for IBM corporation systems, and named CSMP, standing for "Continuous System Modeling Program," Since it is representative of most technical POLs, and is widely known and used, we choose to use it to illustrate the role of POLs in the type of modeling and simulation discussed in this book. The CSMP was designed to be used with IBM computers; but most other major computer manufacturers have similar packages, and many of the simulation features can be found in commercial general-purpose libraries such as IMSL (International Mathematics and Statistics Library).

Continuous System Modeling Program (CSMP)

The brief description presented in the following few pages is intended only as an introduction to this POL. The material presented is that culled from about 12 years of use in treating small simulation problems in both a consulting engineering and an academic setting. Consequently, many of the most sophisticated features are left untouched. For the reader interested in the more advanced features and a more comprehensive description, the book by Speckhart and Green,[8] and the IBM commercial literature, are recommended.

Elementary Rules and Advice

The Continuous System Modeling Program is built around FORTRAN as a source language; so, knowledge of FORTRAN is assumed. However, in contrast to regular FORTRAN, CSMP uses a convenient free-format input scheme, wherein all columns, 1 through 72, are available for placing statements and data. No statement numbers are needed. All identifiers, constants, expressions, operation symbols, and so on, are exactly the same as in FORTRAN, with two exceptions. The exceptions are that subscripting cannot be used, and only real-number treatment exists unless specific integer identifiers are designated with the statement: FIXED IIII. Comment statements are programmed by placing an asterisk in column 1. Continuation of a statement is programmed by putting a blank and three consecutive periods, and then continuing the statement on the next line. Some additional reserved identifiers are used. For example, TIME is the independent variable unless changed with a rename statement, discussed with other reserved words a little later. Obvious from the reserved independent variable name, CSMP is intended primarily as a simulator for initial value problems; all simulations start at TIME = 0.0, unless special programming is done.

Segmentation and Sorting

Following a pattern often used when working with applied differential equations, the typical CSMP program is set up in three distinct parts called the "initial segment," the "dynamic segment," and the terminal segment. Although it is not mandatory that all three segments be labeled, it is recommended to make the program more readable. These segments are described in the following.

Initial Segment. This segment occurs at the beginning of the program and is labeled at its start with the reserved word INITIAL, which is best shown as a one-word statement. All information normally associated with starting a solution for an initial value problem is entered here. For example, initial conditions, constants, parameters, data-set functions that are part of the model, and any preliminary calculations that need to be done only once, are correctly placed here. Both FORTRAN assignment statements and CSMP data statements (description later) can be used. However, the order in which the statements are placed is not important because this segment is automatically sorted to give the most appropriate statement sequence for the simulation to be done. This latter feature is one of the most convenient when programming in CSMP, but it can also be one of the most frustrating for the beginner. It means that no FORTRAN branching statements can be used in this segment unless they are placed in a special section designated to have no automatic sorting done (a NOSORT section; how to invoke such a section is discussed under "Sorting").

Dynamic Segment. A statement with the word DYNAMIC must precede this segment, which in turn follows right after the initial segment. Here any FORTRAN statements that are used to describe the behavior of the model(s) must be placed. Also to be placed here are structure statements containing the special CSMP function used to perform the simulation integrations (the INTGRL function, described under structure statements). Again the arrangement of the statements in this segment is irrelevant to the program operation because the statements are put into a best sequence automatically. Thus FORTRAN branching statements cannot be used here unless they are put into a nosort section.

Terminal Segment. This segment, which starts with a statement with the single word TERMINAL, follows immediately after the dynamic segment. It contains any computations that might have to be done at the end of a simulation. Arrangement of statements is important in this segment because no automatic rearrangement of the statements is done. As a result it is common to find FORTRAN branching statements used in this segment. A number of different CSMP control statements are also correctly placed in this part of the program. For example, statements that control the plotted output, the integration method used, the increment sizes, the simulation span, and the termination-of-run commands are entered here.

Sorting. This feature made CSMP one of the first POLs that attempted to remove some of the aversive aspects of programming. It automatically chooses the correct order in which the statements must be executed regardless of how the programmer placed them in the program. All statements in the initial and dynamic segments of a CSMP program are automatically sorted by this feature unless they are placed within nosort sections. A nosort section is preceded by the single-word statement NOSORT; it is ended with the single-word statement SORT. It can also be ended by the statements DYNAMIC or TERMINAL. Multiple nosort sections can occur in either the initial or dynamic segments.

CSMP Statements

Three types of statements are used in CSMP programs. Abbreviated information on each is presented in the following.

Data Statements. These statements are placed in the initial segment and serve to get information into the program. Each starts with one of the words INCON, CONSTANT, PARAMETER, or FUNCTION, followed by a blank, and then assignments of values to identifiers chosen by the programmer at this time. The first three of these statements can be used interchangeably, but using them to label information appropriate to the name gives a more readable program. An example and description of each follows.

INCON XI=0.0,YI=11.3, ZI=5.2

This statement is used to get initial conditions put into the program. Notice that each initial condition is given a name.

CONSTANT PI=3.141593, K=14.2

Constants are entered with this statement.

PARAMETER R1=6.8, R2=7.6, G=3.E-4

Temporary constants (parameters) are inserted into the program with this statement. Multiple simulation runs, with different parameter values, can be initiated with a special form of this statement as shown below.

PARAMETER EM=(10.92, 30.281, 61.9, 90.11)

Here the simulation run would be repeated four times with EM taking on the values inside the parentheses, EM = 10.92 first, EM = 30.281 second, and so on. But it should be noted that only one parameter may initiate multiple runs, and the number of runs is limited to 50.

FUNCTION FX=(−6.2, 2.1), (0.0,3.5), (103.1, 6.5), (103.8, 26.2), . . .

This statement is used to get tabular-function data into the program. The function is given an identifier (FX in this example), and data pairs are entered sequentially from lowest to highest, as they would be in a typical engineering data table. The first number appearing inside each parenthesis is the value of the independent variable; the second is the matching dependent variable value. More than one function may be used in a program, and there is no limit on the number of data pairs that can be used.

Structure Statements. These statements are the equivalent of the arithmetic assignment statements in FORTRAN and, in fact, are precisely that in many cases, as the examples will show. Most of these statements will occur in the dynamic segment where they are used to describe the model(s) and simulation. In addition to regular FORTRAN assignment statements, special CSMP functions can be used in a manner similar to the way built-in functions are used in FORTRAN. Although more than 30 CSMP functions are available, only 3 will be introduced here. Example statements and descriptions follow.

DYDX=INTGRL(IC, D2YDX2)

The CSMP integrator function, designated by the reserved word INTGRL, is shown. The IC inside the parenthesis is the initial condition, previously assigned its name and value in the initial segment. The D2YDX2 is the integrand, presumably described by a FORTRAN statement in the dynamic segment. A symbolic representation of what INTGRL does is

$$\frac{dy}{dx} = \int_{IC} \frac{d^2y}{dx^2}\,dx$$

Various methods of integration are available and may be specified in the terminal segment. If no method is specified, a variable step size, fourth-order Runge–Kutta method is used.

YE=AFGEN(FX, X)

The linear function generator AFGEN is illustrated. Actually, this is a linear interpolator that uses tabular-function data entered with the FUNCTION statement in the initial segment. The X is a value to be used as the independent variable for this function. In use the FX data is scanned with the independent-variable value X until it is bracketed by known points. A linear interpolation is then done on the dependent data, and the resulting value is stored as YE.

YE=NLFGEN (FX, X)

Where a more refined interpolation is desired, a second-degree (quadratic) inter-polator is available as illustrated above. The meanings of the terms inside the parentheses are the same as described for AFGEN. Both AFGEN and NLFGEN are among the most useful functions available in CSMP for engineering simula-tions. Many of the other functions can be approximated with these functions, and the use of unrefined data, coming directly from the field or laboratory, presents no difficulties.

Control Statements. The statements used to designate each segment, namely INITIAL, DYNAMIC, and TERMINAL, are control statements. Typically, control statements can be mixed with other statements and can occur in any order. Most CSMP programmers prefer to use the LABEL and RENAME control statements at the beginning of the program to enhance readability. Most are used in the terminal segment, however. Examples and descriptions follow.

LABEL EXAMPLE NO. 999

This statement designates a label that will be printed at the top of the trendplot results. When placed at the beginning of the program, it also serves as a program label.

RENAME TIME=X, DELT=DX, FINTIM=SPAN

This statement is used to change the reserved names of some of the run controls, as well as the independent variable. DELT and FINTIM are discussed under "TIMER," and TIME is the default name of the independent variable. Used at the beginning of the program, it improves understandability.

METHOD RKSFX

The method of integration used by the INTGRL function can be specified with this statement. If this statement is omitted, a variable-step size, fourth-order Runge–Kutta method is used. Seven different methods are available, but only four are listed here.

1. *RKSFX* specifies that a fixed-step-size, fourth-order Runge–Kutta method be used.
2. *RKS* specifics that a variable-step-size, fourth-order Runge–Kutta method be used. This is the default method.
3. *ADAMS* specifies that a second-order Adams predictor–corrector method be used.
4. *RECT* specifies that the Euler method be used.

TIMER DELT=0.01, OUTDEL=0.02, FINTIM=5.0

This statement is used to specify the initial computation increment for the inde-pendent variable (DELT), the increment desired for the print-plot output

(OUTDEL), and the span of the independent variable for the complete simulation (FINTIM).

PRTPLOT Y(DYDX,YE), Z

This statement is used to specify that a trendplot be made of the values of Y, followed by a trendplot of Z values. It also specifies that the values of DYDX and YE be printed alongside and to the right of the Y trendplot. The independent variable, and dependent variable, values are always printed alongside and to the immediate left of every trendplot.

CONTINUE

This should not be confused with a FORTRAN continue statement, which will always have a statement number to its left. This statement is used instead of the END statement if it is desired to change another control statement at this point and to continue the simulation without getting an automatic reset of the independent variable to zero.

END

This statement does not indicate the end of program execution as it does in FORTRAN or BASIC. Instead, it simply means that a simulation is ready to accept new data and/or control statements. Thus the independent variable is automatically reset to zero, and the initial conditions are reset. If an END statement is followed by a STOP statement, it means that execution stops.

CALL RERUN

Use of this control statement is another way to recycle the program through another simulation. All data inserted initially in the initial segment will be used again unless FORTRAN statements in the terminal segment are used to assign new values.

ENDJOB

This is the only CSMP statement that must start in column one. It is the last statement in any CSMP program.

Use of FORTRAN Subprograms

Another extremely convenient characteristic of CSMP is that a standard FORTRAN function subprogram can be placed between the STOP and ENDJOB statements. This means that a structure statement can be used in the dynamic segment that uses a non-CSMP special function. If use of a FORTRAN subroutine is desired, the subprogram is, as with the function, placed between the stop and endjob. However, the well-known call statement cannot be used. Instead a special statement, which mimics the function subprogram use, is invoked. In this statement all output identifiers, which would normally occur in the argument list of the call statement, must be placed (separated by commas) on the left-hand side of the equal sign; the right-hand side is treated as though it were a function. An example is

X, Y, K = EXAMP(A, B, C)

The subroutine statement for this example would be

SUBROUTINE EXAMP(X, Y, K, A, B, C)

A standard FORTRAN call statement may be used if it is placed inside a NOSORT section of the dynamic segment.

Some examples of CSMP programs are given in Examples 5.10, 5.11, and 5.12.

EXAMPLE 5.10

Simulation of the x versus t behavior over the span $t=0.0$ to $t=10.0$ is desired for the system described by the following model. Set up a CSMP program to do the simulation using the default method. Let $DF(t)=B \cos Ct/\pi$, $A=100$, $B=10$, and $C=10$. Use a starting computing increment of $\Delta x=0.1$, an output increment of 0.5, and use a FORTRAN subprogram to determine $DF(t)$.

$$\frac{d^2y}{dx^2}+Ax=(B)DF(t) \qquad x(0)=0.0, \quad x'(0)=0.0$$

▶ The CSMP program is shown as Figure 5.22.

```
!
LABEL  CSMP EXAMPLE 5.10
* COMMENTS USE ASTERISK IN COLUMN ONE
* FOR A CONTINUATION STATEMENT, USE A BLANK AND THREE CONSECUTIVE ...
* PERIODS IN THE PRECEDING STATEMENT.
INITIAL
        INCON  XI = 0.0, DXDTI = 0.0
        CONSTANT  A = 100., B = 10.
        PARAMETER  C = 10.
DYNAMIC
        D2XDT2 = A * X - DF(B,C,TIME)
        X = INTGRL(XI,DXDT)
        DXDT = INTGRL(DXDTI,D2XDT2)
TERMINAL
* NOTE THAT DEFAULT METHOD IS VARIABLE-STEP 4TH-ORDER RUNGE KUTTA METHOD
        TIMER  DELT = 0.1, OUTDEL = 0.5, FINTIM = 10.0
        PRTPLOT  X(DXDT)
* TIME AND X VALUES WILL BE PRINTED TO THE IMMEDIATE LEFT OF X TREND PLOT.
* DXDT VALUES WILL BE PRINTED TO THE IMMEDIATE RIGHT OF THE X TREND PLOT.
        END
        STOP
* FORTRAN SUBPROGRAM WHICH FOLLOWS STARTS IN COLUMN 7
        FUNCTION  DF(A,B,X)
        PI = 3.141593
        DF = A * COS(B*X/PI)
        RETURN
        END
* ENDJOB STATEMENT MUST START IN COLUMN 1
ENDJOB
```

Figure 5.22 CSMP program, Example 5.10.

EXAMPLE 5.11

A tabular driving function is found with a third-order model representing a physical system as shown below. Set up a CSMP program to simulate A versus B over the span of $B = 0.0$ to $B = 5.8$, using a fixed step ($\Delta B = 0.05$) fourth-order Runge–Kutta method. Use the rename feature to make the program more readable, and use the quadratic function generator (interpolator) to manipulate the driving function $f(B)$. Let the output interval be 0.1.

$$
\begin{array}{ccccccc}
B = & -2.0 & -1.0 & 0.0 & 1.0 & 3.0 & 10.0 \\
f(B) = & -14.1 & -2.7 & 6.5 & 12.8 & 21.1 & 49.6
\end{array}
$$

$$\frac{d^3 A}{dB^3} - \pi A^{1.87} = f(B) \qquad A(0) = 1.2, \quad A'(0) = 0.0, \quad A''(0) = 0.0$$

► The CSMP program is given as Figure 5.23.

```
LABEL CSMP EXAMPLE 5.11
        RENAME  TIME = B, DELT = DB, FINTIM = MAXB
INITIAL
        INCON  AO = 1.2, DADBO = 0.3, D2ADBO = 0.0
        CONSTANT PI = 3.141593
        FUNCTION FB = (-2.0,-14.1), (-1.0,-2.7), (0.0,6.5), ...
            (1.0,12.8), (3.0,21.1), (10.0,49.6)
DYNAMIC
        A = INTGRL(AO,DADB)
        FD = NLFGEN(FB,B)
        D3ADB3 = FD + PI * A * ABS(A) ** 0.87
        D2ADB2 = INTGRL(D2ADBO,D3ADB3)
        DADB = INTGRL(DADBO,D2ADB2)
TERMINAL
        METHOD RKSFX
        TIMER  DB = 0.05, OUTDEL = 0.1, MAXB = 5.8
        PRTPLOT  A(DADB,D2ADB2), DADB(D2ADB2,FD)
        END
        STOP
ENDJOB
```

Figure 5.23 CSMP program, Example 5.11.

EXAMPLE 5.12

The simulation problem described in Example 5.8 (also Example 5.7) is to be done with a CSMP program. Set up an appropriate CSMP program containing

```
!
LABEL  CSMP EXAMPLE 5.12
       RENAME  TIME = X, DELT = DX, FINTIM = SPAN
INITIAL
       CONSTANT  X1 = 0.0, YB = 0.3, NS = 10., EP = 0.0001, KS = 1.
       INCON  Y1 = 0.0, Z1 = 0.1
       PARAMETER  ZA = 0.1, ZB = 1.0
DYNAMIC
       Y = INTGRL(Y1,Z)
       Z = INTGRL(Z1,DZDX)
       DZDX = XX - Y ** 2
       XX = X + X1
TERMINAL
       METHOD  RECT
* FORTRAN STATEMENTS (TO IMPLEMENT THE SECANT ACCELERATOR) FOLLOW.
          IF(KS .GT. NS) GO TO 40
          DF = Y - YB
          IF(ABS(DF) .LT. EP) GO TO 60
          IF(KS .LT. 1.5) GO TO 20
          IF(KS .LT. 2.5) GO TO 10
          BA = BB
   10     BB = Y
          Z1 = (ZA*(BB-YB) - ZB*(BA-YB))/(BB - BA)
          ZA = ZB
          ZB = Z1
          GO TO 30
   20     BA = Y
          Z1 = ZB
   30     KS = KS + 1.
          CALL RERUN
   40     WRITE(6,50)
   50     FORMAT(20X,'NO CONVERGENCE')
   60     CONTINUE
       TIMER  DX = 0.012, SPAN = 0.6
       END
       TIMER  OUTDEL = 0.012
       PRTPLOT  Y(Z)
       END
       STOP
ENDJOB
```

Figure 5.24   CSMP program, Example 5.12.

FORTRAN logic in the terminal segment to accomplish the secant acceleration and converge the number of shoots as quickly as possible.

▶ The CSMP program is given as Figure 5.24.

Other POL Packages

The Continuous System Modeling Program is by no means the only POL available for simulating from ordinary derivative models. From among the others, some of the better known packages are DYNAMO, CSMP III, MIMIC, and EASY. Also, various differential equation solvers are included within some well-known general-purpose software packages, for example, SAS and IMSL. Each different simulator has advantages and strong features, as well as disadvantages. Undoubtedly, many of the current POLs will be extensively modified in the near future to make them suitable for interactive use on personal computers. The reader interested in more information on these POLs is urged to review the references, particularly the information listed in Shoup,[9] and Jacoby and Kowalik.[10]

Developing Your Own Simulation Package

The very objective of having POLs leads to some conflicts and contradictions when trying to design a marketable software package. Obviously, the more problem-specific the package design becomes, the more efficient it can be. But this generally precludes a product's having any appeal to a large user group. The alternative, found in most of the successful packages to date, veers away from the basic POL concept by providing a large number of methodology options directed at several categories of a problem type. The resulting package is large and requires large computing equipment. Consequently, it is no surprise that the usefulness of most POL packages has been demonstrated in large organizations where the large equipment resides, and where many day-to-day decisions fall within the POLs range. Within smaller organizations, and even for isolated groups within large organizations, access to such POL packages has been limited. Furthermore, it has been the experience of many engineers working on state-of-the-art problems that the problems tend to fall outside the categories and options provided by the POL package available. Finally, as this is written, it is apparent that significant de-centralization of computing equipment is taking place as the personal computer becomes the engineer's desktop companion and compatibility of the large POL packages is in doubt.

Thus a number of reasons exist for an engineer to develop his own simulation package from time to time. With the recognition that the typical engineer is not trained in software development, but that he is reasonably proficient in programming in one or two languages, some recommendations are made in the following paragraphs. A common comment from engineers who have developed their own packages is that the task is very much like designing and making a unique reusable tool to get a tough job done. Although not much of a market exists for such tools, it is known that they are exchanged on a personal and professional level. But it is generally realized that it is not worthwhile to put much effort into making the package understandable to a large, diverse group of users. On the other hand, and particularly if any development is done on an employer's time, the engineer has an obligation to provide a tool with sufficient documentation that another, skilled in the same art, can operate it usefully, or that it can be turned over to a computer science group for further development without excessive verbal explanations. Within this context, some guidelines are offered.

The package should be constructed from program modules if possible. These modules should accomplish specific subtasks with as much independence from other modules as can be arranged. Programming languages like FORTRAN, which have the independent subprogram capability, work well for constructing these modules. The central idea behind using modules is that they can be developed and debugged in relative isolation, thus minimizing logic and programming error propagation. Furthermore, they can usually be arranged in a very clear-to-read sequence, reflecting the start of a simulation and progressing from module-to-module, without too much jumping back, until simulation results are displayed. The foregoing approach is known by various names, but the current computer science jargon refers to it as modular, or structured, top-down program design. Actually, it reflects process, or task, flow chart concepts traditionally used in engineering. Properly implemented, this type of modular program design also tends to minimize computer execution time during the simulation. Most of the examples shown in this book are set up in this fashion.

The subprograms and modules should have as much concise documentation built into them as possible. Typically, comment statements work best here, but sometimes program printouts can be used instead. Not always necessary, but certainly an enhancement for any package, is a short explanatory report related directly to the package through the comments and identifiers (names) used. Such reports may be the source of information on the algorithms used, test cases, symbol definitions, and so on. Among items that should be included in the documenting comments at the beginning of a package are a package or program name reflecting what kind of simulation the package does, the name of the developer with a date, methods used, and possible limitations. Additional comments should head each module and/or subprogram, describing its task.

The input module(s) should have an echo output of some kind. Careful labeling with the intended units should be part of this output. For a larger, or more complicated package, some testing of the input correctness is advisable. Appropriate diagnostic messages should be printed out if errors are detected. Means of correcting a few errors interactively, without shutting down the run, is useful in interactive program packages.

The output module should contain print commands for explanatory headings and labels for all simulation results. A graphical display module is a necessity for most engineering simulations.

Examples against which the package and/or modules within the package can be checked with test runs should be provided. These should be kept available for reuse whenever changes or additions are made to the package, to make sure that extraneous errors have not been introduced.

Generally, almost any technique that makes the package more readable and understandable (at least in an overview fashion) by someone who may not be familiar with the programming language used, is desirable. But this presumes that detailed explanations of the algorithms are not necessary.

It is assumed that the reader interested in setting up his own simulation package (for treating ordinary-derivative model problems) realizes at this stage that he can use the modules presented within the examples as building material.

REFERENCES

1. R. W. Hornbeck, *Numerical Methods*, Quantum Publishers, Inc., New York, 1975.

2. C. F. Gerald, *Applied Numerical Analysis*, Addison-Wesley Publishing Co., Reading, Mass., 1973.

3. R. L. Ketter and S. P. Prawel, Jr., *Modern Methods of Engineering Computation*, McGraw-Hill Book Company, New York, 1969.

4. M. L. James, G. M. Smith, and J. C. Wolford, *Applied Numerical Methods for Digital Computation with Fortran and CSMP*, 2nd ed., IEP, Thomas Y. Crowell Co., Inc., New York, 1977.

5. J. H. Ferziger, *Numerical Methods for Engineering Application*, John Wiley & Sons, Inc., New York, 1981.

6. J. M. Ortega and W. G. Poole, Jr., *An Introduction to Numerical Methods for Differential Equations*, Pitman Publishing Corporation, Belmont, Calif., 1981.

7. K. E. Atkinson, *An Introduction to Numerical Analysis*, John Wiley & Sons, Inc., New York, 1978.

8. F. H. Speckhart and W. L. Green, *A Guide to Using CSMP—The Continuous System Modeling Program*, Prentice-Hall, Inc., Englewood Cliffs, N.J., 1976.

9. T. E. Shoup, *A Practical Guide to Computer Methods for Engineers*, Prentice-Hall, Inc., Englewood Cliffs, N.J., 1979.

10. S. L. S. Jacoby and J. S. Kowalik, *Mathematical Modeling with Computers*, Prentice-Hall, Inc., Englewood Cliffs, N.J., 1980.

PROBLEMS

5.1   Demonstrate an Euler method simulation of T versus t with a tabular algorithm using $\Delta t = 0.1$ and $0.0 \leqslant t \leqslant 0.5$.

$$\frac{dT}{dt} - \left(\frac{1}{240}\right) Q_i - \left(\frac{1}{4.8}\right)(T_A - T) = 0; \qquad T(0) = T_A$$

$$Q_i = 2000 \qquad T_A = 70$$

5.2   Demonstrate an Euler method simulation of P versus S with a tabular algorithm using $\Delta S = 0.1$ and $0.0 \leqslant S \leqslant 0.5$.

$$\frac{dP}{dS} + P^2 = 1; \qquad P(0) = 0.$$

5.3   Demonstrate an Euler method simulation of x versus t with a tabular algorithm using $\Delta t = 0.1$ and $0.0 \leqslant t \leqslant 0.3$.

$$3x - \frac{dx}{dt} + t = 0; \qquad x(0) = 10.$$

5.4   Repeat Problem 5.1, 5.2, or 5.3 except use one of the Runge–Kutta methods instead of the Euler method.

5.5   Verification of a numerical simulation method requires an exact solution of the following model. Demonstrate a "variables separable" solution. (*Hint:* Appendix A1.)

$$0.6 \cos 4y + \frac{dx}{dy} = 0.0; \qquad x(0) = 4.0$$

5.6   Verification of a numerical simulation method requires an exact solution of the following model. Demonstrate a "variables separable" solution. (*Hint:* Appendix A1.)

$$x - 20 - \frac{dx}{dy} = 0; \qquad x(0) = 5.$$

5.7   Demonstrate a third-order Adams predictor–corrector method simulation of y versus x with a tabular algorithm using $\Delta x = 0.5$ and $0.0 \leqslant x \leqslant 3.0$. Use a fourth-order Runge–Kutta starter and only one correction at each step.

$$y - \frac{dy}{dx} + 0.5x = 0; \qquad y(0) = 2.5$$

5.8   Demonstrate a third-order Adams predictor-corrector method simulation of f versus t with a tabular algorithm using $\Delta t = 0.1$ and $0.0 \leqslant t \leqslant 0.5$. Use a fourth-order Runge–Kutta starter and only one correction at each step.

$$\frac{df}{dt} = 1 - f^2; \qquad f(0) = 0.0$$

5.9   Verification of a numerical simulation method requires an exact solution of the following model. Demonstrate a "variables separable" method solution. (*Hint:* Appendix A1.)

$$2y^3 - 5 + \frac{d^2x}{dy^2} = 0.0; \qquad x(0) = 10$$
$$x'(0) = 0.0$$

5.10   Verification of a numerical simulation method requires an exact solution of the following model. Demonstrate a D operator solution. (*Hint:* Appendix A1.)

$$\frac{d^2y}{dx^2} - 16y = 0; \qquad y(0) = 10$$
$$y'(0) = 0.0$$

5.11   A simulation of T versus t from the following model is required, using $\Delta t = 0.5$ and $0.0 \leqslant t \leqslant 5.0$. Demonstrate your preference of a method: an exact solution (see Appendix A1) or a fourth-order Runge–Kutta method tabulation. Be prepared to defend the choice of either method.

$$\frac{dT}{dt} + 0.552T = 0; \qquad T(0) = 70$$

5.12 Demonstrate an Euler method and a second-order Runge–Kutta method simulation of y versus x with tabular algorithms using $\Delta x = 0.2$ and $1.0 \leqslant x \leqslant 1.6$.

$$\frac{dy}{dx} + (\sin y)^{1.5} = \frac{0.5}{x}; \qquad y(1.0) = 13.0$$

5.13 Demonstrate an Euler method simulation of x versus y with a tabular algorithm using $\Delta y = 0.5$ and $0.0 \leqslant y \leqslant 1.0$.

$$\frac{dx}{dy} - y^2 x^{0.3} = g(y); \qquad x(0) = 1.0$$

y	0.0	0.5	1.0	1.5
$g(y)$	3.6	5.8	-2.2	6.8

5.14 Demonstrate a third-order Adams predictor-corrector method simulation of x versus y with a tabular algorithm using $\Delta y = 0.15$ and $0.0 \leqslant y \leqslant 0.75$. Use a fourth-order Runge–Kutta starter and only one correction at each step.

$$\frac{dx}{dy} = 1 \times 10^{-10}(x^4 - 1 \times 10^{12}); \qquad x(0) = 530.$$

5.15 A simulation of y versus x from the following model is required, using $\Delta x = 0.1$ and $0.0 \leqslant x \leqslant 20$.. Demonstrate your preference of a method: an exact solution (*Hint:* Appendix A1), or a third-order Adams predictor–corrector method tabulation. Be prepared to defend the choice of either method.

$$\frac{d^2 y}{dx^2} + 4\frac{dy}{dx} + 8y = 0; \qquad \begin{aligned} y(0) &= 0.0 \\ y'(0) &= 1.0 \end{aligned}$$

5.16 Demonstrate an Euler method simulation of y versus x with a tabular algorithm using $\Delta x = 0.05$ and $0.0 \leqslant x \leqslant 0.2$. Compare the results with an exact solution simulation. (*Hint:* Appendix A1.)

$$\frac{dy}{dx} - y = x^2; \qquad y(0) = 1.0$$

5.17 Demonstrate a fourth-order Runge–Kutta method simulation of y versus x with a tabular algorithm using $\Delta x = 0.1$ and $0.0 \leqslant x \leqslant 0.3$.

$$1.2x^{2.2} - \frac{dy}{dx} = -y^{1.9}; \qquad y(0) = 1.0$$

5.18 Demonstrate a third-order Adams predictor–corrector method simulation of A versus B with a tabular algorithm using $\Delta B = 0.1$ and $0.0 \leqslant B \leqslant 0.5$. Use a fourth-order Runge–Kutta starter and only one correction at each step.

$$1.2B^{2.2} - \frac{dA}{dB} = -A^{1.9}; \qquad A(0) = 1.0$$

5.19 Demonstrate an Euler method simulation of x and t versus y with a tabular algorithm using $\Delta y = 0.5$ and $0.0 \leqslant y \leqslant 2.0$.

$$\frac{dx}{dy} - \frac{x^2}{t} = 0; \qquad x(0) = 1.0$$

$$\frac{dt}{dy} + t - x = 1; \qquad t(0) = 1.0$$

5.20 Demonstrate a fourth-order Runge–Kutta method simulation of p and q versus t with a tabular algorithm using $\Delta t = 0.5$ and $0.0 \leqslant t \leqslant 0.5$.

$$\frac{dp}{dt} - \frac{p^2}{q} = 0; \qquad p(0) = 1.0$$

$$\frac{dq}{dt} + q - p = 1; \qquad q(0) = 1.0$$

5.21 Demonstrate a third-order Adams predictor–corrector method simulation of y and x versus t with a tabular algorithm using $\Delta t = 0.1$ and $0.0 \leqslant t \leqslant 0.4$. Use a fourth-order Runge–Kutta starter and only one correction at each step.

$$\frac{dy}{dt} - \frac{10.5}{y^2} + 3.8 \cos x = t; \qquad y(0) = 2.0$$

$$\frac{dx}{dt} + 5x - 2.3y^{1.2} = 10; \qquad x(0) = -2.0$$

5.22 Demonstrate an Euler method simulation of w, y, and z versus x with a tabular algorithm using $\Delta x = 0.1$ and $0.0 \leqslant x \leqslant 0.4$.

$$\frac{dy}{dx} - 2z + y^2 - w = 0.; \qquad y(0) = 1.0$$

$$w = z/y - x^2; \qquad w(0) = 0.0$$

$$\frac{dz}{dx} = 0.5z + y; \qquad z(0) = 0.0$$

5.23 Demonstrate an Euler method simulation of y_1 and y_2 versus t with a tabular algorithm using $\Delta t = 0.1$ and $0.0 \leqslant t \leqslant 0.3$.

$$\frac{dy_1}{dt} - y_2 t - y_1 = 0.; \qquad y_1(0) = 0.0$$

$$\frac{dy_2}{dt} - y_1 = 0.; \qquad y_2(0) = 1.0$$

5.24   Demonstrate an Euler method simulation of y and $dy/dx = z$ versus x with a tabular algorithm using $\Delta x = 0.1$ and $0.0 \leqslant x \leqslant 0.4$.

$$\frac{d^2 y}{dx^2} - \frac{dy}{dx} + y^{1.5} = 5.; \qquad y(0) = 1.0$$
$$y'(0) = 0.0$$

5.25   Demonstrate a third-order Adams predictor–corrector method simulation of p and $dp/dy = w$ versus y with a tabular algorithm using $\Delta y = 0.5$ and $0.0 \leqslant y \leqslant 2.0$. Use a fourth-order Runge–Kutta starter and only one correction per step.

$$\frac{d^2 p}{dy^2} + \ln p^2 = 3.^{-y}; \qquad p(0) = 10.$$
$$p'(0) = 0.0$$

5.26   Demonstrate an Euler method simulation of y versus x with a tabular algorithm using $\Delta x = 0.5$ and $0.0 \leqslant x \leqslant 1.5$. Compare results with an exact solution. (*Hint:* Appendix A1.)

$$\frac{d^2 y}{dx^2} + y = 0.; \qquad y(0) = 0.0$$
$$y'(0) = 3.0$$

5.27   Demonstrate a third-order Adams predictor–corrector method simulation of y and $dy/dx = z$ versus x with a tabular algorithm using $\Delta x = 0.2$ and $0.0 \leqslant x \leqslant 0.8$. Use a fourth-order Runge–Kutta starter and only one correction per step.

$$\frac{d^2 y}{dx^2} - \frac{dy}{dx} + y^{1.5} = 5.; \qquad y(0) = 1.0$$
$$y'(0) = 0.0$$

5.28   Verification of a numerical simulation method requires an exact solution of the following model. Demonstrate a general solution using a D operator. (*Hint:* Appendix A1.)

$$\frac{d^2 h}{dt^2} + 20 \frac{dh}{dt} + 6h = 0.0; \qquad \text{with two conditions}$$

5.29   Demonstrate a fourth-order Runge–Kutta method simulation of y versus t with a tabular algorithm using $\Delta t = 0.1$ and $0.0 \leqslant t \leqslant 0.1$. Compare the results with some other numerical simulation.

$$\frac{d^2y}{dt^2} - y[\exp(t) + 1.]^{-1} = 0.; \qquad \begin{array}{l} y(0) = 1.0 \\ y'(0) = 0.0 \end{array}$$

5.30 Demonstrate an exact solution simulation of A versus B from the following model using a D operator. (*Hint:* Appendix A1.) Compare results with a numerical simulation.

$$\frac{d^2A}{dB^2} + A = 0.0; \qquad \begin{array}{l} A(0) = 0.0 \\ A'(0) = 3.0 \end{array}$$

5.31 Demonstrate an exact solution simulation of y versus x from the following model, using the undetermined coefficient method. (*Hint:* Appendix A1.) Compare results with a numerical simulation.

$$\frac{d^2y}{dx^2} + \frac{3dy}{dx} + 2y = 2.8; \qquad \begin{array}{l} y(0) = 0.0 \\ y'(0) = 1.0 \end{array}$$

5.32 Demonstrate an exact solution simulation of x versus t and dx/dt versus t, from the following model, using the undetermined coefficient method. (*Hint:* Appendix A1.) Compare your results with a numerical simulation.

$$\frac{d^2x}{dt^2} + 16x = 10 \cos 4t; \qquad \begin{array}{l} x(0) = 0.0 \\ x'(0) = 8.0 \end{array}$$

5.33 Verification of a numerical simulation method requires an exact solution of the following model. Demonstrate an undetermined coefficient method solution. (*Hint:* Appendix A1.) Compare results with a numerical simulation.

$$\frac{d^2x}{dt^2} + 16x = 10e^t; \qquad \begin{array}{l} x(0) = 0.0 \\ x'(0) = 8.0 \end{array}$$

5.34 Demonstrate an exact general solution that can be used to simulate y versus x from the following model, using the variation of parameters method. (*Hint:* Appendix A1.)

$$\frac{d^2y}{dx^2} - y = \ln x; \qquad \text{with two conditions}$$

5.35 Demonstrate an exact general solution that can be used to simulate y versus x from the following model, using the variation-of-parameters method. (*Hint:* Appendix A1.)

$$\frac{d^2y}{dx^2} - 4y = e^{x^2}, \qquad \text{with two conditions}$$

5.36 Demonstrate an Euler method shooting method simulation of x and $dx/dy=z$ versus y with a tabular algorithm using $\Delta y=0.5$ and (1) $z(0.)=0.0$, and (2) $z(0.)=2.0$. Limit the number of shoots to three. [*Hint:* Use $z(0.)=1.0$ for the third shoot.]

$$\frac{d^2x}{dy^2}-x=-y^2; \qquad x(0.0)=1.0$$
$$x(1.0)=2.25$$

5.37 Demonstrate an Euler shooting method simulation of y and $dy/dt=z$ versus t with a tabular algorithm using $\Delta t=0.5$ and (1) $z(0.)=0.0$, and (2) $z(0.)=1.0$. Limit the number of shoots to three. [*Hint:* Use $z(0.)=0.5$ for the third shoot.]

$$\frac{d^2y}{dt^2}-y=t; \qquad y(0.)=1.0$$
$$y'(1.5)=3.3$$

5.38 Demonstrate an Euler shooting method simulation of y versus x with a tabular algorithm using $\Delta x=0.5$ and (1) $y'(0)=-0.5$, and (2) $y'(0)=0.5$. Limit the number of shoots to three. [*Hint:* Use $y'(0)=0.0$ for the third shoot.]

$$\frac{d^2y}{dx^2}-\frac{\cos(x/2)}{y^2}=0.; \qquad y(0)=1.0$$
$$y(1.5)=1.74$$

5.39 Demonstrate an Euler shooting method simulation of a versus b with a tabular algorithm using $\Delta b=0.2$ (1) $a'(5.0)=0.0$, and (2) $a'(5.0)=2.0$. Limit the number of shoots to three. [*Hint:* Use $a'(5.0)=1.0$ for the third shoot.

$$\frac{d^2a}{db^2}-\cos a=b; \qquad a(5.0)=0.0$$
$$a'(6.0)=6.8770$$

5.40 Demonstrate an Euler shooting method simulation of y versus x with a tabular algorithm using $\Delta x=0.5$ and (1) $P=0.5$, and (2) $P=1.5$. Limit the number of shoots to three. Notice that a trial-and-error parameter determination (P value) is being done directly from the differential model, but it is apparent that a complete set of boundary and initial conditions must be available in such a case. (*Hint:* Use $P=1.0$ for the third shoot.)

$$\frac{d^2y}{dx^2}-\frac{\cos Px}{y^2}=0,; \qquad y(0.)=1.0$$
$$y'(0.)=-0.5$$
$$y(1.5)=1.14$$

5.41   Demonstrate an Euler shooting method simulation of y versus x with a tabular algorithm using $\Delta x = 0.5$ and (1) $y'(0.) = 2.0$, and (2) $y'(0.) = 4.0$. Apply the secant acceleration scheme to get the $y'(0.)$ values for subsequent shoots, but limit the total number of shoots to 4.

$$\frac{d^2 y}{dx^2} - y^{1.1} = 0.; \qquad y(0.) = 0.0$$
$$y(1.5) = 5.0$$

5.42   Demonstrate an Euler shooting method simulation of a versus b with a tabular algorithm using $\Delta b = 0.2$ and (1) $a'(5.0) = 0.0$, and (2) $a'(5.0) = 2.0$. Apply the secant acceleration scheme to get the $a'(5.0)$ values for subsequent shoots, but limit the total number of shoots to three. Compare the results with Problem 5.39.

$$\frac{d^2 a}{db^2} - \cos a = b; \qquad a(5.0) = 0.0$$
$$a'(6.0) = 6.8770$$

5.43   Set up the matrix equation required for a finite-difference matrix method simulation of y versus x. Use $\Delta x = 0.2$.

$$x^2 y + x \frac{dy}{dx} + x^2 \frac{d^2 y}{dx^2} = 0.0; \qquad y(0.1) = 1.0$$
$$y(0.9) = 0.0$$

5.44   (a) Set up the matrix equation required for a finite-difference matrix method simulation of y versus x. Use $\Delta x = 1.0$.
(b) Complete the simulation by solving with the Gauss–Seidel method, using starters of $y_i = 1.0$, and $\varepsilon = 0.2$. (*Hint:* Appendix A2.)

$$\frac{d^2 y}{dx^2} - y = 10.; \qquad y(0) = 1.0$$
$$y(4.0) = 2.0$$

5.45   (a) Set up the matrix equation required for a finite-difference matrix method simulation of P versus W. Use $\Delta W = 0.2$.
(b) Complete the simulation by solving with the Gauss–Jordan method. (*Hint:* Appendix A2.)

$$\frac{d^2 P}{dW^2} + W^2 P = S(W); \qquad P(0) = 1.0$$
$$P'(0.8) = 5.0$$

where $\qquad\qquad S = 10.0 \qquad$ at $W = 0.2$
$$S = 0.0 \text{ for all other values of } W$$

5.46   (a) Set up the matrix equation required for a finite-difference matrix method simulation of a versus b. Use $\Delta b = 1.0$.

(b) Complete the simulation by solving with the Gauss elimination method. (*Hint*: Appendix A2.)

$$\frac{d^2a}{db^2} + b^{1.2}a = 10, +0.5b; \qquad a(0.0) = 10.0$$
$$a(4.0) = 12.0$$

5.47   (a) Set up the matrix equation required for a finite-difference matrix method simulation of *y* versus *x*. Use $\Delta x = 1.0$.
(b) Complete the simulation by solving with the Gauss–Jordan method. (*Hint*: Appendix A2.)
(c) Complete the simulation by solving with the Gauss–Seidel method, using starters of $y_i = 0.0$, and $\varepsilon = 1.0$. (*Hint*: Appendix A2.)

$$\frac{d^2y}{dx^2} + 0.25y = 8.0; \qquad y(0.0) = 0.0$$
$$y(4.0) = 66.$$

5.48   Write a CSMP program for simulating *x*, dx/dy, and d^2x/dy^2 versus *y*, using a variable-step-size, fourth-order Runge–Kutta method, an initial step size of $\Delta y = 0.1$, a print step of $\Delta y_p = 0.2$, and a span of $0.0 \leqslant y \leqslant 15.0$. (*Hint*: Let *y* be represented by TIME in the program.

$$3x^{1.2}\frac{dx}{dy} + 5\cos x = 5 + 5.2y - \frac{d^3x}{dy^3}; \qquad x(0) = 0.0$$
$$x'(0) = 5.0$$
$$x''(0) = 1.35$$

5.49   Write a CSMP program for simulating *y* versus *x*, using the default method, a step size of $\Delta x = 0.2$, and a span of $0.0 \leqslant x \leqslant 100.0$. Use the RENAME feature to get *x* instead of TIME and XSPAN instead of FINTIM. [*Hint*: Beware of $0.5/x$; use a small ε in $0.5/(x + \varepsilon)$.]

$$\frac{dy}{dx} + [\sin y]^{1.5} = \frac{0.5}{x}; \qquad y(0) = 13.0$$

5.50   Write a CSMP program for simulating *x* versus *t*, using the default method, a step size of $\Delta t = 0.01$, and a span of $0.0 \leqslant t \leqslant 2.0$. Print-plot both *x* and *z*.

$$\frac{dx}{dt} + x^{1.5} - z = 5.; \qquad x(0.) = 1.0$$

$$\frac{dz}{dt} - z^2 = 2t; \qquad z(0.) = 0.0$$

5.51   Write a CSMP program for simulating *z* versus *x*, using a fixed-step fourth-order Runge–Kutta method, a step size of $\Delta x = 0.01$, a print step of $\Delta x_p = 0.02$, and a span of $0.0 \leqslant x \leqslant 5.0$. Apply a quadratic interpolator (function generator) to $G(z)$ and $F(x)$ and the RENAME feature to get X instead of TIME, MAXX instead of FINTIM, and DX instead of DELT.

$$\frac{dz}{dx} - G(z) = F(x); \qquad z(0) = 20.$$

x	0.0	10.0	500.	z	-60.0	-10.0	100.
$F(x)$	-2.0	23.0	1.0	$G(z)$	2.0	0.0	3.0

5.52 Write a CSMP program for simulating A versus t, and B versus t, using a fixed-step fourth-order Runge–Kutta method, a step size of $\Delta t = 0.1$, and a span of $0.0 \leqslant t \leqslant 50.0$.

$$\frac{dA}{dt} - \frac{10.5}{A^2} + 3.8 \cos B = t; \qquad A(0) = 2.0$$

$$\frac{dB}{dt} + 5B - 2.3 A^{1.2} = 10; \qquad B(0) = -2.0$$

5.53 Write a CSMP program for simulating y versus x and dy/dx versus x, using a fixed-step fourth-order Runge–Kutta method, a step size of $\Delta x = 0.01$, and a span of $0.0 \leqslant x \leqslant 2.0$. Use the RENAME feature to get X instead of TIME, DX instead of DELT, and MX instead of FINTIM.

$$\frac{d^2 y}{dx^2} - \frac{dy}{dx} + y^{1.5} = 5.0; \qquad \begin{aligned} y(0) &= 1.0 \\ y'(0) &= 0.0 \end{aligned}$$

5.54 Write a CSMP program for simulating y versus t, using a fixed-step fourth-order Runge–Kutta method, a step size of $\Delta t = 0.01$, a print step of $\Delta t_p = 0.02$, and a span of $0.0 \leqslant t \leqslant 5.0$. Use the RENAME feature to get T for TIME, and the nonlinear function generator to get interpolated values from $f(y)$.

$$5y^2 \mid \frac{dy}{dt} = -f(y); \qquad y(0) - 10.0$$

y	$-10.$	-3.0	0.0	16.	43.
$f(y)$	0.6	0.2	-0.01	1.1	5.81

5.55 Write a CSMP program for simulating y versus t, and dy/dt versus t, using the Euler method, a step size of $\Delta t = 0.002$, and a span of $0.0 \leqslant t \leqslant 0.4$.

$$\frac{d^2 y}{dt^2} + A \left(\frac{dy}{dt} \right)^n + B[1 + C(1 - y)^m] y = 0.; \qquad y(0) = 0.1$$

The parameters are

$$A = 0.9022 \qquad B = 6.434 \qquad y'(0) = 0.0$$
$$C = 609.23 \qquad n = 1.5 \qquad m = -2.32$$

5.56 Write a CSMP program for simulating Q versus A, and dQ/dA versus A, using the default method, a step size of $\Delta A = 0.1$, a print step of $\Delta A_p = 0.2$, and a span of $0.0 \leqslant A \leqslant 40$. Use the RENAME feature to get A instead of TIME, and ASPAN instead of FINTIM. Use a linear function generator for interpolation.

$$\frac{d^2Q}{dA^2} + 2.1\left(\frac{dQ}{dA}\right)^{1.3} + f(x) = 0; \qquad Q(0) = 1.0$$

$$x = -0.1\frac{dQ}{dA}; \qquad Q'(0) = 0.0$$

x	$-20.$	$-15.$	0.0	$30.$	$35.$
$f(x)$	0.6	-5.0	-6.3	-2.0	3.8

5.57 Write a CSMP program for simulating x versus t and y versus t, using a fixed-step fourth-order Runge–Kutta method, a step size of $\Delta t = 0.001$, a print step of $\Delta t_p = 0.002$, and a span of $0.0 \leqslant t \leqslant 0.2$. Print-plot both x and y.

$$\frac{d^2y}{dt^2} + CD\left(\frac{dy}{dt}\right)^m - E\left(\frac{dx}{dt} - \frac{dy}{dt}\right)^n + Fy - G(x-y) = 0.0; \qquad \begin{array}{l} y(0) = 0.0 \\ y'(0) = 10.0 \end{array}$$

$$\frac{d^2x}{dt^2} + A\left(\frac{dx}{dt} - \frac{dy}{dt}\right)^n + B(x-y) = 0.; \qquad x(0) = 0.0; \qquad x'(0) = 10.0$$

The parameters are

$$A = 30. \qquad B = 502. \qquad C = 3.0$$
$$D = 12. \qquad E = 7.9 \qquad F = 4080.$$
$$G = 500. \qquad n = 1.25 \qquad m = 1.1$$

The following problems require significant use of a computer.

5.58(L) The behavior of the liquid level in an accumulator for a pumped-fluid system can be shown to be governed by the following differential model, where h represents the level and t represents the time.

$$\frac{dh}{dt} + C_1\left(\gamma h + \frac{H}{H-h}\right) - C_2(A + A\sin wt) = C_3; \qquad h(0) = h_0$$

Develop a FORTRAN or BASIC program using an Euler method for simulating h versus t. Assume consistent units and demonstrate your program using the parameters given below. Present the results with a trend print-plot.

$$H = 10.3 \qquad \gamma = 52.1 \qquad C_1 = 0.002 \qquad C_2 = 1.5$$
$$C_3 = 0.0308 \qquad A = 0.78 \qquad w = 3.0 \qquad h_0 = 5.0$$
$$0.0 \leqslant t \leqslant 10.0 \qquad\qquad \Delta t = 0.1$$

5.59(L) The temperature, T, of the compressed air remaining in a storage tank varies with the rate at which air is bled off and the amount of heat transferred through (or from) the wall of the tank to the air remaining inside. A model that can be set up to simulate the air temperature follows, where t represents time.

$$\frac{dT}{dt} + \left(\frac{A_1}{B}\right) T = \left(\frac{A_2}{B}\right); \qquad T(0) = T_0$$

$$A_1 = M_r(C_p - C_v)/C_v$$

$$A_2 = Q/C_v$$

$$Q = HA(T_\infty - T)$$

$$B = M_i + M_r$$

Develop a FORTRAN or BASIC program, using an Euler method, for simulating T versus t. Assume consistent units and demonstrate your program using the parameters given below. Present the results with a trend print-plot or other graphical display.

$$C_p = 0.24 \qquad C_v = 0.171 \qquad h = 1.6 \qquad A = 10.0$$
$$T_\infty = 65. \qquad M_r = 19.7 \qquad M_i = 11.8 \qquad T_0 = 65.$$
$$0.0 \leqslant t \leqslant 0.5 \qquad\qquad \Delta t = 0.005$$

5.60(L) The instantaneous velocity, V, of a shuttle moving on a relatively friction-free surface can be shown to be

$$\frac{dV}{dt} + AV = BF; \qquad V(0) = V_0$$

Develop a FORTRAN or BASIC program, using an Euler method, for simulating V with respect to time, t. Assume consistent units and demonstrate your program using the following parameters. Present the results with a trend print-plot or other graphical display.

$$A = 2.0 \qquad\qquad B = 0.4 \qquad F = 73. \qquad V_0 = 0.0$$
$$0.0 \leqslant t \leqslant 3.0 \qquad\qquad \Delta t = 0.02$$

5.61(L) The following differential model can be set up to simulate the liquid level in a tank that has a constant flow rate in, but a cyclic flow rate out, where h represents the level and t the time.

$$16\sqrt{5h-h^2}\,\frac{dh}{dt}=-\sin\frac{\pi t}{60}; \qquad h(0)=h_0$$

Develop a FORTRAN or BASIC program, using an Euler method for simulating h versus time, t. Assume consistent units and demonstrate your program using the following parameters. Present your results with a trend print-plot or other suitable graphical display.

$$h_0=2. \qquad 0.0\leqslant t\leqslant 120. \qquad \Delta t=1.2$$

5.62(L) The instantaneous pressure, P, in a compressed air tank is to be simulated with the following model, where t represents time.

$$\frac{dP}{dt}-\left(\frac{P_1}{V}\right)\left[C_0-C_1\left(\frac{P}{P_i}\right)^n-S(t)\right]=0. \qquad P(0)=P_0$$

t	$\leqslant 10.$	$\leqslant 13.5$	$\leqslant 21.5$	$\leqslant 25.$	$\leqslant 33.$	$\leqslant 36.5$, etc.
$S(t)$	0.0	0.97	0.0	0.97	0.0	0.97, etc.

Develop a FORTRAN or BASIC program, using an Euler method, for simulating P versus time, t. Assume consistent units and demonstrate your program using the following parameters. Present your results with a suitable printout and graphical display.

$$P_1=2116.8 \qquad V=46. \qquad C_0=0.49 \qquad C_1=0.05$$
$$P_i=14.7 \qquad n=0.714 \qquad P_0=14.7$$
$$0.0\leqslant t\leqslant 60. \qquad \Delta t=0.5$$

5.63(L) The differential model for simulating the water temperature, T_s, in a solar heating system, where t is time, is

$$\frac{dT_s}{dt}+C_1T_s-C_2(S+T_a)=C_3; \qquad T_s(6.)=T_0$$

Develop a FORTRAN or BASIC program, using an Euler method, for simulating the temperature, T_s, versus time. Apply a linear interpolation subprogram to give the required T_a and S values. Assume consistent units and demonstrate your program using the following parameters. Present your results with a professional-looking printout and graphical display.

$$C_1=0.17044 \qquad C_2=0.003933 \qquad C_3=0.928998$$
$$T_0=60 \qquad 0.0\leqslant t\leqslant 20 \qquad \Delta t=0.1$$

t	T_a	S	t	T_a	S
6.0	0.0	0.0	14.0	8.0	2190
7.0	1.0	0.0	15.0	6.0	1420
8.0	0.0	850	16.0	4.0	0.0
9.0	2.0	1380	17.0	1.0	0.0
10.0	4.0	2170	18.0	0.0	0.0
11.0	10.0	3060	19.0	0.0	0.0
12.0	10.0	3130	20.0	0.0	0.0
13.0	8.0	2760	—	—	—

5.64(L) The behavior of a compressed-air flow system driven by a wind turbine is studied with the following model, where P represents the receiver pressure and t represents time.

$$\frac{dP}{dt} - (0.95 - 0.05P^{1.4})F + C(P+1.3)\sqrt{Q} = 0.; \qquad P(0) = P_0$$

$P = 1.$ if P determined from above is less than 1.

$$F = \frac{1}{\pi} + 0.5 \sin wt - \frac{2}{\pi}((\tfrac{1}{3}) \cos 2wt + (\tfrac{1}{15}) \cos 4wt + (\tfrac{1}{35})\cos 6wt$$

$$+ (\tfrac{1}{63}) \cos 8wt)$$

Develop a FORTRAN or BASIC program, using an Euler method, for simulating P versus time. Apply a subprogram to evaluate F. Assume consistent units and demonstrate your program using the following parameters. Present your results with a professional-looking printout and graphical display.

$C = 0.01008$ $\qquad$ $w = 0.035$ $\qquad$ $Q = 0.$ unless $P > 1.3$; then $Q = P - 1.3$

$0.0 \leqslant t \leqslant 360$ $\qquad$ $\Delta t = 3.0$

5.65(L) The tube side fluid temperature, T, in a shell-and-tube heat exchanger can be simulated from the following model, where x represents position along the tube side.

$$\frac{dT}{dx} + AT = AT_s; \qquad T(0) = T_0$$

$A = \pi Dh/(\dot{m}c)$

$h = 0.023(k/D)[4m/(\pi D\mu)]^{0.8}(c\mu/k)^{0.4}$

$c = 0.251 + 3.46 \times 10^{-5}T - 14400./(T+460)^2$

$\mu = 0.032(T/460. + 1.)^{0.935}$

$k = k(T)$, tabular function given below

Develop a FORTRAN or BASIC program, using an Euler method, for simulating T with respect to position, x. Apply a linear interpolation subprogram to get k values. Assume consistent units and demonstrate your program using the following parameters. Present your results with a trend print-plot or other graphical display.

$$D = 0.05 \qquad \dot{m} = 25. \qquad T_0 = 80. \qquad T_s = 587.$$
$$0.0 \leqslant x \leqslant 5.0 \qquad \Delta x = 0.05$$

T	32.	212.	392.	572.
$k(T)$	0.0085	0.0133	0.0181	0.0228

5.66(L) Repeat any of the preceding problems [from Problem 5.58(L)] except use a fourth-order Runge–Kutta method or a third-order Adams predictor–corrector method.

5.67(L) The short-time variation of air temperature in a room is often very complicated and difficult to simulate. One simplistic model, which has been used successfully where forced-air heating systems are involved, is

$$\frac{dT}{dt} - \left(\frac{78.1}{V}\right)(q_1 + q_2 + q_3 + q_4 + q_5) = 0.0; \qquad T(0) = T_0$$

$$q_1 = 220.5(60 - T)$$
$$q_2 = 164.0(62 - T)$$
$$q_3 = 14.7(40 - T)$$
$$q_4 = 0. \text{ if } T_H > T$$
$$= 96.(120 - T) \text{ if } T_L < T, \text{ or}$$
$$\text{if } T_H < T \text{ and } 0.0 < \frac{dT}{dt}$$

$$q_5 = 9.2(-20. - T)$$

Develop a FORTRAN or BASIC program, using a fourth-order Runge–Kutta method, for simulating temperature, T, versus time, t. Assume consistent units and demonstrate your program using the following parameters. Present your results with an appropriate graphical display and printout.

$$V = 1000. \qquad T_H = 70. \qquad T_L = 65.$$
$$T_0 = 50. \qquad 0.0 \leqslant t \leqslant 0.5 \qquad \Delta t = 0.005$$

5.68(L) In the design of a wastewater treatment system, the oxygen concentration in the water owing to a special aerator operation can be simulated from the following model.

$$\frac{dC}{dt} - \left(\frac{N\theta Q^a H^{0.78}}{V}\right)(C_1 - C) = 0; \qquad C(0) = C_0$$

t	0.0	1.0	3.0	4.0	6.0	8.0	9.0
$Q(t)$	31.0	20.0	5.6	10.9	17.0	15.2	13.5

Develop a FORTRAN or BASIC program for simulating the concentration, C, versus time, t, using a fourth-order Runge–Kutta method. Apply a subprogram to get linearly interpolated values of Q. Assume consistent units and demonstrate your program using the following parameter values. Present your results with a trend print-plot or other graphical display.

$$N = 12. \qquad \theta = 3.1672 \qquad a = 1.377 \qquad H = 15.$$
$$V = 2 \times 10^5 \qquad C_1 = 11.35 \qquad C_0 = 2.$$
$$0.0 \leqslant t \leqslant 9.0 \qquad\qquad\qquad \Delta t = 0.05$$

5.69(L) The capacities and pressurization times of a pneumatic accumulator on a large shock-absorbing system can be simulated from the following model.

$$\frac{dP}{dt} = C_1 \dot{m} + C_2 Q; \qquad P(0) = P_0$$

$$\dot{m} = C_3 \left(\frac{P}{P_i}\right)^n \sqrt{1 - \left(\frac{P}{P_i}\right)^m}$$

Develop a FORTRAN or BASIC program for simulating the pressure, P, with respect to time, t, using a fourth-order Runge–Kutta method. Assume consistent units and demonstrate your program using the following parameters. Present the results with trend print-plot or other graphical display.

$$C_1 = 20909.28, \quad C_2 = 155.9649, \quad C_3 = 0.0360213; \qquad P_0 = 110.$$
$$n = 0.71429 \qquad P_i = 200. \qquad m = 0.28571$$
$$Q = -0.1 \qquad 0.0 \leqslant t \leqslant 1.0 \qquad \Delta t = 0.01$$

5.70(L) The temperature transducer response in a food processing system can be simulated from the following differential model.

$$\frac{d\theta}{dt} + C_1 [C_2 - C_3(\theta - \theta_1)](\theta - \theta_1) = 0.; \qquad \theta(0) = \theta_0$$

$$\theta_1 = C_4 + C_5 \sin wt$$

Develop a FORTRAN or BASIC program for simulating θ with respect to time, t, using a fourth-order Runge–Kutta method. Assume consistent units and demonstrate your program using the following parameters. Present the results with a trend print-plot or other graphical display.

$$C_1 = 1.41 \qquad C_2 = 5.2 \qquad C_3 = 0.003$$
$$C_4 = 200. \qquad C_5 = 5.3 \qquad w = 24.881$$
$$\theta_0 = 70. \qquad 0.0 \leqslant t \leqslant 0.24 \qquad \Delta t = 0.002$$

5.71(L) The liquid level, Z, in a reservoir connected to another reservoir through a pipeline system can be simulated from the following model.

$$\frac{d^2Z}{dt^2} + \left(\frac{A_1}{A}\right)\left(\frac{0.5f}{D}\right)\left(\frac{L_e}{L}\right)\left|\frac{dZ}{dt}\right|\frac{dZ}{dt} + \left(\frac{gA_1}{L}\right)\left(\frac{1}{A_1} + \frac{1}{A_2}\right)Z = 0.0$$

$$Z(0) = 40.0 \qquad\qquad Z'(0) = 0.0$$

Develop a FORTRAN or BASIC program for simulating Z and Z' versus time, t, using a fourth-order Runge–Kutta method. Assume consistent units and demonstrate your program using the following parameters. Present your results with a trend print-plot or other graphical display of both Z and Z'.

$A_1 = 200.$	$A = 7.0686$	$f = 0.024$	$D = 3.0$
$L_e = 2437.5$	$L = 2000.$	$g = 115812.$	$A_2 = 300.$
$0.0 \leqslant t \leqslant 2.0$	$\Delta t = 0.02$		

5.72(L) The swaying movement of a traffic-light mounting system in a high wind is estimated from the following model.

$$\frac{d^2\theta}{dt^2} + S|\theta|^2 - C_1 \left|\left(V - P\frac{d\theta}{dt}\right)\cos\theta\right|^n = 0.0$$

$$V = C_2\left(1 + \sin\frac{\pi t}{2}\right)$$

$$\theta(0) = 0.0 \qquad\qquad \theta'(0) = 0.0$$

Develop a FORTRAN or BASIC program, using a fourth-order Runge–Kutta method, for simulating the angular displacement, θ, and velocity, θ', versus time, t. Assume consistent units and demonstrate your program using the following parameters. Present the results with a printout and computer-graphics display of both displacement and velocity.

$S = 40.$	$C_1 = 1 \times 10^{-6}$	$C_2 = 200.$	$P = 240.$
$n = 1.5$	$0.0 \leqslant t \leqslant 20.$	$\Delta t = 0.2$	

5.73(L) The temperature, T, in an internally heated porous plate with one-dimensional flow-through cooling can be simulated from the following model.

$$\frac{d^2T}{dx^2} - A\frac{dT}{dx} = S$$

$$T(0) = 0.045 \qquad T'(0) = 3.1$$

Develop a FORTRAN or BASIC program, using a fourth-order Runge–Kutta method, for simulating T and T' versus position, x. Assume consistent units and demonstrate your program using the following parameters. Present the results with a trend print-plot of both T and T'.

$$A = 2.5 \qquad S = -10.75$$
$$0.0 \leqslant x \leqslant 0.5 \qquad \Delta x = 0.005$$

5.74(L) A lumped spring dashpot model is used to study certain suspension systems. Displacements, x, are simulated by the following model.

$$\frac{d^2x}{dt^2} + \left(\frac{D}{C_1}\right)\frac{dx}{dt} + \left(\frac{k}{C_1}\right)x = \left(\frac{D}{C_1}\right)\frac{dx_0}{dt} + \left(\frac{k}{C_1}\right)x_0$$

$$x(0) = 0.0 \qquad x'(0) = 0.0$$

$$x_0 = A(1 - \cos wt)$$

$$D = C_2\left[1 + C_3\left(\frac{dx}{dt} - \frac{dx_0}{dt}\right)^n\right]$$

Develop a FORTRAN or BASIC program, using a fourth-order Runge–Kutta method, for simulating x and x' versus t. Assume consistent units and demonstrate your program using the following parameters. Present your results with a trend print-plot of both x and x'.

$$C_1 = 3500. \qquad C_2 = 160. \qquad C_3 = 10. \qquad n = 0.5$$
$$k = 500. \qquad A = 2.0 \qquad w = 2.0$$
$$0.0 \leqslant t \leqslant 10. \qquad \Delta t = 0.1$$

5.75(L) The internal damping behavior of an energy storage flywheel system is being studied by simulations of relative angular displacements, θ, from the following model.

$$\frac{d^2\theta}{dt^2} + \frac{A}{J}\left(\frac{d\theta}{dt}\right)^{1.4} + \frac{k}{J}\theta = 0.$$

$$\theta(0) = 10. \qquad \theta'(0) = 0.0$$

Develop a FORTRAN or BASIC program, using a fourth-order Runge–Kutta method, for simulating θ and θ' versus time, t. Assume consistent units and demonstrate your program using the following parameters. Present the results with an appropriate graphical display and printout.

$$J = 4.1 \qquad A = 0.17 \qquad k = 20.8$$
$$0.0 \leqslant t \leqslant 15 \qquad \Delta t = 0.1$$

5.76(L) In the design of an aerator system for treatment of a process effluent, the residence time of air bubbles is simulated from the following model.

$$\frac{d^2x}{dt^2} + AB^{-0.333}\left(\frac{dx}{dt}\right)^2 - \frac{C}{B} = -1.0$$

$$x(0) = 0.0 \qquad x'(0) = 0.0$$

$$B = \gamma(10 - x) + 2116.8$$

Develop a FORTRAN or BASIC program, using a fourth-order Runge–Kutta method, for simulating displacement, x, and velocity, x', versus time, t. Assume consistent units and demonstrate your program using the following parameters. Present the results with an appropriate graphical display and printout.

$$\gamma = 62.4 \qquad A = 2.9277 \times 10^4 \qquad C = 1.76 \times 10^6$$
$$0.0 \leqslant t \leqslant 0.005 \qquad\qquad\qquad \Delta t = 5 \times 10^{-5}$$

5.77(L) A gas spring used to absorb valve-opening shocks in a liquid-pumping system is to be studied by simulating displacements, y, from the following model.

$$\frac{d^2 y}{dt^2} + \left[\frac{Cg}{(aL\gamma)} \right] \left(\frac{dy}{dt} \right)^{1.5} + \left(\frac{2g}{L} \right) \left[1 + \frac{nP_0(1 - y/L_0)^{-(n+1)}}{\gamma L_0} \right] y = 0$$
$$y(0) = 0.1 \qquad\qquad y'(0) = 0.0$$

Develop a FORTRAN or BASIC program, using a fourth-order Runge–Kutta method, for simulating y and y' versus time, t. Assume consistent units and demonstrate your program using the following parameters. Present your results with a trend print-plot or other graphical display.

$$C = 0.35 \qquad g = 32.17 \qquad a = 0.02 \qquad L = 10.$$
$$\gamma = 62.4 \qquad n = 1.32 \qquad P_0 = 28800 \qquad L_0 = 1.0$$
$$0.0 \leqslant t \leqslant 0.4 \qquad\qquad \Delta t = 0.002$$

5.78(L) Certain leg movements of a human runner are being studied by simulations from the following model.

$$\frac{d^2 x}{dt^2} + R \left(\frac{dx}{dt} \right)^n + Sx^m = F$$
$$x(0) = 1.0 \qquad\qquad x'(0) = 0.0$$

Develop a FORTRAN or BASIC program, using a fourth-order Runge–Kutta method, for simulating displacements, x, and velocities, x', versus time, t. Assume consistent units and demonstrate your program using the following parameters. Present the results with suitable graphical display and compare with an exact-solution simulation.

$$R = 2. \qquad n = 1.0 \qquad S = 3. \qquad m = 1.0$$
$$F = 0.0 \qquad 0.0 \leqslant t \leqslant 6.0 \qquad \Delta t = 0.05$$

5.79(L) The motion, y, of a nonlinear spring system can be simulated from the following model.

$$\frac{d^2 y}{dt^2} + (A - 1)y + 2y^3 = 0.0$$

$$y(0) = 0.0 \qquad\qquad y'(0) = 0.04$$

Develop a FORTRAN or BASIC program, using a fourth-order Runge–Kutta method, for simulating y and y' versus time, t. Assume consistent units and demonstrate your program using the following parameters. Present the results with a trend print-plot or other graphical display.

$$A = 0.04 \qquad 0.0 \leqslant t \leqslant 10. \qquad \Delta t = 0.1$$

5.80(L) The vibrational behavior of a slightly unbalanced motor during a startup is being simulated from the following model.

$$\frac{d^2x}{dt^2} + \frac{C}{M}\frac{dx}{dt} + \frac{k}{M}x = \frac{m}{M}w^2 \sin wt$$

$$x(0) = 0.0 \qquad\qquad x'(0) = 0.0$$

$$w = 120\pi\left(1 - \exp\frac{-t}{4}\right)$$

$$C = 1.2\left|\frac{dx}{dt}\right|^n$$

Develop a FORTRAN or BASIC program, using a fourth-order Runge–Kutta method, for simulating centerline displacements, x, and velocities, x', versus time, t. Assume consistent units and demonstrate your program using the following parameters. Present your results with a trend print-plot or other graphical display.

$$M = 0.1 \qquad n = 0.2 \qquad k = 800. \qquad m = 0.009$$
$$0.0 \leqslant t \leqslant 0.5 \qquad\qquad \Delta t = 0.005$$

5.81(L) The efficiency of a shredder system is being analyzed from the following model, where $\theta' = d\theta/dt$ represents angular velocity.

$$\frac{d^2\theta}{dt^2} + \frac{C}{I}\left(\frac{d\theta}{dt}\right)^n + \frac{B}{I}\frac{d\theta}{dt} - \frac{T(\theta')}{I} = 0$$

$$\theta(0) = 0.0 \qquad\qquad \theta'(0) = 500.$$

θ'	$T(\theta')$	θ'	$T(\theta')$
0	0	450	96
150	0	500	65
200	4	550	3
250	21	600	0
400	98	1000	0

Develop a FORTRAN or BASIC program, using a fourth-order Runge–Kutta method for simulating θ and θ' versus time, t. Use a linear interpolator to get T values. Assume consistent units and demonstrate your program using the following parameters. Present your results with a suitable graphical display.

$$C = 42.7 \qquad \text{for } 1.0 \leqslant t \leqslant 3.0, \text{ otherwise } C = 0.0$$

$$B = 0.5 \qquad I = 86. \qquad n = 0.7$$

$$0.0 \leqslant t \leqslant 6.0 \qquad \Delta t = 0.05$$

5.82(L) The following model is being used to approximate the nonlinear damped-spring behavior of a heavy equipment tire-wheel combination.

$$\frac{d^2y}{dt^2} + a_1 \left(\frac{dy}{dt}\right)^{1.6} + a_2 S = a_3$$

$$y(0) = 0.5 \qquad\qquad y'(0) = 0.0$$

$$S = C_1 [C_2 + (C_3 - C_4 y)(C_5 + C_6 y - C_7 y^2)^{0.5}]^{-1}$$

Develop a FORTRAN or BASIC program, using a fourth-order Runge–Kutta method, for simulating y and y' versus time, t. Assume consistent units and demonstrate your program using the following parameters. Present the results with a trend print-plot.

$a_1 = 1.508562$	$a_2 = 0.26466$	$a_2 = 386.4$	$C_1 = 7200$
$C_2 = 4.0$	$C_3 = 0.8$	$C_4 = 0.33333$	$C_5 = 1.36$
$C_6 = 0.53333$	$C_7 = 0.11111$	$0.0 \leqslant t \leqslant 3.0$	$\Delta t = 0.02$

5.83(L) A feasibility study of a new plastic housing for a delicate electronic system uses the following model to simulate movements and velocities after a drop.

$$\frac{d^2x}{dt^2} + \left(\frac{gC}{W}\right)\left(\frac{dx}{dt} - \frac{dy}{dt}\right)^n + \left(\frac{gK}{W}\right)(x - y) = 0.0$$

$$x(0) = 0.0 \qquad\qquad x'(0) = 25.0$$

$$\frac{d^2y}{dt^2} + \left(\frac{g}{w}\right)\left[B\left(\frac{dy}{dt}\right)^m - C\left(\frac{dx}{dt} - \frac{dy}{dt}\right)^n + Sy - K(x - y)\right] = 0.0$$

$$y(0) = 0.0 \qquad\qquad y'(0) = 25.0$$

Develop a FORTRAN or BASIC program, using a fourth-order Runge–Kutta method, for simulating x, x', y, and y' versus time, t. Assume consistent units and demonstrate your program using the following parameters. Present your results with a suitable graphical display.

$g = 32.17$	$W = 10.6$	$w = 16.0$	$n = 1.26$
$m = 1.10$	$K = 20.2$	$S = 2340$	$C = 13.2$
$B = 7.9$	$0.0 \leqslant t \leqslant 0.2$	$\Delta t = 0.001$	

5.84(L) The heat loss from a cylindrical pipe system is being studied with temperature simulations from the following model.

$$\frac{d^2T}{dr^2} + \left(\frac{1}{r}\right)\frac{dT}{dr} + \frac{b}{a+bT}\left(\frac{dT}{dr}\right)^2 = 0.0$$

$$T(0.1) = 150. \qquad T'(0.1) = -314.54$$

Develop a FORTRAN or BASIC program, using a fourth-order Runge–Kutta method, for simulating T and T' versus radial position, r. Assume consistent units and demonstrate your program using the following parameters. Present your results with a graphical display.

$$a = -1.54 \times 10^{-2} \qquad b = 1.87 \times 10^{-4}$$
$$0.1 \leqslant r \leqslant 0.28 \qquad \Delta r = 9 \times 10^{-4}$$

5.85(L) The deformations, y, in a heavily loaded cantilever beam are being simulated from the following model.

$$\frac{d^2y}{dx^2} - \frac{M}{EI}\left[1+\left(\frac{dy}{dx}\right)^2\right]^{1.5} = 0.0$$

$$y(0) = 0.0 \qquad y'(0) = 0.0$$

$$M = wLx - \frac{wL^2}{2} - \frac{wx^2}{2}$$

Develop a FORTRAN or BASIC program, using a fourth-order Runge–Kutta method, for simulating y and y' versus x. Assume consistent units and demonstrate your program using the following parameters. Present your results with a graphical display.

$$w = 8.0 \qquad 0.0 \leqslant x \leqslant L$$
$$L = 10.0$$
$$\Delta x = \frac{L}{100} \qquad EI = 1.37816 \times 10^{-6}$$

5.86(L) Movements, x, of a spring-mounted mechanical system can be simulated from the following model.

$$\frac{d^2x}{dt^2} + m(x^2-1)\frac{dx}{dt} + x = 0.0$$

$$x(0) = 0.75 \qquad x'(0) = 0.0$$

Develop a FORTRAN or BASIC program, using a fourth-order Runge–Kutta method, for simulating x and x' versus time, t. Assume consistent units and demonstrate your program using the following parameters. Present your results with a trend print-plot.

$$m = 4.0 \qquad 0.0 \leqslant t \leqslant 20.0 \qquad \Delta t = 0.1$$

5.87(L) The performance of a small sounding rocket in still air can be simulated from the following model.

$$\frac{d^2x}{dt^2}+\left(\frac{Cg}{W}\right)\left(\frac{dx}{dt}\right)^{1.8}=\left(\frac{F}{W}-1\right)g$$

$$x(0)=0.0 \qquad\qquad x'(0)=0.0$$

t	F	W	t	F	W
0.00	0.00	0.0324	0.40	—	0.0292
0.10	—	0.0318	0.50	—	0.0284
0.12	1.73	—	0.60	—	0.0273
0.13	1.77	—	0.70	—	0.0261
0.15	1.60	—	0.80	0.72	0.0250
0.17	1.10	—	0.83	0.66	0.0247
0.20	0.90	0.0304	0.85	0.50	—
0.22	0.82	—	0.87	0.00	—
0.24	0.76	—	12.	0.00	0.0247
0.30	0.72	0.0299	—	—	—

Develop a FORTRAN or BASIC program, using a fourth-order Runge–Kutta method, for simulating elevation, x, and velocity, x', versus time, t. Apply linear interpolation to get F and W values. Assume consistent units and demonstrate your program using the following parameters. Present results with a graphical display.

$$g=32.17 \qquad C=1.77\times10^{-6}$$
$$0.0\leqslant t\leqslant8.0 \qquad \Delta t=0.05$$

5.88(L) The cool-down behavior of a small container and its liquid contents is being studied using lumped analysis and the following model.

$$\frac{dL}{dt}+A_1(L-C)=0; \qquad L(0)=70$$

$$\frac{dC}{dt}+A_2(C-T_s)+A_3(C-L)=0; \quad C(0)=70$$

$$A_1=\frac{h_iA_i}{M_LS_L}$$

$$A_2=\frac{h_oA_o}{M_CS_C}$$

$$A_3=\frac{h_iA_i}{M_CS_C}$$

Develop a FORTRAN or BASIC program, using a third-order Adams predictor–corrector method with three corrections per step with a fourth-order Runge–Kutta starter, for simulating L and C versus time, t. Assume consistent units and demonstrate your program using the parameters given below. Present the results graphically.

$$h_i = 6.0 \qquad h_o = 3.0 \qquad A_i = 0.8 \qquad A_o = 0.85$$
$$M_L = 0.55 \qquad M_C = 0.11 \qquad S_C = 0.21 \qquad S_L = 1.0$$
$$T_S = 41.0 \qquad 0.0 \leqslant t \leqslant 0.5 \qquad \qquad \Delta t = 0.005$$

5.89(L) Repeat any problem from 5.58(L) through 5.88(L), except develop the program in CSMP or other appropriate POL. Compare results using different simulation methods specified in CSMP by the METHOD statement.

5.90(L) Deflections, y, of a compression-loaded strut can be simulated from the following nonlinear model.

$$\frac{d^2y}{dx^2} + \left[\frac{Py+M}{EI}\right]\left[1 + \left(\frac{dy}{dx}\right)^2\right]^{1.5} = 0.0$$

$$y(0) = 0.0 \qquad\qquad y(L) = 0.0$$

$$M = 5x \text{ for } x \leqslant \frac{L}{2}, \qquad \text{otherwise } M = 5(L-x)$$

Develop a FORTRAN or BASIC program, using a fourth-order Runge–Kutta shooting method, for simulating y and y' versus x. Use starters of (1) $y'(0) = -0.01$, and (2) $y'(0) = -0.50$, and apply a secant acceleration scheme ($\varepsilon = 0.01$ for the convergence test but limit the number of shoots to 20) through a subprogram. Assume consistent units and demonstrate your program using the following parameters. Present results graphically.

$$P = 500 \qquad\qquad EI = 5 \times 10^5 \qquad L = 100.$$

$$0.0 \leqslant x \leqslant L \qquad \Delta x = \frac{L}{100}$$

(See Chapter 7 topics also.)

5.91(L) The temperatures within an extruded plastic cover for a space application can be simulated from the following model.

$$\frac{d^2T}{dx^2} + H^{-1}\frac{dH}{dx}\frac{dT}{dx} - \left[\frac{\sigma e}{kH}\right]T^4 = 0.0$$

$$T(0) = 350. \qquad\qquad T(L) = 350.$$

$$H = A + Bx$$

Develop a FORTRAN or BASIC program, using a fourth-order Runge–Kutta shooting method, for simulating T and T' versus position, x. Use starters of (1) $T'(0) = -2500$, and (2) $T'(0) = -5000.$, and apply a secant acceleration scheme ($\varepsilon = 0.01$ for the convergence test but limit the number of shoots to 20) through a subprogram. Assume consistent units and demonstrate your program using the following parameters. Present your results graphically.

$$A = 0.02 \qquad B = 0.1 \qquad k = 1.5 \qquad e = 0.9$$

$$\sigma = 5.67 \times 10^{-8} \qquad L = 0.2 \qquad 0.0 \leqslant x \leqslant L \qquad \Delta x = \frac{L}{100}$$

5.92(L) The deflections, y, of a simply supported composite beam can be simulated from the following linearized model.

$$\frac{d^2 y}{dx^2} = \frac{M}{EI} \qquad y(0) = 0.0, \quad y(L) = 0.0$$

$$M = \frac{wx(L-x)}{2} + P(L - L_p) \frac{x}{L} - M_r$$

$M_r = P(x - L_p)$ for $x \geqslant L_p$, otherwise $M_r = 0.0$

$EI = 36.9 \times 10^6$ for $A \leqslant x \leqslant B$, otherwise $EI = 9.3 \times 10^6$

$w = 13.7$ for $A \leqslant x \leqslant B$, otherwise $w = 6.0$

Develop a FORTRAN or BASIC program, using a finite-difference matrix method and a Gauss–Jordan equation-set solver, for simulating y versus x. Assume consistent units and demonstrate your program using the following parameters. Present your results graphically.

$$L = 14. \qquad P = 500. \qquad L_p = 10. \qquad A = 5.5$$

$$B = 8.5 \qquad 0.0 \leqslant x \leqslant L \qquad \Delta x = \frac{L}{50}$$

(See Problem 5.93(L) and Chapter 7 topics also.)

5.93(L) The deflections, y, of a simply supported composite beam can be simulated from the following nonlinear model.

$$\frac{d^2 y}{dx^2} - \frac{M}{EI} \left[1 + \left(\frac{dy}{dx} \right)^2 \right]^{1.5} = 0.0$$

$$y(0) = 0.0 \qquad\qquad y(L) = 0.0$$

$$M = \frac{wx(L-x)}{2} + P(L - 585L) \frac{x}{L} - M_r$$

$M_r = P(x - L_p)$ for $x \geqslant L_p$, otherwise $M_r = 0.0$

$EI = 36.9 \times 10^6$ for $A \leqslant x \leqslant B$, otherwise $EI = 9.3 \times 10^6$

$w = 13.7$ for $A \leqslant x \leqslant B$, otherwise $w = 6.0$

Develop a FORTRAN or BASIC program, using a fourth-order Runge–Kutta shooting method, for simulating y and y' versus position, x. Use starters of (1) $y'(0) = -0.01$, and (2) $y'(0) = -0.05$, and apply a secant acceleration scheme ($\varepsilon = 0.00001$ for the convergence test but limit the number of shoots to 20) through a subprogram. Assume consistent units and demonstrate your program using the following parameters. Present results graphically.

$$L=14. \qquad P=500. \qquad L_p=10. \qquad A=5.5$$

$$B=8.5 \qquad 0.0 \leqslant x \leqslant L \qquad \Delta x = \frac{L}{100.}$$

(See Chapter 7 topics also.)

5.94(L) The deflections, y, in a restrained cantilever beam arm of a remote manipulator system under a load, F, can be simulated from the following linearized model.

$$\frac{d^2y}{dx^2} - \frac{M_a + M_b}{EI} = 0.$$

$$y(0)=0.0 \qquad y'(0)=0.0 \qquad y(L)=1.0$$
$$M_a=2400(192-x) \text{ for } A_1 \leqslant x \leqslant A_2, \quad \text{otherwise } M_a=0.0$$
$$M_b=F(x-228) \text{ for } A_3 \leqslant x \leqslant A_4, \quad \text{otherwise } M_b=0.0$$
$$EI=20 \times 10^6 \text{ for } B_1 \leqslant x \leqslant B_2$$
$$EI=15 \times 10^6 \text{ for } B_3 \leqslant x \leqslant B_4, \quad \text{otherwise } EI=10 \times 10^6$$

Develop a FORTRAN or BASIC program, using a finite-difference matrix method and Gauss–Jordan equation-set solver, for simulating y versus x. Assume consistent units and demonstrate your program using the following parameters. Present your results graphically.

$$L=240. \qquad A_1=0.0 \qquad A_2=192. \qquad A_3=0.0 \qquad A_4=228.$$
$$B_1=0.0 \qquad B_2=100 \qquad B_3=200. \qquad B_4=L$$
$$F=1756.04 \qquad 0.0 \leqslant x \leqslant L \qquad \Delta x = 2.0$$

(See Chapter 7 topics also.)

5.95(L) Repeat the preceding problem, except use a fourth-order Runge–Kutta shooting method and consider the problem to be an inverse problem, where F has to be determined. Use starters of (1) $F=1000.$ and (2) $F=2000.$, and apply a secant acceleration scheme ($\varepsilon=0.1$ for the convergence test but limit the number of shoots to 20) through a subprogram.

5.96(L) The steady temperatures, T, in a heat transfer fin can be simulated from the following linear model.

$$\frac{d^2T}{dx^2} - PT = PT_a$$

$$T(0)=T_0, \qquad T'(L)=\left(\frac{h}{k}\right)(T_a-T(L))$$

$$P=\frac{h(2t+2w)}{12kwt}$$

Develop a FORTRAN or BASIC program, using a finite-difference matrix method and Gauss–Jordan equation-set solver, for simulating T versus position, x. Assume consistent units and demonstrate your program using the following parameters. Present your results graphically. Compare your results with a simulation from an exact solution and/or a shooting method simulation [see Problem 5.97(L)].

$$L = 9.126 \qquad h = 3.9 \qquad k = 30.0 \qquad w = 1.465$$
$$t = 0.228 \qquad T_0 = 503. \qquad T_a = 91.$$
$$0.0 \leqslant x \leqslant L \qquad \Delta x = 0.3042$$

5.97(L) Repeat Problem 5.96(L), except use a fourth-order Runge–Kutta shooting method. Use starters of (1) $T'(0) = -100.$ and (2) $T'(0) = -200.$, and apply a secant acceleration scheme ($\varepsilon = 0.1$ for the convergence test but limit the number of shoots to 20) through a subprogram.

Some Case-Study Projects

5.98(C) Set up an appropriate differential model and perform some simulations of a thrown or batted object, like a tennis ball or baseball. Try to account for such influences as wind and the spinning aerodynamic effects.

5.99(C) Set up an appropriate differential model and perform some simulations of the thermal energy flows from, or to, a small residential building. Include sufficient time in the simulation span to account for some seasonal climate changes.

5.100(C) Set up an appropriate differential model and perform some simulations of water levels in a small natural watershed region. Include at least one lake and stream.

5.101(C) Set up an appropriate differential model and perform some simulations of a vehicle traveling over a rough road surface. Include as much detail as possible about the individual wheels and suspension components.

5.102(C) Set up an appropriate differential model and perform some simulations of deflections of a composite beam or strut system under various loadings. Include as much detail as possible about the composite interactions, within the restrictions of one-dimensional modeling.

6
Numerical Simulations from Partial-Derivative Models[1]

ELLIPTIC-MODEL SIMULATIONS WITH FINITE-DIFFERENCE METHODS[2]

The domain of the simulation is determined from the boundary conditions. We concentrate on plane x, y (rectangular Cartesian coordinates) domains, but the methods will be general and extendable to other coordinate systems and three dimensions. The central ideas have already been discussed in Chapter 5 (Matrix Methods for Linear Models), but most of the information will be repeated here in the context of multidimensional simulations.

The first step is to discretize (divide up into discrete parts) the domain. Basically, two different procedures can be used. In the first, we superimpose a grid of uniformly spaced, intersecting lines (whose directions coincide with the coordinates of the model) over the domain, taking care to place intersections on the boundaries or as near to the boundaries as possible. Obviously, for highly irregular boundaries, we need to use a relatively fine grid to get the desired intersections on or near the boundaries. Each intersection then becomes a nodal point for our discretization process. Furthermore, we use an integer index scheme to identify the grid lines, typically an i value for each y direction line, and a j value for each x direction line. Thus each nodal point can be located physically, just as on a map of the domain, by using a double i, j subscript. In the second, we divide the domain up into cells. It is generally convenient to use uniform-size rectangular cells, except at the boundaries, where half-size cells are used. However, it should be noted that *we do not have to use a uniform size*. Each cell is then marked, or labeled, with a single symbol; typically, a single number that can be used as a subscript is best. Then we locate a point in each cell where we feel that "lumped" storage can be best represented (ordinarily this is at the geometric center, except for a boundary cell, where it is advantageous to place it right on the boundary). These points are called nodal points also, but the reader is cautioned against assuming that they are the same as the nodal points from the grid discretization procedure. It should be evident that

---

[1] For additional background information, consult the references.[1 − 7]
[2] Finite-element methods are treated in Chapter 7.

for highly irregular domains, the cell discretization procedure has some setup advantages over the grid scheme; however, this procedure should not be mistaken for a "finite-element" method (see Chapter 7).

The next step is to get a general algebraic difference equation, or the equivalent, identified as a replacement for the model differential equation. If the grid discretization procedure has been used, this is a simple matter of substituting finite-difference approximations for the derivatives, using central differences around each node i, j wherever possible. If a Neumann (slope) condition exists at a boundary, then a finite-difference equation (formed from forward or backward difference approximations) will have to be written for the boundary also. If the cell scheme has been used for discretizing, then a more basic equation, using the universal flow concept (Equation 2.6), will have to be set up. A common trick used here is to assume that the flow-driving potential (force) at the cell of interest is less than the potentials of all the adjacent cells and/or boundary effects. Then we simply sum all flows into the cell of interest and set the result to zero. In effect, when we do this, we are applying the basic conservation principle directly to the cells without setting up a differential-equation model first. It is interesting that we need to know at least four flow resistances and that we can write this equation out in only a very general fashion because we have not established any definite pattern for the adjacent cell markers.

To continue, specific algebraic equations have to be written for each node or cell, using the generic equations from the foregoing. In doing this, we note a significant difference between the Neumann boundary equations we get from the two different discretization procedures. The equations coming from the cell approach are more complicated and show influences from adjacent boundary cells; whereas the grid approach equations are very simple two-term relationships, showing no influences from adjacent boundary nodes. It is well known that the former give more accurate simulations when compared with actual measurements. Consequently, it is common practice to mix the two approaches, using the grid procedure for interior nodes and the cell procedure for the boundary nodes. Some obvious differences have to occur in the subscripting.

Some elliptic problems will be given with redundant conditions. If available, the Dirichlet conditions should always be chosen over any coinciding Neumann conditions. Sometimes, however, only Neumann conditions are given. Care must be exercised in this case because the resulting simulation will not be unique. The recommended procedure is to introduce a one-point artificial Dirichlet condition (usually a zero or one is preferred) superimposed on the given Neumann condition at that point. Also, so-called "interior boundaries" of the Neumann type (i.e., regions completely surrounded by the domain) are best approximated by source, or sink, values if the boundary-region areas are not too large.

The resulting set of equations is simplified and re-posed into a linear matrix equation. It should be apparent that the matrix equation originating from the cell discretization approach is much easier to read and has subscripts consistent with the mathematical representation. For either case, the coefficient matrix of the matrix equation is sparse; that is, it consists mostly of zeros and probably has a dominant diagonal. The bandwidth (overall span of the diagonals on each side of the main diagonal that contain nonzero elements), may be small relative to the size of the matrix, but we observe that the size of this bandwidth is dependent upon the pattern used to number the cells if the cell discretization was used. Generally speaking, a small bandwidth is desirable. The driving vector is made up from Dirichlet boundary values and any local source/sink constants.

Many methods are available for solving a linear matrix equation. We shall remain with the two introduced in Chapter 5, namely, the elementary Gauss elimination method (an exact method), and the elementary Gauss–Seidel method (an iterative method). However, the reader is urged to check his or her own software sources for other more sophisticated solvers, especially if a large matrix equation (over about 100 unknowns) is the basis of a simulation. For those not familiar with the foregoing methods, brief explanations are given in Appendix A, and their use is demonstrated in the examples. If the matrix equation originated from the grid discretization procedure, and the Gauss–Seidel solver is to be used, significant array storage can be saved by posing the equations in Liebmann, or explicit, form, and simply cycling the computation process through the physical i, j subscripts. This is illustrated in the examples. Occasionally, nonlinearities show up, for example when the resistances in the cell approach are dependent upon the driving potentials. In such cases the exact solvers cannot be used. The iterative solvers can be used, but the simulation values upon which they converge, if convergence is ever obtained, may not be unique. If simulation results do not seem reasonable in such a case, the recommendation is to repose the explicit equations and try again. The alternative is model linearization.

Examples 6.1, 6.2, and 6.3 illustrate some of the foregoing items.

EXAMPLE 6.1

Temperatures within a long, square cross section beam are to be simulated using the following Laplace equation model. (1) Using $\Delta x = \Delta y = 0.2$, the grid discretization procedure, and a Gauss–Seidel equation-set solver (use a calculator and tabular algorithm), demonstrate a T versus x and y finite-difference simulation. (2) Change the grid spacing to $\Delta x = \Delta y = 0.06$ and set up a BASIC computer program to do the same simulation.

$$\frac{\partial^2 T}{\partial x^2} + \frac{\partial^2 T}{\partial y^2} = 0; \qquad T(0.0, y) = 80., \quad T(0.6, y) = 100.$$
$$T(x, 0.0) = 60., \quad T(x, 0.6) = 150.$$

▶ **1.** The superimposed grid is given i values in the x direction, for example, $i=1$ at $x=0.0$, $i=2$ at $x=0.2$, and so on, and j values in the y direction, for example, $j=1$ at $y=0.0$, and so on. Thus we have four interior points. The general difference equation is written around the arbitrary i, j nodal point by substituting first central difference approximations for the derivatives in the model, giving the following algebraic equation.

$$\frac{T_{i+1,j} - 2T_{i,j} + T_{i-1,j}}{(\Delta x)^2} + \frac{T_{i,j+1} - 2T_{i,j} + T_{i,j-1}}{(\Delta y)^2} = 0$$

But simplification gives us our general finite-difference equation.

$$T_{i-1,j} + T_{i,j-1} - 4T_{i,j} + T_{i+1,j} + T_{i,j+1} = 0$$

The unknowns are $T_{2,2}$, $T_{2,3}$, $T_{3,2}$, $T_{3,3}$. Specific equations can be written for each node using the foregoing equation. We will substitute the Dirichlet values directly for any boundary points in the equations.

Node $i=2, j=2$: $\qquad$ $80 + 60 - 4T_{2,2} + T_{3,2} + T_{2,3} = 0$
Node $i=2, j=3$: $\qquad$ $80 + T_{2,2} - 4T_{2,3} + T_{3,3} + 150 = 0$
Node $i=3, j=2$: $\qquad$ $T_{2,2} + 60 - 4T_{3,2} + 100 + T_{3,3} = 0$
Node $i=3, j=3$: $\qquad$ $T_{2,3} + T_{3,2} - 4T_{3,3} + 100 + 150 = 0$

These equations can be re-posed into the following matrix equation. Note that the bandwidth of the coefficient matrix is five.

$$\begin{bmatrix} -4.0 & 1.0 & 1.0 & 0.0 \\ 1.0 & -4.0 & 0.0 & 1.0 \\ 1.0 & 0.0 & -4.0 & 1.0 \\ 0.0 & 1.0 & 1.0 & -4.0 \end{bmatrix} \begin{Bmatrix} T_{2,2} \\ T_{2,3} \\ T_{3,2} \\ T_{3,3} \end{Bmatrix} = \begin{Bmatrix} -140.0 \\ -230.0 \\ -160.0 \\ -250.0 \end{Bmatrix}$$

However, the explicit (Liebmann) form of writing the equation set works better with the Gauss–Seidel method:

$$T_{2,2} = \frac{140.0 + T_{2,3} + T_{3,2}}{4.0}$$

$$T_{2,3} = \frac{230.0 + T_{2,2} + T_{3,3}}{4.0}$$

$$T_{3,2} = \frac{160.0 + T_{2,2} + T_{3,3}}{4.0}$$

$$T_{3,3} = \frac{250.0 + T_{2,3} + T_{3,2}}{4.0}$$

We choose starters of $T_{i,j} = 100.0$, and a convergence parameter of $\varepsilon = 0.5$, for our manual computations. The results are shown in Table 6.1.

A sample calculation is given for iteration no. 3, and for node $i=3$ and $j=2$. Using the third Liebmann equation, we get

$$T_{3,2} = \frac{160.0 + 83.44 + 110.94}{4.0} = 88.60$$

$$\Delta T_{3,2} = 88.60 - 88.13 = 0.47$$

2. We can choose a much smaller convergence parameter when using the

Table 6.1 Computational Results, Example 6.1(1)

Iteration	$T_{2,2}$	$\Delta T_{2,2}$	$T_{2,3}$	$\Delta T_{2,3}$	$T_{3,2}$	$\Delta T_{3,2}$	$T_{3,3}$	$\Delta T_{3,3}$	F.Pts.
—	100[a]	—[b]	100[a]	—[b]	100[a]	—[b]	100[a]	—[b]	4[c]
1	85.0	15.0	103.75	3.75	86.25	13.75	110.0	10.0	4
2	82.5	2.5	105.63	1.88	88.13	1.88	110.94	0.94	4
3	83.44	0.94	106.1	0.47	88.6	0.47	111.18	0.24	1
4	83.68	0.24	106.22	0.12	88.72	0.12	111.24	0.06	0

[a]Initial guesses.
[b]Difference between current value and one from preceding iteration.
[c]Number of points that do not satisfy $|\Delta T_{i,j}| \leqslant \varepsilon$.

computer; so we choose $\varepsilon = 0.01$. However, we keep the starting values of $T_{i,j} = 100.0$. The BASIC program setup is shown in Figure 6.1.

```
                              Main Program
100    REM EXAMPLE 6.1 BASIC PROGRAM FOR ELLIPTIC-MODEL SIMULATION
110    REM FINITE DIFFERENCE SIMULATION WITH GAUSS-SEIDEL SOLVER
120    CLS! CLEAR20! DEFINT I-N
130    DIM T(50,50)
140    READ NI, NJ, MX, EP, SV
150      DATA 11, 11, 100, .01, 100.
160    REM SET STARTER AND BOUNDARY VALUES
170      FOR I = 1 TO NI! FOR J = 1 TO NJ
180        T(I,J) = SV
190        IF I = 1 THEN T(I,J) = 80.
200        IF I = NI THEN T(I,J) = 100.
210        IF J = 1 THEN T(I,J) = 60.
220        IF J = NJ THEN T(I,J) = 150.
230      NEXT J,I
240    REM ITERATION LOOP
250    FOR IT = 1 TO MX! M = 0
260    REM COMPUTATION LOOPS, CYCLING THROUGH NODES-- THE I, J  VALUES
270        FOR I = 2 TO NI-1! FOR J = 2 TO NJ-1
280          SA = T(I,J)
290          T(I,J) = ( T(I-1,J) + T(I+1,J) + T(I,J-1) + T(I,J-1) )/4.
300          DF = SA - T(I,J)
310    REM TEST FOR CONVERGENCE
320          IF ABS(DF) > EP THEN M = M + 1
330        NEXT J,I
340        IF M <= 0 GOTO 380
350        PRINT"ITERATION NO.="!IT!" NO. OF POINTS FAILING TO CONVERGE ="!M
360    NEXT IT
370    LPRINT "******* NO CONVERGENCE !!!"
380    REM PRINT RESULTS
390    LPRINT TAB(20)"GAUSS-SEIDEL RESULTS"
400    LPRINT "CONVERGENCE PARAMETER ="!EP
410    LPRINT "NO. OF ITERATIONS COMPLETED ="!IT
420      FOR I = 1 TO NI! FOR J = 1 TO NJ
430        LPRINT "I ="!I!" J ="!J!" T ="!T(I,J)
440      NEXT J,I
450    END
```

Figure 6.1   BASIC program using a Gauss–Seidel solver, Example 6.1.

EXAMPLE 6.2

Repeat the problem described in Example 6.1, except use a cell-discretization procedure and a Gauss elimination method. (1) Assume a cell size of 0.2×0.2, with a flow resistance across the cell, in either the x or y direction, of $R = 0.05$. Show a sketch of the cells and perform a manual simulation (calculator and tabular algorithm) of the cell temperatures. (2) Change the left-hand boundary (i.e., at $x = 0$.) condition to a convective-type condition having a cell-area flow resistance of $R_c = 0.01$, and having an adjacent environment temperature of $T_c = 80$. Note that this is equivalent to specifying a Neumann condition. Set up a FORTRAN program to do the simulation.

▶ **1.** The sketch of the cell layout is shown in Figure 6.2. We choose a single-number marker (label) for each cell, "lump" the storage of each interior cell in the geometric center of the cell, and designate it with an "O". The storage in a boundary cell is placed on the boundary but is centered otherwise.·

In using the cell-discretization procedure, we have to recognize how the original model was set up. In this case we recognize this as an energy-flow problem and go back to the steady state conservation principle, Equation 2.3,

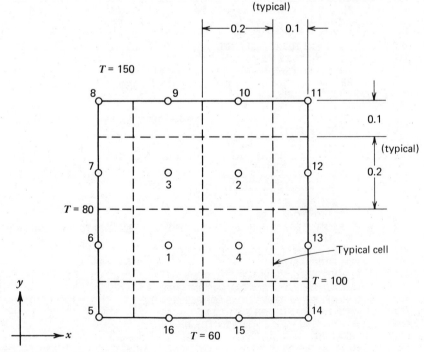

Figure 6.2 Cell layout, Example 6.2.

and the universal flow formula, Equation 2.6. Each cell becomes a control volume. To each cell we apply the trick of setting its lumped temperature lower than the adjacent cells, applying Equation 2.6 for the flow from each adjacent cell and summing all flows to zero. We use a hyphenated subscript to indicate the flow direction, for example, the subscript i–j means the flow from cell i to cell j.

Cell 1: $q_{6-1} + q_{4-1} + q_{16-1} + q_{3-1} = 0.$

$$\frac{T_6 - T_1}{R_{6-1}} + \frac{T_4 - T_1}{R_{4-1}} + \frac{T_{16} - T_1}{R_{16-1}} + \frac{T_3 - T_1}{R_{3-1}} = 0.$$

Cell 2: $q_{3-2} + q_{12-2} + q_{4-2} + q_{10-2} = 0.$

$$\frac{T_3 - T_2}{R_{3-2}} + \frac{T_{12} - T_2}{R_{12-2}} + \frac{T_4 - T_2}{R_{4-2}} + \frac{T_{10} - T_2}{R_{10-2}} = 0.$$

Cell 3: $q_{7-3} + q_{2-3} + q_{1-3} + q_{9-3} = 0.$

$$\frac{T_7 - T_3}{R_{7-3}} + \frac{T_2 - T_3}{R_{2-3}} + \frac{T_1 - T_3}{R_{1-3}} + \frac{T_9 - T_3}{R_{9-3}} = 0.$$

Cell 4: $q_{1-4} + q_{13-4} + q_{15-4} + q_{2-4} = 0.$

$$\frac{T_1 - T_4}{R_{1-4}} + \frac{T_{13} - T_4}{R_{13-4}} + \frac{T_{15} - T_4}{R_{15-4}} + \frac{T_2 - T_4}{R_{2-4}} = 0.$$

Observing that all the flow resistances are equal, we get rid of them by division. Then, substituting the boundary-cell temperatures into the flow equations, we get the following equation set.

Cell 1: $80. + T_4 + 60. + T_3 - 4T_1 = 0.$
Cell 2: $T_3 + 100. + T_4 + 150. - 4T_2 = 0.$
Cell 3: $80. + T_2 + T_1 + 150. - 4T_3 = 0.$
Cell 4: $T_1 + 100. + 60. + T_2 - 4T_4 = 0.$

We then re-pose this set into a matrix equation. The similarities with the matrix equation from Example 6.1 are evident.

$$\begin{bmatrix} -4.0 & 0.0 & 1.0 & 1.0 \\ 0.0 & -4.0 & 1.0 & 1.0 \\ 1.0 & 1.0 & -4.0 & 0.0 \\ 1.0 & 1.0 & 0.0 & -4.0 \end{bmatrix} \begin{bmatrix} T_1 \\ T_2 \\ T_3 \\ T_4 \end{bmatrix} = \begin{bmatrix} -140.0 \\ -250.0 \\ -230.0 \\ -160.0 \end{bmatrix}$$

The bandwidth of the coefficient matrix is surprisingly large (7), but we quickly realize that this could be decreased by changing the marking (numbering) scheme for the cells. For this small matrix, we choose not to. The simulation is completed by solving the matrix equation using a Gauss elimination method. The calculator-based computations are shown as Table 6.2. Results can be compared with the Example 6.1 results in Table 6.1.

Table 6.2 Gauss Elimination Calculator Results, Example 6.2

Stage	Reduced Matrices					Auxiliary Equations				
1	-4.0	0.0	1.0	1.0	-140.0	1.0	0.0	-0.25	-0.25	35.0
	0.0	-4.0	1.0	1.0	-250.0	0.0	0.0	0.0	0.0	0.0
	1.0	1.0	-4.0	0.0	-230.0	1.0	0.0	-0.25	-0.25	35.0
	1.0	1.0	0.0	-4.0	-160.0	1.0	0.0	-0.25	-0.25	35.0
2	-4.0	0.0	1.0	1.0	-140.0					
	0.0	-4.0	1.0	1.0	-250.0		1.0	-0.25	-0.25	62.5
	0.0	1.0	-3.75	0.25	-265.0		1.0	-0.25	-0.25	62.5
	0.0	1.0	0.25	-3.75	-195.0		1.0	-0.25	-0.25	62.5
3	-4.0	0.0	1.0	1.0	-140.0					
	0.0	-4.0	1.0	1.0	-250.0					
	0.0	0.0	-3.5	0.5	-327.5			1.0	-0.14286	93.571
	0.0	0.0	0.5	-3.5	-257.5			0.5	-0.07143	46.786
4	-4.0	0.0	1.0	1.0	-140.0					
	0.0	-4.0	1.0	1.0	-250.0					
	0.0	0.0	-3.5	0.5	-327.5					
	0.0	0.0	0.0	-3.4286	-304.29					

Back Substitution

$$T_4 = -304.29/(-3.4286) = 88.751.$$
$$T_3 = (-327.5 - 0.5T_4)/(-3.5) = 106.25.$$
$$T_2 = (-250.0 - T_4 - T_3)/(-4.0) = 111.25.$$
$$T_1 = (-140.0 - T_4 - T_3)/(-4.0) = 83.750.$$

2. We have to develop additional equations for the cells on the left-hand boundary for this case. We do this by applying Equations 2.3 and 2.6 again, but now to cells 6 and 7 on the left-hand boundary.

Cell 6: $\quad q_{c-6} + q_{1-6} + q_{5-6} + q_{7-6} = 0.$

$$\frac{T_c - T_6}{R_{c-6}} + \frac{T_1 - T_6}{R_{1-6}} + \frac{T_5 - T_6}{R_{5-6}} + \frac{T_7 - T_6}{R_{7-6}} = 0.$$

Cell 7: $\quad q_{c-7} + q_{3-7} + q_{6-7} + q_{8-7} = 0.$

$$\frac{T_c - T_7}{R_{c-7}} + \frac{T_3 - T_7}{R_{3-7}} + \frac{T_6 - T_7}{R_{6-7}} + \frac{T_8 - T_7}{R_{8-7}} = 0.$$

We observe that all the resistances are *not* the same in this case, in contrast to the previous case; but from the sizes of the cells, we can get good estimates.

$$R_{5-6} = R_{7-6} = 0.5R_{1-6} = 0.025$$
$$R_{6-7} = R_{8-7} = 0.5R_{3-7} = 0.025$$

Using $R_{c-6} = R_{c-7} = 0.01$, and $T_c = 80.$, we can simplify the equations for cell 6 and 7 and add them to the previous set.

```
                7
C EXAMPLE 6.2 FORTRAN PROGRAM FOR ELLIPTIC-MODEL SIMULATION
C FINITE DIFFERENCE SIMULATION WITH GAUSS-ELIMINATION SOLVER
          DIMENSION  A(6,7), T(6)
C READ NUMBER OF UNKNOWNS
C USES LIST DIRECTED INPUT-OUTPUT OF FORTRAN 77
C FORMATTED READ/WRITE STATEMENTS MAY BE REQUIRED ON OTHER COMPILERS
          READ*, N
          PRINT*,'NO. OF UNKNOWNS =', N
          M = N + 1
          L = N - 1
C READ AND ECHOPRINT ELEMENTS OF MATRIX EQUATION; DRIVING VECTOR IS  A(I,M)
          DO 10 I = 1,N
          READ*, ( A(I,J), J = 1,M )
   10     PRINT*,( A(I,J), J = 1,M )
C MATRIX REDUCTION
          DO 100 K = 1,L
            KP1 = K + 1
              DO 50 I = KP1,N
                QUOT = A(I,K)/A(K,K)
                DO 30 J = KP1,M
   30             A(I,J) = A(I,J) - QUOT * A(K,J)
   50         CONTINUE
              DO 60 I = KP1,N
   60           A(I,K) = 0.
  100       CONTINUE
C BACK SUBSTITUTION
          T(N) = A(N,M)/A(N,N)
          DO 200 NN = 1,L
            SUM = 0.
            I = N - NN
            IP1 = I + 1
            DO 150 J = IP1,N
  150         SUM = SUM + A(I,J) * T(J)
  200       T(I) = ( A(I,M) - SUM )/A(I,I)
C PRINT RESULTS
          PRINT*,'               GAUSS-ELIMINATION RESULTS'
          DO 300 I = 1,N
            IF( I .GT. 4 ) THEN
              II = I + 1
              ELSE
            ENDIF
  300       PRINT*,'I =',II,' T =',T(I)
          STOP
          END
```

```
  6
  -4., 0., 1., 1., 1., 0., -60.
  0., -4., 1., 1., 0., 0., -250.
  1., 1., -4., 0., 0., 1., -150.
  1., 1., 0., -4., 0., 0., -160.
  1., 0., 0., 0., -10., 2., -520.
  0., 0., 1., 0., 2., -10., -700.
```

Figure 6.3 FORTRAN program using a Gauss elimination solver, Example 6.2.

Cell 1: $T_6 + T_4 + 60 + T_3 - 4T_1 = 0.$

Cell 2: $T_3 + 100. + T_4 + 150. - 4T_2 = 0.$

Cell 3: $T_7 + T_2 + T_1 + 150. - 4T_3 = 0.$

Cell 4: $T_1 + 100. + 60. + T_2 - 4T_4 = 0.$

Cell 6: $520. + T_1 + 2T_7 - 10T_6 = 0.$

Cell 7: $700. + T_3 + 2T_6 - 10T_7 = 0.$

Re-posing gives us the following matrix equation.

$$\begin{bmatrix} -4.0 & 0.0 & 1.0 & 1.0 & 1.0 & 0.0 \\ 0.0 & -4.0 & 1.0 & 1.0 & 0.0 & 0.0 \\ 1.0 & 1.0 & -4.0 & 0.0 & 0.0 & 1.0 \\ 1.0 & 1.0 & 0.0 & -4.0 & 0.0 & 0.0 \\ 1.0 & 0.0 & 0.0 & 0.0 & -10.0 & 2.0 \\ 0.0 & 0.0 & 1.0 & 0.0 & 2.0 & -10.0 \end{bmatrix} \begin{bmatrix} T_1 \\ T_2 \\ T_3 \\ T_4 \\ T_6 \\ T_7 \end{bmatrix} = \begin{Bmatrix} -60.0 \\ -250.0 \\ -150.0 \\ -160.0 \\ -520.0 \\ -700.0 \end{Bmatrix}$$

The FORTRAN program to solve this matrix equation, giving us the desired simulation, is shown in Figure 6.3. The simulation results are $T_1 = 85.158$, $T_2 = 112.67$, $T_3 = 111.24$, $T_4 = 89.458$, $T_6 = 79.938$, and $T_7 = 97.111$.

EXAMPLE 6.3

Repeat the simulation asked for in Example 6.1, except change the left-hand boundary condition to the following Neumann condition.

$$\frac{\partial T}{\partial x} = 25(T - T_c) \qquad \text{at } x = 0.0 \quad \text{and where } T_c = 80.0$$

Also, use the Gauss elimination program developed in Example 6.2 to solve the resulting matrix equation and complete the simulation.

▶ The same grid indexing scheme as described in Example 6.1 is used again. However, we have to add two equations to the set developed in Example 6.1, to take care of the Neumann condition at nodes $i=1$, $j=2$; and $i=1$, $j=3$. The additional finite-difference formulation required is generated by substituting a forward finite-difference approximation for the derivative in the given boundary condition, as follows.

$$\frac{T_{i+1,j} - T_{i,j}}{\Delta x} = 25(T_{i,j} - T_c)$$

This simplifies to

$$T_{i+1,j} - 6T_{i,j} + 5T_c = 0$$

From this we get the two new equations.

Node $i=1, j=2$: $T_{2,2}-6T_{1,2}+400.0=0.0$

Node $i=1, j=3$: $T_{2,3}-6T_{1,3}+400.0=0.0$

The new matrix equation follows; we observe that it differs significantly from the matrix equation in part 2 of Example 6.2.

$$
\begin{bmatrix}
-6.0 & 0.0 & 1.0 & 0.0 & 0.0 & 0.0 \\
0.0 & -6.0 & 0.0 & 1.0 & 0.0 & 0.0 \\
1.0 & 0.0 & -4.0 & 1.0 & 1.0 & 0.0 \\
0.0 & 1.0 & 1.0 & -4.0 & 0.0 & 1.0 \\
0.0 & 0.0 & 1.0 & 0.0 & -4.0 & 1.0 \\
0.0 & 0.0 & 0.0 & 1.0 & 1.0 & -4.0
\end{bmatrix}
\begin{Bmatrix}
T_{1,2} \\ T_{1,3} \\ T_{2,2} \\ T_{2,3} \\ T_{3,2} \\ T_{3,3}
\end{Bmatrix}
=
\begin{Bmatrix}
-400.0 \\ -400.0 \\ -60.0 \\ -150.0 \\ -160.0 \\ -250.0
\end{Bmatrix}
$$

Using the program from Example 6.2 (see Figure 6.3) to solve this matrix equation, gives us the desired simulation results.

$$T_{1,2}=80.724$$
$$T_{1,3}=84.609$$
$$T_{2,2}=84.345$$
$$T_{2,3}=107.66$$
$$T_{3,2}=89.002$$
$$T_{3,3}=111.66$$

Comparing these results with those obtained from part 2 of Example 6.2, we note that some temperatures (those along the left boundary, in particular) are considerably different. In general, those using the cell discretization approach match experimental information better.

PARABOLIC MODEL SIMULATIONS WITH FINITE-DIFFERENCE METHODS

Most parabolic models of interest to engineers have time as one of the independent variables. We will concentrate our discussion of simulation methods on the simplest of these time-dependent models, namely, the one having a position co-ordinate as the only other independent variable. For more complicated models the reader is directed to the references. The typical linear model, also known as the "heat equation," has a parameter, α, known as a "diffusivity" in many applications. The magnitude of α will influence the choice of simulation method in most situations. Using the standard symbols, x for the spatial independent variable and t for time, we repeat the model equation here, for convenience in referencing, as Equation 6.1.

$$\frac{\partial^2 y}{\partial x^2}=(\alpha)^{-1}\frac{\partial y}{\partial t}; \quad \text{with three (initial and boundary) conditions} \qquad (6.1)$$

A variety of finite-difference simulation methods exists for such models. We will consider three that have distinctly different attributes and application advantages. The fully explicit method is by far the easiest to program, but it has a severe constraint that couples the space and time increments to each other and it is only first-order accurate in time. The fully implicit method does not have the time-increment constraint, but it requires a matrix equation solver and it is also only first-order accurate in time. The Crank–Nicolson method is similar to the fully implicit method in that it has no time-increment constraint and needs a matrix equation solver, but it is more complicated to set up and is second-order accurate in time. Descriptions follow.

The Fully Explicit Method

In this method a forward finite-difference approximation is substituted for the time-derivative term in Equation 6.1. Along with a central-difference approximation for the space derivative term, this permits an algebraic formulation to be set up for a forward-in-time value of the dependent variable, y. Letting the subscript i represent spatial nodal points and k subscripts stand for discrete points in time, we get the following equation.

$$\frac{y_{i+1,k}-2y_{i,k}+y_{i-1,k}}{(\Delta x)^2}=\frac{y_{i,k+1}-y_{i,k}}{\alpha\,\Delta t}$$

This can be re-posed into the following explicit form.

$$y_{i,k+1}=\left[\frac{\alpha\,\Delta t}{(\Delta x)^2}\right](y_{i+1,k}+y_{i-1,k})+\left[1-\frac{2\alpha\,\Delta t}{(\Delta x)^2}\right]y_{i,k}$$

Since all the k subscripts on the right-hand side are the same, it becomes convenient to drop the k subscript altogether, as shown in Equation 6.2.

$$y_i^*=Fo(y_{i+1}+y_{i-1})+(1-2Fo)y_i \tag{6.2}$$

where
$$Fo=\frac{\alpha\,\Delta t}{(\Delta x)^2}$$

y_i^* = new value of y_i, at time $t_k+\Delta t$

But a severe constraint exists in the use of Equation 6.2. It can be observed that the term $(1-2Fo)$ must not be allowed to go negative; otherwise certain basic physical system laws (e.g., the second law of thermodynamics) will be violated. Also, aside from the physical law violations, from the numerical viewpoint, large instabilities or "blow ups" can occur. Consequently, Equation 6.2 must always be used in conjunction with the constraint given in Equation 6.3.

$$\Delta t\leqslant 0.5\,\frac{(\Delta x)^2}{\alpha} \tag{6.3}$$

This means that the analyst doing the simulation cannot choose Δt and Δx independently. Either Δt or Δx can be chosen arbitrarily; but once one is chosen, the other must conform to Equation 6.3. A particularly simple form of Equation 6.2 evolves if the equality limit in Equation 6.3 is used, and is shown as Equation 6.4.

$$y_i^* = 0.5(y_{i+1} + y_{i-1}) \tag{6.4}$$

Equation 6.4 is the basis for a traditional engineering-graphics method called the Schmidt-plot method, once widely used to simulate certain transient behaviors in heat transfer systems.

Because the explicit stepping forward in time, which occurs when applying the above equations, is similar to a description of marching, this method is also commonly referred to as the "marching" method. The parabolic problems, as a result, are also known as marching problems.

Applications of the marching method are illustrated in Example 6.4.

EXAMPLE 6.4

A long slab of porous insulating material is sandwiched between two impermeable walls, as illustrated in Figure 6.4. Initially, the moisture concentration throughout the slab is $C = 1.0$; but at time $t = 0.0$, the concentration on the left-hand boundary at $x = 0.0$ is raised to $C = 5.0$ and held steady, while at the same time the concentration on the right-hand boundary at $x = 0.4$ is held steady at $C = 1.0$. Assume Equation 6.1 has been determined to be the appropriate model for simulating transient concentrations, C values, within the porous slab. (1) If the diffusivity for the moisture travel through the slab is $\alpha = 1.0$, set up and perform a simulation of the concentration profiles $C(x, t)$, using the finite-difference marching method with an x increment of 0.1, and a $\Delta t = 0.005$ from Equation 6.3. Use a calculator and

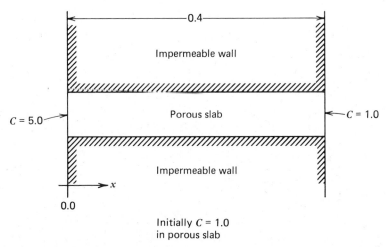

Figure 6.4   Porous slab, Example 6.4.

tabular algorithm to demonstrate the methodology at five successive times from $t=0.0$. (2) Set up a FORTRAN program based on Equations 6.2 and 6.3 to give simulations for the general marching problem. Demonstrate the program by applying it to the problem herein, using $\Delta x=0.05$ and $\Delta t=0.001$. Have it give a $C(x, t)$ profile at $t=0.02$.

▶ **1.** We replace the y's with C's in Equation 6.1 and add the following conditions.

$$C(x, 0)=1.0, \quad C(0, t)=5.0, \quad C(0.4, t)=1.0$$

Just as with the elliptic model methods, our first step is to discretize the spatial domain. We use a grid scheme with i subscripts, starting with $i=1$ at the left-hand boundary $(x=0.0)$, and ending with $i=5$ at the right-hand boundary. Next we compute the maximum allowable Δt from Equation 6.3. We find it to be 0.005; so the specified Δt is acceptable and we will not have to decrease its size. Furthermore, we observe that we can use the simple formulation of Equation 6.4 because $Fo=0.5$. Substituting the boundary values immediately, we get the following set of explicit equations for the internal nodes.

$$C_2^*=0.5C_3+2.5$$
$$C_3^*=0.5(C_2+C_4)$$
$$C_4^*=0.5C_3+0.5$$

From these equations we can compute successive C values, as shown in Table 6.3. We handle the step change in the boundary conditions at $t=0.0$ by enclosing them in parentheses in the initial condition column at $t=0.0$.

Sample calculations are given for $t=0.015$.

$$C_1^*=C_1=5.0$$
$$C_2^*=0.5(5.0+2.0)=3.5$$
$$C_3^*=0.5(3.0+1.0)=2.0$$
$$C_4^*=0.5(2.0+1.0)=1.5$$
$$C_5^*=C_5=1.0$$

Table 6.3 Calculator Results For C values, Example 6.4

		C_i Values Marching→					
i	$x\downarrow$	$t=0.0$	0.005	0.010	0.015	0.020	0.025
1	0.0	1.0(5.0)	5.0	5.0	5.0	5.0	5.0
2	0.1	1.0	3.0	3.0	3.5	3.5	3.75
3	0.2	1.0	1.0	2.0	2.0	2.5	2.5
4	0.3	1.0	1.0	1.0	1.5	1.5	1.75
5	0.4	1.0(1.0)	1.0	1.0	1.0	1.0	1.0

7

```
C EXAMPLE 6.4 FORTRAN PROGRAM FOR PARABOLIC-MODEL SIMUALTIONS
C FINITE-DIFFERNCE EXPLICIT (MARCHING) METHOD
C MODEL IS D2Y/DX2 = (1/DF) DY/DT
C DX = X INCREMENT ; DT = TIME INCREMENT  ; DF = DIFFUSIVITY VALUE
C NX = NO. OF X NODES ; NT = NO. OF TIME STEPS
C YI = INITIAL CONDITION ; YL = LEFT BOUND. COND. ; YR = RIGHT BOUND. COND.
C T2, T3, T4, T5 = TIMES AT WHICH SIMULATIONS TO BE DISPLAYED
C PUT PARAMETER VALUES INTO DATA STATEMENTS 111 AND 112
              DIMENSION Y(101), YT(101), YS(101,5), TP(5)
111           DATA DX, DT, DF, YI, YL, YR / 0.05, 0.001, 1.0, 1.0, 5.0, 1.0 /
112           DATA T2, T3, T4, T5, NX / 0.005, 0.01, 0.02, 0.025, 9 /
              DM = 0.5*DX**2/DF
              IF( DT .LE. DM ) GO TO 10
              PRINT*,'SPECIFIED DT WAS TOO LARGE. MAXIMUM DT =', DT
              PRINT*,'SIMULATION CONTINUED WITH MAXIMUM DT.'
              DT = DM
10            FM = DF*DT/DX**2
              NT = T5/DT + 1
C INITIALIZATIONS
              DO 20 I = 1,NX
              Y(I) = YI
20            YS(I,1) = YI
              TP(1) = 0.0
              N=NX-1
C MARCHING LOOP
              DO 100 IT = 1,NT
              T = IT*DT
              Y(1) = YL
              Y(NX) = YR
C NODAL POINT LOOP
              DO 30 I = 2,N
30            YT(I) = FM*(Y(I+1) + Y(I-1)) + (1 - 2*FM)*Y(I)
              YT(1) = Y(1)
              YT(NX) = Y(NX)
C CHECK IF Y PROFILE IS TO BE SAVED FOR DISPLAY LATER
              K = 0
              IF(ABS(T-T2) .LT. DT/2) K = 2
              IF(ABS(T-T3) .LT. DT/2) K = 3
              IF(ABS(T-T4) .LT. DT/2) K = 4
              IF(ABS(T-T5) .LT. DT/2) K = 5
C PLACE TEMPORARY ARRAY YT INTO Y
C SAVE PROFILE INTO YS IF FLAGGED ( K .GT. 0 ) FOR DISPLAY
              DO 50 I = 1,NX
              Y(I) = YT(I)
              IF( K .GT. 0 ) YS(I,K) = YT(I)
50            CONTINUE
              TP(K) = T
100           CONTINUE
C DISPLAY RESULTS
              DO 150 K = 1,5
              PRINT*,'------- Y PROFILE AT TIME =', TP(K)
              DO 90 I = 1,NX
              X = (I-1)*DX
              PRINT*,'I=',I,' X=',X,' Y=',YS(I,K)
90            CONTINUE
150           CONTINUE
              STOP
              END
```

Figure 6.5 FORTRAN program, Example 6.4.

189

2. The FORTRAN computer program for the general marching problem is given in Figure 6.5. Results of the demonstration run follow.

$$C_2 = 4.2031$$
$$C_3 = 3.4537$$
$$C_4 = 2.7908$$
$$C_5 = 2.2382$$
$$C_6 = 1.8015$$
$$C_7 = 1.4689$$
$$C_8 = 1.2138$$

The Fully Implicit Method

In this method we use the same grid discretization as before but substitute a backward-difference approximation for the order-one (time) derivative term in Equation 6.1. Along with a central-difference approximation for the order-two term, this gives the algebraic formulation shown in Equation 6.5.

$$y_{i-1} - \left(2 + \frac{1}{Fo}\right) y_i + y_{i+1} = -\frac{1}{Fo} y_i^o \tag{6.5}$$

where
$$Fo = \alpha \frac{\Delta t}{(\Delta x)^2}$$

y_i^o = the "old," or preceding, value of y_i

Here, as with the explicit method, the subscript for time has been suppressed; and the $y_{i,k-1}$ term is shown as y_i^o, where the superscript simply means that the existing value of y_i, determined from the preceding time step, must be used.

With this method the choice of Δx and Δt is constrained only by accuracy considerations, and the use of a constraint equation like Equation 6.3 is not necessary. The method is said to be "unconditionally stable" because an answer can always be computed without any blowups. This offers some advantages in simulating some engineering systems, where the value of α may be large and small Δx's are desired. However, a subtle disadvantage is that the method is only first-order accurate in time, and the fact that a large Δt is used can lead to relatively inaccurate simulations without any indication of trouble. A common recommendation is to use a Δt no larger than about twice the size of the Δt allowed by Equation 6.3. Another disadvantage is that a complete matrix equation has to be solved for every Δt step. Consequently, a good, fast equation-set solver is a necessity. Exact method solvers have been preferred. Alternately, if the coefficient matrix is not too large, it can be inverted once at the beginning of a simulation; and then a fast matrix-multiplier using the inverted matrix and an updated driving vector can be invoked at each time step. If this method can be used, it usually competes favorably with the explicit method in computer run time. Another approach is to reduce the coefficient into an upper and lower consolidated matrix, and then simply apply a back substitution process, using the updated driving vector, for each subsequent time step. A refined version of such a process, known as the Crout method, is found in much software. We choose to look at the inverse matrix procedure in Example 6.5.

EXAMPLE 6.5

Repeat the simulation asked for in Example 6.4, except make the following changes. (1) Use a time step of $\Delta t = 0.01$, $\Delta x = 0.1$, and a fully implicit method with an inverted coefficient matrix. (2) Set up a BASIC program around a repetitive use of the Gauss elimination method to treat the general implicit case. Demonstrate the program by repeating the simulation asked for in part (1), but using $\Delta x = 0.05$ and $\Delta t = 0.0025$.

▶ **1.** In this case $Fo = (1)(0.01)/(0.1)^2 = 1.0$, and Equation 6.5 becomes

$$C_{i-1} - 3C_i + C_{i+1} = -C_i^o$$

Applying this equation to each interior node, we get the following set of equations:

$$C_1 - 3C_2 + C_3 = -C_2^o$$
$$C_2 - 3C_3 + C_4 = -C_3^o$$
$$C_3 - 3C_4 + C_5 = -C_4^o$$

Noting that C_1 and C_5 are boundary conditions, we can re-pose this set into the matrix equation that must be solved at each successive time step of the simulation.

$$[A] \begin{Bmatrix} C_2 \\ C_3 \\ C_4 \end{Bmatrix} = - \begin{Bmatrix} C_2^o + C_1 \\ C_3^o \\ C_4^o + C_5 \end{Bmatrix}$$

where the coefficient matrix $[A]$ is given by the following.

$$[A] = \begin{bmatrix} -3.0 & 1.0 & 0.0 \\ 1.0 & -3.0 & 1.0 \\ 0.0 & 1.0 & -3.0 \end{bmatrix}$$

Although a number of different methods exist for inverting matrices, one of the simplest is a modified Gauss elimination method, called the Gauss–Jordan method. A brief description of this method is given in Appendix A. Here we assume that the inverted coefficient matrix has already been found by applica-

```
100         REM EXAMPLE 6.5 BASIC PROGRAM FOR PARABOLIC-MODEL SIMULATION
110         REM FINITE-DIFFERENCE FULLY IMPLICIT METHOD
120         REM USING A GAUSS-ELIMINATION EQUATION-SET SOLVER
130         REM MODEL IS D2Y/DX2 = (1/DF) DY/DT
140         REM DX = X INCREMENT;  DT = TIME INCREMENT;  DF = DIFFUSIVITY VALUE
150         REM NX = NO. OF X NODES;  NT = NO. OF TIME INCREMENTS
160         REM YI = INITIAL CONDITION; YL = LEFT B. C.; YR = RIGHT B. C.
170         REM T2,T3,T4,T5 = TIMES AT WHICH SIMULATIONS ARE TO BE DISPLAYED
180         CLS; CLEAR20; DEFINT I-N
190         READ DX, DT, DF, YI, YL, YR
200            DATA 0.05, 0.0025, 1.0, 1.0, 5.0, 1.0
210         READ T2, T3, T4, T5, NX
220            DATA 0.005, 0.01, 0.02, 0.025, 9
230         DIM Y(NX), YS(NX,5), TP(5), A(NX,NX+1), AT(NX,NX)
240         NT = T5/DT + 1 ;  FM = DF*DT/DX!2
250         REM INITIALIZATIONS
260            FOR I = 1 TO NX
270               Y(I) = YI; YS(I,1) = YI
280            NEXT I
290               N = NX-2; M = N+1; L = N-1
300         REM SETTING UP THE PERMANENT COEFF. MATRIX (AT)
310            FOR I = 1 TO N; FOR J = 1 TO N
320               AT(I,J) = 0.0
330            NEXT J,I
340               FOR I = 1 TO N
350                  IF I = 1 THEN AT(I,I) = -(2+1/FM); AT(I,I+1) = 1.0; GOTO 380
360                  IF I = N THEN AT(I,I) = -(2+1/FM); AT(I,I-1) = 1.0; GOTO 380
370                  AT(I,I-1) = 1.0; AT(I,I) = -(2+1/FM); AT(I,I+1) = 1.0
380               NEXT I
390         TP(1) = 0.0;  Y(1) = YL;  Y(NX) = YR
400         REM TIME-INCREMENTING LOOP
410         FOR IT = 1 TO NT
420            T = IT*DT
430            PRINT"COMPUTATION UNDERWAY AT T=";T
440         REM RECOVERING ORIGINAL COEFF. MATRIX
450            FOR I = 1 TO N; FOR J = 1 TO N
460               A(I,J) = AT(I,J)
470            NEXT J,I
480         REM SETTING DRIVING VECTOR
490            FOR I = 1 TO N
500               A(I,M) = -1/FM*Y(I+1)
510               IF I = 1 THEN A(I,M) = A(I,M) - Y(1);  GOTO 530
520               IF I = N THEN A(I,M) = A(I,M) - Y(M+1)
530            NEXT I
540         REM USES GAUSS-ELIMINATION SUBROUTINE TO SOLVE SET
550                     GOSUB 1000
560         REM CHECK IF Y VECTOR IS TO BE SAVED FOR DISPLAY
570            K = 0
580            IF ABS(T-T2) < DT/2 THEN K = 2;  GOTO 620
590            IF ABS(T-T3) < DT/2 THEN K = 3;  GOTO 620
600            IF ABS(T-T4) < DT/2 THEN K = 4;  GOTO 620
610            IF ABS(T-T5) < DT/2 THEN K = 5
620         REM RECOVERS Y VECTOR FROM A(I,M);  SAVES DISPLAY VECTOR YS
```

Figure 6.6*a*   BASIC program, Example 6.5.

```
630              FOR I = 1 TO N
640                 Y(I+1) = A(I,M)
650                 IF K > 0 THEN YS(I+1,K) = A(I,M)
660              NEXT I
670           TP(K) = T; YS(1,K) = Y(1); YS(NX,K) = Y(NX)
680           NEXT IT
690           REM DISPLAY RESULTS
700             FOR K = 1 TO 5
710                PRINT " ****** Y PROFILE AT TIME=";TP(K)
720                FOR I = 1 TO NX
730                   X = (I-1)*DX
740                   PRINT"I=";I;" X=";X;" Y=";YS(I,K)
750                NEXT I
760                INPUT"FOR NEXT DISPLAY, PRESS 'ENTER'"; ZZ
770             NEXT K
780           INPUT "REPEAT DISPLAY? -- Y/N"; A$
790           IF A$ = "Y" GOTO 690
800           END
                              Gauss Elimination Subroutine
1000          REM GAUSS-ELIMINATION SUBROUTINE
1010          REM N = NO. OF UNKNOWNS; ASSUMES; M = N+1  AND  L = N-1
1020          REM A(I,J) = COEFFICIENT MATRIX; I = 1 TO N,  J = 1 TO M
1030          REM A(I,M) = DRIVING VECTOR
1040          REM SOLUTION RETURNED AS A(I,M)
1050          REM ASSUMES WELL-POSED, NONSINGULAR MATRIX
1060          FOR K = 1 TO L; K1 = K + 1
1070            FOR I = K1 TO N
1080              Q = A(I,K)/A(K,K)
1090                FOR J = K1 TO M
1100                  A(I,J) = A(I,J) - Q*A(K,J)
1110                NEXT J
1120            NEXT I
1130            FOR I = K1 TO N
1140              A(I,K) = 0.0
1150            NEXT I
1160          NEXT K
1170          FOR NN = 1 TO L
1180            SU = 0.0; I = N-NN; I1 = I+1
1190              FOR J = I1 TO N
1200                SU = SU + A(I,J)*A(J,M)
1220              NEXT J
1230            A(I,M) = (A(I,M) - SU)/A(I,I)
1240          NEXT NN
1250          RETURN
1260          END
```

Figure 6.6b   BASIC program, Example 6.5 *(continued)*.

tion of the Gauss–Jordan method, as follows.

$$[A]^{-1} = \begin{bmatrix} -0.38095 & -0.14286 & -0.04762 \\ -0.14286 & -0.42857 & -0.14286 \\ -0.04762 & -0.14286 & -0.38095 \end{bmatrix}$$

To get one profile simulated, we perform the following matrix multiplication.

$$
\begin{Bmatrix} C_2 \\ C_3 \\ C_4 \end{Bmatrix} = [A]^{-1} \begin{Bmatrix} -C_2^o - C_1 \\ -C_3^o \\ -C_4^o - C_5 \end{Bmatrix}
$$

Thus, at time $t = 0.01$, we get the following:

$$
\begin{Bmatrix} C_2 \\ C_3 \\ C_4 \end{Bmatrix} = [A]^{-1} \begin{Bmatrix} -(1.0)-(5.0) \\ -(1.0) \\ -(1.0)-(1.0) \end{Bmatrix} = \begin{Bmatrix} 2.52 \\ 1.57 \\ 1.19 \end{Bmatrix}
$$

Updating the driving vector, we get for the next profile, at $t = 0.02$:

$$
\begin{Bmatrix} C_2 \\ C_3 \\ C_4 \end{Bmatrix} = [A]^{-1} \begin{Bmatrix} -7.52 \\ -1.57 \\ -2.19 \end{Bmatrix} = \begin{Bmatrix} 3.20 \\ 2.06 \\ 1.42 \end{Bmatrix}
$$

We can compare these C profiles with the results from Example 6.4 at the same times. Significant differences are noted. In this case the implicit values are probably not as good as the explicit values because of the large time increment used in the former. We will look at another method shortly that does better, yet retains the stability of the implicit method.

2. A BASIC program to treat the general implicit case using successive applications of the Gauss elimination solver is shown as Figure 6.6. It should be apparent that this approach is not very efficient if large sets, or many time steps, are involved. In such cases a good, repetitive Crout-method solver (using one reduction of the coefficient matrix at the beginning of a run) is probably the most efficient. The results from the BASIC program demonstration run at $t = 0.02$ follow.

$$
\begin{aligned}
C_2 &= 4.1679 \\
C_3 &= 3.3969 \\
C_4 &= 2.7318 \\
C_5 &= 2.1922 \\
C_6 &= 1.7743 \\
C_7 &= 1.4573 \\
C_8 &= 1.2107
\end{aligned}
$$

The Crank-Nicolson Method

At best, with either the fully explicit or fully implicit method just described, we get only the time-accuracy equivalent of the Euler method in ordinary-derivative simulations. Consequently, for extensive extrapolations into time, we would prefer some method giving extrapolation accuracy closer to the Runge–Kutta, or predictor–corrector methods. One such method, which retains the relative simplicity of the implicit method, is the Crank–Nicolson method. An order more accurate in time than the fully implicit method, it is also unconditionally stable. Its main disadvantages are a slightly more complicated matrix equation, and the need, like the implicit method, to solve this matrix equation at each time step. We get the appropriate finite-difference equation by substituting a forward difference for the time derivative and the average of two central differences in x, one at the forward time, for the x derivative in Equation 6.1. This is shown in Equation 6.6, where the time-variable subscript has been suppressed as before; instead we use an asterisk superscript to indicate forward in time.

$$\frac{[(y_{i+1}-2y_i+y_{i-1})^*/2+(y_{i+1}-2y_i+y_{i-1})/2]}{(\Delta x)^2}=\frac{(\alpha)^{-1}(y_i^*-y_i)}{\Delta t} \tag{6.6}$$

Using the previously used grouping, $Fo=\alpha\,\Delta t/(\Delta x)^2$, inverted, $B\equiv1/Fo$, we can re-pose Equation 6.6 into the form shown by Equation 6.7.

$$y_{i-1}^*-2(1+B)y_i^*+y_{i+1}^*=-y_{i-1}-2(1-B)y_i-y_{i+1} \tag{6.7}$$

From this equation we can form a matrix equation. However, this is best shown with an example (Example 6.6).

EXAMPLE 6.6

Repeat the simulation asked for in Example 6.4, except modify it as follows. (1) Use a time step of $\Delta t=0.01$, $\Delta x=0.1$, and the Crank–Nicolson method with a Gauss elimination solver. (2) Set up a BASIC program for the general application of the Crank–Nicolson method. Demonstrate the program by repeating the simulation asked for in part (1), but using $\Delta x=0.05$ and $\Delta t=0.0025$.

▶ **1.** In this case $B=(0.1)^2/(1\times0.01)=1.0$. Applying Equation 6.7 to each node, we get the following set of equations:

$$(C_1-4C_2+C_3)^*=-(C_1+C_3)$$
$$(C_2-4C_3+C_4)^*=-(C_2+C_4)$$
$$(C_3-4C_4+C_5)^*=-(C_3+C_5)$$

Thus the matrix equation that must be solved at each time step (after the first step) is

$$\begin{bmatrix} -4.0 & 1.0 & 0.0 \\ 1.0 & -4.0 & 1.0 \\ 0.0 & 1.0 & -4.0 \end{bmatrix}\begin{Bmatrix} C_2 \\ C_3 \\ C_4 \end{Bmatrix}^* = -\begin{bmatrix} 0.0 & 1.0 & 0.0 \\ 1.0 & 0.0 & 1.0 \\ 0.0 & 1.0 & 0.0 \end{bmatrix}\begin{Bmatrix} C_2 \\ C_3 \\ C_4 \end{Bmatrix}-\begin{Bmatrix} 10.0 \\ 0.0 \\ 2.0 \end{Bmatrix}$$

The boundary conditions are kept in a separate vector so that it is easier to modify the right-hand side from time step to time step. For time $t=0.01$, we must solve the following matrix equation.

$$\begin{bmatrix} -4.0 & 1.0 & 0.0 \\ 1.0 & -4.0 & 1.0 \\ 0.0 & 1.0 & -4.0 \end{bmatrix} \begin{Bmatrix} C_2 \\ C_3 \\ C_4 \end{Bmatrix}^* = - \begin{Bmatrix} 11.0 \\ 2.0 \\ 3.0 \end{Bmatrix}$$

Using a calculator and a tabular algorithm (we do not show the computation table here; see Table 6.2 for the technique), we get the following simulation values.

$$C_2^* = 3.1429$$
$$C_3^* = 1.5714$$
$$C_4^* = 1.1429$$

For the next set of simulated values, we have to update the right-hand side of the matrix equation. The resulting matrix equation has the same left-hand side, but the driving vector becomes

$$\begin{Bmatrix} -C_3 - 10.0 \\ -C_2 - C_4 \\ -C_3 - 2.0 \end{Bmatrix} = - \begin{Bmatrix} 11.5714 \\ 4.2858 \\ 3.5714 \end{Bmatrix}$$

Again using a calculator, we get the following simulation values at $t = 0.02$.

$$C_2^* = 3.4694$$
$$C_3^* = 2.3061$$
$$C_4^* = 1.4694$$

Comparing the foregoing solutions with those from Example 6.4 (Table 6.3), and Example 6.5 (implicit values), we observe that the Crank–Nicolson values fall on either side of the other simulation values, indicating the direction of extrapolation inaccuracies with respect to boundary changes.

2. A Basic program to treat the general, one-dimensional, parabolic model case with the Crank–Nicolson method is shown as Figure 6.7. The results of the demonstration run, at $t = 0.02$, follow.

$$C_2 = 4.1830$$
$$C_3 = 3.4185$$
$$C_4 = 2.7485$$
$$C_5 = 2.1970$$
$$C_6 = 1.7678$$
$$C_7 = 1.4458$$
$$C_8 = 1.2023$$

```
100    REM EXAMPLE 6.6 BASIC PROGRAM FOR PARABOLIC-MODEL SIMULATION
110    REM FINITE-DIFFERENCE CRANK NICOLSON METHOD
120    REM USING A GAUSS-ELIMINATION EQUATION-SET SOLVER
130    REM MODEL IS D2Y/DX2 = (1/DF) DY/DT
140    REM DX = X INCREMENT; DT = TIME INCREMENT; DF = DIFFUSIVITY VALUE
150    REM NX = NO. OF X NODES; NT = NO. OF TIME STEPS
160    REM YI = INITIAL CONDITION; YL = LEFT B.C. ; YR = RIGHT B.C.
170    REM T2,T3,T4,T5 = TIMES AT WHICH SIMULATIONS ARE TO BE DISPLAYED.
180    CLS! CLEAR20! DEFINT I-N
190    READ DX, DT, DF, YI, YL, YR
200       DATA 0.05, 0.0025, 1.0, 1.0, 5.0, 1.0
210    READ T2, T3, T4, T5, NX
220       DATA 0.005, 0.01, 0.02, 0.025, 9
230    DIM Y(NX), YS(NX,5), TP(5), A(NX,NX+1), BL(NX,NX), BR(NX,NX)
240    NT = T5/DT + 1!  B = DX!2/(DF*DT)
250    REM INITIALIZATIONS
260       FOR I = 1 TO NX
270          Y(I) = YI! YS(I,1) = YI
280       NEXT I
290          N = NX-2! M = N+1! L = N-1
300    REM SETTING UP THE PERMANENT COEFFICIENT MATRICES! <BL> AND <BR>
310       FOR I = 1 TO N! FOR J = 1 TO N
320          BL(I,J) = 0.0! BR(I,J) = 0.0
330       NEXT J,I
340       FOR I = 1 TO N
350          IF I = 1 THEN BL(I,I) = -2*(1 + B)! BL(I,I+1) = B!
                          BR(I,I) = -2*(1 - B)! BR(I,I+1) = -B! GOTO 380
360          IF I = N THEN BL(I,I) = -2*(1 + B)! BL(I,I-1) = B!
                          BR(I,I) = -2*(1 - B)! BR(I,I-1) = -B! GOTO 380
370          BL(I,I-1) = B! BL(I,I) = -2*(1 + B)! BL(I,I+1) = B!
             BR(I,I-1) = -B! BR(I,I) = -2*(1-B)! BR(I,I+1) = -B
380       NEXT I
390    TP(1) = 0.0
400    REM TIME-INCREMENTING LOOP
410       FOR IT = 1 TO NT
420          T = IT*DT
430          PRINT"COMPUTING UNDERWAY AT T=";T
440    REM RECOVERING ORIGINAL COEFFICIENT MATRIX
450       FOR I = 1 TO N! FOR J = 1 TO N
460          A(I,J) = BL(I,J)
470       NEXT J,I
480    REM SETTING THE DRIVING VECTOR
490       FOR I = 1 TO N! A(I,M) = 0.0
500          FOR J - 1 TO N
510             A(I,M) = A(I,M) + BR(I,J)*Y(J+1)
520          NEXT J
530          IF I = 1 THEN A(I,M) = A(I,M) - B*(Y(1) + YL)! GOTO 550
540          IF I = N THEN A(I,M) = A(I,M) - B*(Y(M+1) + YR)
550       NEXT I
560    REM USE GAUSS-ELIMINATION SUBROUTINE TO SOLVE SET
570                     GOSUB 1000
580    REM CHECK IF Y VECTOR IS TO BE SAVED FOR DISPLAY
590       K = 0
```

Figure 6.7a   BASIC program, Example 6.6.

```
600        IF ABS(T-T2) < DT/2 THEN K = 2: GOTO 640
610        IF ABS(T-T3) < DT/2 THEN K = 3: GOTO 640
620        IF ABS(T-T4) < DT/2 THEN K = 4: GOTO 640
630        IF ABS(T-T5) < DT/2 THEN K = 5
640    REM RECOVERS Y VECTOR FROM A(I,M): SAVES DISPLAY VECTOR YS
650    FOR I = 1 TO N
660        Y(I+1) = A(I,M)
670        IF K > 0 THEN YS(I+1,K) = A(I,M)
680    NEXT I
690    Y(1) = YL: Y(NX) = YR: TP(K) = 0.0: YS(1,K) = Y(1): YS(NX,K) = Y(NX)
700    NEXT IT
710    REM DISPLAY RESULTS
720    FOR K = 1 TO 5
730        PRINT"****** Y PROFILE AT TIME=";TP(K)
740        FOR I = 1 TO NX
750            X = (I-1)*DX
760            PRINT"I=";I;" X=";X;" Y=";YS(I,K)
770        NEXT I
780    INPUT"FOR NEXT DISPLAY, PRESS 'ENTER'"; ZZ
790    NEXT K
800    INPUT"REPEAT DISPLAY?--Y/N";A$
810        IF A$ = "Y" GOTO 710
820    END
```

<u>Place Gauss-Elimination Subroutine 1000 Here</u>
<u>(see Figure 6.6 for Source Listing)</u>

Figure 6.7b   BASIC program, Example 6.6 (continued).

HYPERBOLIC MODEL SIMULATIONS WITH FINITE-DIFFERENCE METHODS

These simulations are much like the parabolic model simulations in that several different finite-difference methods are applicable. However, some methods are significantly less complicated than others. We choose to confine our attention to only the simplest explicit method. We also limit our interest to the most basic of the higher-order hyperbolic models—namely, the one-dimensional linear wave equation—in order to illustrate the simulation method. The model equation is shown in Equation 6.5.

$$\frac{\partial^2 y}{\partial x^2} = C^{-2} \frac{\partial^2 y}{\partial t^2}; \quad \text{with four conditions} \tag{6.5}$$

It should be noted that the parameter C represents a propagation speed of the wave y through the domain designated by the span of x. The simulation method is called the "direct explicit" method herein to distinguish it from other more-complicated methods that exist, which are described in the literature.

The Direct Explicit Method

As with other finite-difference procedures, our first task is to discretize the x domain. This is best done with a grid discretization process, which has been described before and is not repeated here. The time discretization is visualized in the same manner as for the parabolic models. Then central difference approximations are substituted for both the derivatives in Equation 6.5. After some simplification, we get the following algebraic equation.

$$y_{i,k+1} = P(y_{i+1,k} + y_{i-1,k}) + 2(1-P)y_{i,k} - y_{i,k-1}$$

where

$$P = \left(C\frac{\Delta t}{\Delta x}\right)^2$$

$$i = x\text{-domain subscript}$$

$$k = \text{time-domain subscript}$$

Next, the time increment is specified as $\Delta t = \Delta x/C$. This choice of Δt not only simplifies the equation, but it can also be shown to be the best choice for simulation accuracy. In any case, Δt should not exceed $\Delta x/C$ for stability. Putting the resulting $P = 1$ into the equation gives the recommended simulator: Equations 6.6 and 6.7.

$$y_{i,k+1} = y_{i+1,k} + y_{i-1,k} - y_{i,k-1} \tag{6.6}$$

$$\Delta t = \frac{\Delta x}{C} \tag{6.7}$$

From Equation 6.6 it is evident that forward-in-time values of y are influenced by surrounding values immediately preceding in time, as well as by y values at the same positions but two steps back in time. The general form of the equation, however, is almost as simple as the parabolic model Equation 6.4 and permits a short program to be used with fast computation times. Aside from the restriction on the time increment, the only other complication is what to do about the $y_{i,k-1}$ value at the start of the simulation. In view of the relative simplicity of the approach to this point, the following procedure is recommended. If initial $\partial y/\partial t$ values are zero, use Equation 6.8 to get started. If initially $\partial y/\partial t = g(x)$, use Equation 6.9 to get started. The justification for the foregoing procedure is described by Gerald.[2]

$$y_{i,2} = 0.5(y_{i+1,1} + y_{i-1,1}) \tag{6.8}$$

$$y_{i,2} = 0.5(y_{i+1,1} + y_{i-1,1}) + \left(\frac{\Delta x}{C}\right)\frac{g_{i-1} + 4g_i + g_{i+1}}{6} \tag{6.9}$$

An application of the foregoing method is illustrated in Example 6.7.

EXAMPLE 6.7

A stretched wire, 0.6 units long, is initially displaced laterally 0.15 units at its center. The initial shape of the wire forms a triangle with the base at the wire's equilibrium position. The model equation is given below. (1) Use $\Delta x = 0.1$, and demonstrate a finite-difference simulation of y displacements manually with a calculator and tabular algorithm for four time steps. (2) Set up a BASIC program to handle the general hyperbolic simulation. Demonstrate the program by repeating the preceding simulation, but use $\Delta x = 0.05$ and let the time span be $t = 0.0$ to $t = 3.0$.

$$\frac{\partial^2 y}{\partial x^2} = 4\frac{\partial^2 y}{\partial t^2}; \qquad y(x, 0) = f(x), \quad y'(x, 0) = 0.0$$
$$y(0, t) = 0.0, \quad y(0.6, t) = 0.0$$

▶ **1.** The x coordinate nodes will take on the values $i = 1$ at $x = 0.0$ to $i = 7$ at $x = 0.6$. The pertinent equations for the internal nodes are extracted from Equations 6.6, 6.7, and 6.8.

$$y_{i,2} = 0.5(y_{i+1,1} + y_{i-1,1})$$

$$y_{i,k+1} = y_{i+1,k} + y_{i-1,k} - y_{i,k-1}$$

$$\Delta t = \frac{\Delta x}{C} = \frac{0.1}{0.5} = 0.2$$

The results of the calculations are shown in Table 6.4.

```
                          Main Program
100      REM EXAMPLE 6.7 BASIC PROGRAM FOR HYPERBOLIC-MODEL SIMULATIONS
110      REM FINITE-DIFFERENCE DIRECT EXPLICIT METHOD
120      REM MODEL IS D2Y/DX2 = (1/C!2) D2Y/DT2
130      REM C IS PROPAGATION SPEED
140      REM DX = X INCREMENT;  DT = TIME INCREMENT
150      REM NX = NO, OF X NODES;  NT = NO, OF TIME STEPS
160      REM YI = INITIAL CONDITION; ENTER YI(X) INTO I,C, LOOP
170      REM G(X) = DY/DT INITIAL CONDITION; ENTER INTO I,C, LOOP
180      REM YL = LEFT BOUNDARY CONDITION;  YR = RIGHT BOUNDARY CONDITION
190      REM T3,T4,T5,T6 = TIMES AT WHICH SIMULATIONS ARE TO BE DISPLAYED,
200      CLS! CLEAR20! DEFINT I-N
210      READ DX, DT, C, YI, YL, YR
220         DATA 0.05, 0.5, 0.5, 0.15, 0.0, 0.0
230      READ T3, T4, T5, T6, NX
240         DATA 0.2, 0.8, 2.4, 3.0, 13
250      DIM Y(NX), G(NX), YT(NX), YS(NX,6), TP(6)
260         DM = DX/C
270         IF ABS(DT-DM) < 0.001*DM GOTO 320
280         PRINT"SPECIFIED DT IS TOO LARGE OR SMALL, OPTIMUM DT = "; DM
290         INPUT"DO YOU WISH TO CONTINUE WITH OPTIMUM DT?--Y/N";A$
300         IF A$ = "Y" THEN DT = DM! GOTO 320
310         PRINT"TO CONTINUE RUN, CHANGE 'DT' IN DATA STATEMENT!!"! END
320      NT = T6/DT + 1
330      REM INITIALIZATIONS
340         FOR I = 1 TO NX
350           YS(I,1) = 0.5*(I-1)*DX
360           IF I > 7 THEN YS(I,1) = 0.05*(0.6 - (I-1)*DX)
370           G(I) = 0.0! YS(I,2) = YS(I,1)! Y(I) = YS(I,1)
380         NEXT I
390      TP(1) = 0.0! Y(1) = YL! Y(NX) = YR
```

Figure 6.8a Basic program, Example 6.7.

```
400        REM TIME INCREMENTING LOOP
410        FOR IT = 1 TO NT
420          T = IT*DT
430          PRINT"COMPUTING UNDERWAY AT TIME =";T
440        REM NODAL POINT LOOP
450          FOR I = 2 TO NX-1
460            IF IT > 1 GOTO 480
470            YT(I) = 0.5*(Y(I+1) + Y(I-1)) + DX/C*(G(I-1)+4*G(I)+G(I-1))/6: GOTO 490
480            YT(I) = Y(I+1) + Y(I-1) - YS(I,1)
490          NEXT I
500        REM CHECK IF Y PROFILE IS TO BE SAVED FOR DISPLAY
510          K = 0
520          IF ABS(T-T3) < DT/2 THEN K = 3: GOTO 560
530          IF ABS(T-T4) < DT/2 THEN K = 4: GOTO 560
540          IF ABS(T-T5) < DT/2 THEN K = 5: GOTO 560
550          IF ABS(T-T6) < DT/2 THEN K = 6
560        REM PLACE TEMPORARY ARRAY YT INTO Y
570        REM IF K > 0 , SAVE PROFILE INTO YS FOR LATER DISPLAY
580        REM SAVE CURRENT Y VALUES IN YS(I,1)
590          FOR I = 1 TO NX
600            YS(I,1) = Y(I): Y(I) = YT(I)
610            IF K > 0 THEN YS(I,K) = YT(I)
620          NEXT I
630        TP(K) = T: Y(1) = YL: Y(NX) = YR
640        NEXT IT
650        REM DISPLAY RESULTS
660          FOR K = 2 TO 6
670            PRINT"****** Y PROFILE AT TIME =";TP(K)
680            FOR I = 1 TO NX
690              X = (I-1)*DX
700              PRINT"I=";I;" X=";X;"  Y=";YS(I,K)
710            NEXT I
720          INPUT"FOR NEXT DISPLAY, PRESS 'ENTER'"; ZZ
730          NEXT K
740        INPUT"REPEAT DISPLAY?--Y/N";B$
750          IF B$ = "Y" GOTO 650
760        END
```

Figure 6.8*b* BASIC program, Example 6.7 *(continued)*.

Table 6.4 Calculator Results For *y* Values, Example 6.7

		y_i Values Progressing in Time$\rightarrow$				
i	$x_i\downarrow$	$t=0.0$	$t=0.2$	$t=0.4$	$t=0.6$	$t=0.8$
1	0.0	0.0	0.0	0.0	0.0	0.0
2	0.1	0.05	0.05	0.05	0.0	-0.05
3	0.2	0.1	0.1	0.05	0.0	-0.05
4	0.3	0.15	0.1	0.05	0.0	-0.05
5	0.4	0.1	0.1	0.05	0.0	-0.05
6	0.5	0.05	0.05	0.05	0.0	-0.05
7	0.6	0.0	0.0	0.0	0.0	0.0

A sample calculation is given at node 3, for $t=0.4$,

$$y_{3,3} = 0.1 + 0.05 - 0.1 = 0.05$$

2. The BASIC computer program is shown in Figure 6.8. Values from the demonstration run are given at $t=0.2$ and 2.4, and at every other node.

Results, Example 6.7

At $t=0.2$	At $t=2.4$
$y_3 = 0.05$	$y_3 = 0.05$
$y_5 = 0.1$	$y_5 = 0.1$
$y_7 = 0.1$	$y_7 = 0.15$
$y_9 = 0.1$	$y_9 = 0.1$
$y_{11} = 0.05$	$y_{11} = 0.05$

REFERENCES

1. R. W. Hornbeck, *Numerical Methods*, Quantum Publishers, Inc., New York, 1975.
2. C. F. Gerald, *Applied Numerical Analysis*, Addison-Wesley Publishing Co., Reading, Mass., 1973.
3. M. L. James, G. M. Smith, and J. C. Wolford, *Applied Numerical Methods for Digital Computation with Fortran and CSMP, 2nd ed.*, IEP, Thomas Y. Crowell Co., Inc., New York, 1977.
4. J. H. Ferziger, *Numerical Methods for Engineering Application*, John Wiley & Sons, Inc., New York, 1981.
5. W. F. Ames, *Numerical Methods for Partial Differential Equations, 2nd ed.*, Academic Press, Inc., New York, 1977.
6. V. Vemuri and W. J. Karplus, *Digital Computer Treatment of Partial Differential Equations*, Prentice-Hall, Inc., Englewood Cliffs, N.J., 1981.
7. P. J. Roache, *Computational Fluid Mechanics*, Hermosa Publishers, Albuquerque, N.M., 1972.

PROBLEMS

6.1 The u values on a 1.5×1.5 square cross section, where $u = 200, 100, 200,$ and 300 on the left, upper, right, and lower boundaries, respectively, can be simulated from a Laplace equation model.

(a) Use a grid-discretization procedure with $\Delta x = \Delta y = 0.5$ and set up an appropriate finite-difference matrix equation for simulating $u_{i,j}$ values.

(b) Demonstrate a simulation by applying a Gauss–Seidel equation-set solver (Liebmann form), with starters of $u_{i,j}=200$, and $\varepsilon=1.0$, using a tabular algorithm.

(c) Repeat part (b), except use a Gauss–Jordan equation-set solver. Compare results with part (b).

6.2 Repeat Problem 6.1, except use a cell discretization procedure. Use a single-number subscript to identify each cell.

6.3 The temperatures, T, on a 3.0×1.5 rectangular cross section, where $T=$ 100, 200, 300, and 400 on the left $(x=0)$, top, right, and bottom $(y=0)$ boundaries, respectively, can be simulated from a Laplace equation model.

(a) Use a grid discretization process, with $\Delta x=1.0$ and $\Delta y=0.5$, and set up an appropriate finite-difference matrix equation for simulating the $T_{i,j}$ values.

(b) Demonstrate a simulation by applying a Gauss–Seidel equation-set solver (starters $T_{i,j}=200$, $\varepsilon=0.5$), using a tabular algorithm.

(c) Repeat part (b), except use a Gauss elimination equation-set solver. Compare results with part (b).

6.4 The pressures, P, in a two-dimensional aquifer can be simulated from the following elliptic model.

$$\frac{\partial^2 P}{\partial x^2}+\frac{\partial^2 P}{\partial y^2}=1.5$$

$$P(x, 0)=30. \qquad P(x, 0.6)=10.$$

$$\frac{\partial P}{\partial x}=0.0 \qquad \text{at } (0, y) \text{ and } (0.3, y)$$

(a) Use a grid discretization procedure, with $\Delta x=0.1$ and $\Delta y=0.2$, and set up a finite-difference matrix equation for simulating the $P_{i,j}$ values.

(b) Demonstrate a simulation by applying a Gauss–Jordan equation-set solver to find all the interior node P values.

(c) How will this simulation differ if a cell discretization procedure is used?

6.5 The concentrations, C, of a substance in a two-dimensional field are simulated from the following model.

$$\frac{\partial^2 C}{\partial x^2}+\left(\frac{0.5}{2.0+0.5x}\right)\frac{\partial C}{\partial x}+\frac{\partial^2 C}{\partial y^2}=0.0$$

$$C(0, x)=5.0, \quad C(1.5, x)=2.0, \quad C(y, 0)=5.0-2.0y, \quad C(y, 1.5)=7.0$$

(a) Use a grid discretization procedure, with $\Delta x=\Delta y=0.5$, and set up a finite-difference matrix equation for simulating the interior node C values.

(b) Demonstrate a simulation by applying a Gauss–Jordan equation-set solver.

(c) Repeat part (b), except use a Gauss–Seidel equation set solver (Liebmann form), using as starters $C_{i,j}=4.0$, a convergence test parameter $\varepsilon=0.1$, and a tabular algorithm. Compare results with part (b).

6.6 The B values on the cross section shown in Figure P6.6 can be simulated from the following model.

$$\frac{\partial^2 B}{\partial x^2} + \frac{\partial^2 B}{\partial y^2} = 0.0$$

with conditions shown on the sketch.

(a) Set up a finite-difference matrix equation for simulating the $B_{i,j}$ values at the internal nodes.
(b) Demonstrate a simulation by applying a Gauss–Seidel solver (starters: $B=0.0$, and convergence tester: $\varepsilon=0.5$), using a tabular algorithm.

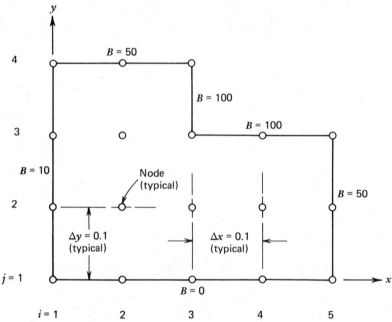

Figure P6.6 Cross section, Problem 6.6.

6.7 Temperature values, T, on the cross section shown in Figure P6.7 can be simulated from the following Poisson equation model.

$$\frac{\partial^2 T}{\partial x^2} + \frac{\partial^2 T}{\partial y^2} = S$$

with conditions shown on the sketch.

$$S_{2,2} = -10.0, \quad \text{otherwise } S_{i,j}=0.0$$

(a) Set up a finite-difference matrix equation for simulating the interior node $T_{i,j}$ values, and $T_{4,2}$.
(b) Demonstrate a simulation by applying a Gauss–Seidel solver (starters: $T=0.0$, and convergence tester: $\varepsilon=0.1$), using a tabular algorithm.

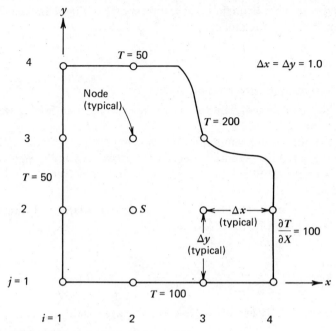

Figure P6.7 Cross section, problem 6.7.

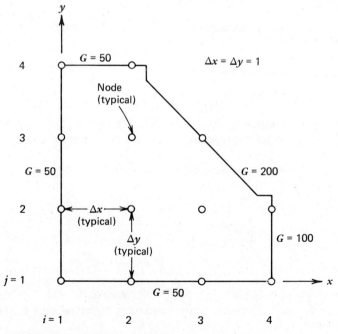

Figure P6.8 Cross section, Problem 6.8.

6.8 Field potentials, G, on the cross section shown in Figure P6.8 can be simulated from the following model.

$$\frac{\partial^2 G}{\partial x^2} + \frac{\partial^2 G}{\partial y^2} = 16.8$$

with conditions shown on the sketch.

(a) Set up a finite-difference matrix equation for simulating the $G_{i,j}$ values at the internal nodes.

(b) Demonstrate a simulation by applying a Gauss–Seidel solver (starters: $G=0.0$, and convergence tester: $\varepsilon=0.5$), using a tabular algorithm.

(c) Repeat part (b), except use a Gauss–Jordan solver. Compare the results with part (b).

6.9 Transient concentrations, C, near a surface of a large slab can be simulated from the following parabolic model.

$$\frac{\partial^2 C}{\partial x^2} = a \frac{\partial C}{\partial t}$$

$$C(x, 0) = 0.0, \qquad C(0, t) = 100., \qquad C(10, t) = 300.$$

(a) Using a fully explicit finite-difference approach, with $\Delta x = 1.0$ and $a = 1$, set up an appropriate equation for simulating C_i values at successive Δt steps. Choose the maximum Δt.

(b) Demonstrate a simulation of C_i values for five Δt steps with a tabular algorithm.

6.10 The following parabolic model can be used for simulating time-varying P values.

$$\frac{\partial^2 P}{\partial x^2} = 10.0 \frac{\partial P}{\partial t}$$

$$P(x, 0) = 0.0, \qquad P(0, t) = 10.0, \qquad P(1.0, t) = 0.0$$

(a) Using a fully implicit finite-difference approach, with $\Delta x = 0.2$ and $\Delta t = 0.4$, set up an appropriate matrix equation for simulating P_i values at successive Δt steps.

(b) Demonstrate a simulation for two Δt steps by applying a Gauss–Jordan equation-set solver.

6.11 The following model can be used to simulate transient T values.

$$\frac{\partial^2 T}{\partial x^2} = 10.0 \frac{\partial T}{\partial t}$$

$$T(x, 0) = 0.0, \qquad T(0, t) = 5 + 102t, \qquad T(1, t) = 0.1 + 10.5t$$

(a) Using a fully implicit finite-difference approach, with $\Delta x = 0.2$ and $\Delta t = 0.4$, set up an appropriate matrix equation for simulating the T_i values at successive time steps.

(b) Demonstrate a simulation for two Δt steps by applying a Gauss–Jordan solver.

6.12 (a) Refer to the model in Problem 6.11, and using a fully explicit finite-difference approach, with $\Delta x = 0.2$ and $\Delta t = 0.2$, set up an appropriate equation for simulating T values at successive Δt steps.
(b) Demonstrate a simulation of T_i values for four Δt steps using a tabular algorithm marching method.

6.13 The simulation of transient temperatures, T, near the surface of a large slab can be based on the following model.

$$\frac{\partial^2 T}{\partial x^2} + \left(\frac{B}{A+BT}\right)\left(\frac{\partial T}{\partial x}\right)^2 = \left(\frac{C}{A+BT}\right)\frac{\partial T}{\partial t}$$

$$T(x, 0) = -20, \qquad T(0, t) = -5, \qquad T(1, t) = -20$$

$$A = 0.4, \qquad B = -0.1, \qquad C = 0.01$$

(a) Using a fully explicit finite-difference approach, with $\Delta x = 0.1$ and the largest Δt possible, set up an appropriate equation for simulating the T_i values at successive time steps.
(b) Demonstrate a simulation with a marching method (tabular algorithm) for five Δt steps.

6.14 Repeat Problem 6.10, except use the Crank–Nicolson method instead of the fully implicit method.

6.15 Repeat Problem 6.11, except use the Crank–Nicolson method instead of the fully implicit method

6.16 The following model is being used to simulate transient C values in a long packed tube.

$$\frac{\partial^2 C}{\partial x^2} = 2.0\,\frac{\partial C}{\partial t}$$

$$C(x, 0) = 0.0, \qquad C(0, t) = 100, \qquad C(7, t) = 20$$

(a) Using a fully explicit finite-difference approach, with $\Delta x = 1.$, and a maximum allowable Δt, set up an appropriate equation for simulating C_i values at successive time steps.
(b) Demonstrate a simulation with a marching method (tabular algorithm) for five Δt steps.

6.17 Repeat Problem 6.16, except change the conditions to the following.

$$C(x, 0) = 0.0, \qquad C(0, t) = 100 - 10t, \quad C(7, t) = 2 + 10\sin 2t$$

6.18 Repeat Problem 6.17, except use the Crank–Nicolson method, with $\Delta t = 2$, and demonstrate the simulation for only two Δt steps.

6.19 The transient temperatures, T, in a large homogeneous wall can be simulated from the following model.

$$\frac{\partial^2 T}{\partial x^2} = 1 \times 10^6\,\frac{\partial T}{\partial t}$$

$$T(x, 0) = 25., \qquad T(0, t) = -20., \qquad T(0.01, t) = 25.$$

(a) Using a fully implicit finite-difference approach, with $\Delta x = 0.002$ and $\Delta t = 4.0$, set up the appropriate matrix equation for simulating T_i values at successive Δt steps.

(b) Demonstrate a simulation by applying a Gauss elimination equation-set solver for two Δt steps.

6.20 Repeat Problem 6.19, except use a fully explicit approach with $\Delta t = 2.0$, and march the simulation out for four Δt steps. Compare the results with Problem 6.19.

6.21 Repeat Problem 6.19, except use a Crank–Nicolson method. Compare the results with Problems 6.19 and 6.20.

6.22 The transient transverse displacements, y, of a stretched cable can be simulated from the following model.

$$\frac{\partial^2 y}{\partial x^2} = 0.01 \frac{\partial^2 y}{\partial t^2}$$

$$y(0, t) = 0.0, \qquad y(6, t) = 0.0$$

x	0.0	1.0	2.0	3.0	4.0	5.0	6.0
$y(x, 0)$	0.0	0.5	0.6	0.4	0.2	0.1	0.0

(a) Using a fully explicit finite-difference approach, with $\Delta x = 1.0$ and $\Delta t = 0.1$, set up an appropriate equation for simulating y_i values at successive Δt steps.

(b) Demonstrate a simulation of y_i values for six Δt steps using a tabular algorithm.

6.23 The compression wave (local displacement $= u$) in a bar that has been impact-loaded at one end, can be simulated from the following model.

$$\frac{\partial^2 u}{\partial x^2} = 0.01 \frac{\partial^2 u}{\partial t^2}$$

$$u(x, 0) = 0.0, \qquad u(0, t > 0) = 0.0, \qquad u(0, 0) = 0.0, \qquad u(6, t) = 0.0$$

(a) Using a fully explicit finite-difference approach, set up an equation for simulating u_i values at successive Δt steps. Use $\Delta x = 1.0$ and $\Delta t = 0.1$.

(b) Demonstrate a simulation of u_i values for six Δt steps using a tabular algorithm.

6.24 The vibrational displacements in a metal strip can be simulated from the following model.

$$\frac{\partial^2 v}{\partial x^2} = 4.0 \frac{\partial^2 v}{\partial t^2}$$

$$v(0, t) = 0.7 \sin 3t, \qquad v(0.6, t) = 0.0$$

x	0.0	0.1	0.2	0.3	0.4	0.5	0.6
$v(x, 0)$	0.05	0.1	0.15	0.1	0.15	0.05	0.0

(a) Using a fully explicit finite-difference approach, with $\Delta x = 0.1$ and $\Delta t = 0.2$, set up an appropriate equation for simulating the v_i values at successive Δt steps.

(b) Demonstrate a simulation of the v_i values for six time steps by use of a tabular algorithm.

The following problems require significant use of a computer.

6.25(L) The temperatures, T, on a rectangular $L \times W$ cross section, where $T = 100$, 50, 50, and 50 on the left ($x = 0$), upper, right ($x = L$), and lower boundaries, respectively, can be simulated from a Laplace equation model. Use a finite-difference grid discretization and set up the appropriate simulation equations. Develop a FORTRAN or BASIC program, using a Gauss–Seidel solver (starters: $T = 50$, convergence tester: $\varepsilon = 0.01$), for simulating the $T_{i,j}$ values. Assume consistent units and demonstrate your program using the following parameters. Present the results in a physically meaningful pattern.

$$L = 3.0, \qquad W = 2.0, \qquad \Delta x = 0.25, \qquad \Delta y = 0.25$$

6.26(L) The deflections, z, of a flat rectangular ($L \times W$) plate under a pressure load can be simulated from the following model.

$$\frac{\partial^2 \theta}{\partial x^2} + \frac{\partial^2 \theta}{\partial y^2} = \frac{P}{R}$$

$\theta = 0$ on all the edges.

$$\frac{\partial^2 z}{\partial x^2} + \frac{\partial^2 z}{\partial y^2} = \theta$$

$z = 0$ on all the edges.

Using a finite-difference grid discretization approach, set up the necessary simulation equations. Develop a FORTRAN or BASIC program, using a Gauss–Seidel solver (starters $= 0.0$, $\varepsilon = 0.0001$), for simulating the $z_{i,j}$ values. Assume consistent units and demonstrate your program using the following parameters. Present the results in a physically meaningful pattern.

$$P = 50 \qquad R = \frac{Et^3}{12(1 - \mu^2)} \qquad \mu = 0.3 \qquad t = 0.04$$

$$L = 1.6 \qquad W = 1.2, \quad E = 15 \times 10^6 \qquad \Delta x = 0.2 \qquad \Delta y = 0.2$$

6.27(L) The pressures, P, in a confined aquifer having both an injection and withdrawal well, as shown in Figure P6.27(L), can be simulated from the following model.

$$\frac{\partial^2 P}{\partial x^2} + \frac{\partial^2 P}{\partial y^2} = S(x, y)$$

with conditions shown on the sketch.

$$S(0.9, 0.3) = -20.0, \qquad S(1.3, 0.7) = 20.0$$

Using a finite-difference grid approach, set up the equations needed for the simulation. Develop a FORTRAN or BASIC program, using a Gauss–Seidel solver (starters $=0.0$, $\varepsilon=0.001$), for simulating the $P_{i,j}$ values. Assume consistent units and demonstrate your program using the following parameters. Present the results in a physically meaningful pattern.

$$\Delta x = 0.1, \qquad \Delta y = 0.1$$

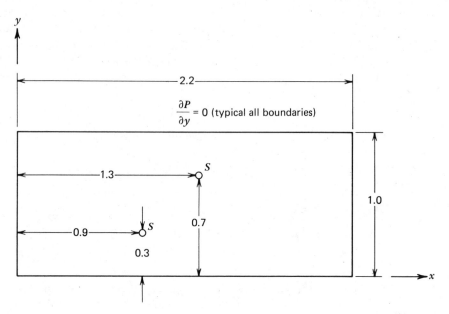

Figure P6.27L Aquifer section, Problem 6.27L.

6.28(L) The temperatures, T, on the cross section shown in Figure P6.28(L) can be simulated from a Laplace equation model. Using a finite-difference grid approach, set up the simulation equations. Develop a FORTRAN or BASIC program, using a Gauss Seidel solver (starters $= 30.$, $\varepsilon=0.5$), for simulating the $T_{i,j}$ values. Assume consistent units and demonstrate your program using the following parameters. Present the results graphically if possible.

$$TP = 70. \qquad TT = 0.0 \qquad TR = 25. \qquad TB = 30.$$
$$W = 15. \qquad H = 15. \qquad \Delta x = 1.0 \qquad \Delta y = 1.0$$

6.29(L) A solar collector has a thin sheet of plastic for a cover. If this sheet is treated as a membrane sealed to a rectangular ($W \times L$) frame, transverse deflections, D, owing to wind pressure can be simulated from the following

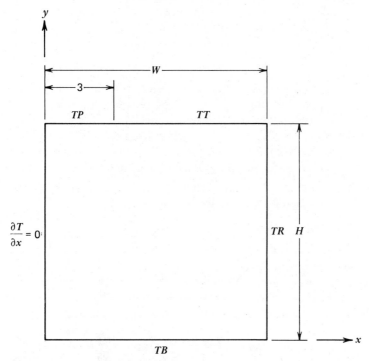

Figure P6.28L Cross section, Problem 6.28L.

model.

$$\frac{\partial^2 D}{\partial x^2} + \frac{\partial^2 D}{\partial y^2} = \frac{-P}{T}$$

$$D = 0.0 \text{ on all edges}$$

Using a finite-difference grid approach, set up the required simulation equations for D. Develop a FORTRAN or BASIC program, using a Gauss–Seidel solver (starters $=0.0$, $\varepsilon = 0.001$), for simulating the $D_{i,j}$ values. Assume consistent units and demonstrate your program using the following parameters. Present the results graphically if possible.

$$W = 8 \qquad L = 4 \qquad P = 0.5 \qquad T = 3.0$$
$$\Delta x = 0.5 \qquad \Delta y = 0.5$$

6.30(L) The temperatures, T, within a ceramic component of a special energy recovery system, can be simulated from a Laplace equation model. Finite-difference approximations for the boundary conditions are given here, where $\Delta x = \Delta y = 1.0$ has been used.

$$T_{i,j} = 65 \text{ for } i = 5 \text{ and } 6, \text{ and } j > 5$$

$$T_{1,j} = \frac{T_{2,j} + T_{1,j+1} + T_{1,j-1}}{3.3}$$

$$T_{10,j} = \frac{10 + T_{9,j} + T_{10,j+1} + T_{10,j-1}}{3.3}$$

$$T_{i,1} = T_{i,2} \quad \text{and} \quad T_{i,12} = T_{i,11}$$

The index spans for the grid scheme are $1 \leqslant i \leqslant 10$ for the x direction, and $1 \leqslant j \leqslant 12$ for the y direction. Continue the finite-difference discretization and set up the simulation equations for the interior $T_{i,j}$ values. Develop a FORTRAN or BASIC program, using a Gauss–Seidel solver (starters = 0.0, $\varepsilon = 0.05$), for simulating the $T_{i,j}$ values. Assume consistent units and demonstrate your program. Present the results graphically if possible.

6.31(L) The temperatures in a ceiling panel, as shown in Figure P6.31(L), can be simulated from the following model.

$$\frac{\partial^2 T}{\partial x^2} + \frac{\partial^2 T}{\partial y^2} = 23.1$$

with conditions shown on the sketch.

Using a finite-difference grid approach, set up the simulation equations for the T's. Develop a FORTRAN or BASIC program, using a Gauss–Seidel solver (starters = 100, $\varepsilon = 0.01$), for simulating the $T_{i,j}$ values. Assume consistent units and demonstrate your program using the following parameters. Present the results graphically if possible.

$$S = 115. \qquad W = 8.0 \qquad H = 4.0 \qquad A_1 = 1.0 \qquad A_2 = 1.0$$
$$B_1 = 0.67 \qquad B_2 = 0.67 \qquad \Delta x = 0.33 \qquad \Delta y = 0.33$$

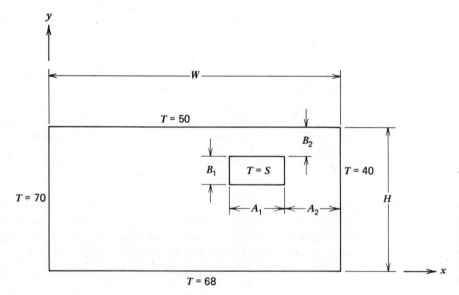

Figure P6.31 L Panel, Problem 6.31 L.

6.32(L) The temperatures, T, in the cross section shown in Figure P6.32(L), can be simulated from a Laplace equation model. Using a finite-difference grid approach, set up the appropriate simulation equations. Develop a FORTRAN or BASIC program, using a Gauss–Seidel solver (starters = 0.0, $\varepsilon = 0.01$), for simulating the $T_{i,j}$ values. Assume consistent units and demonstrate your program using the following parameters. Present the results graphically if possible.

$$A_1 = 0.6 \qquad A_2 = 0.6 \qquad B_1 = 0.4 \qquad B_2 = 0.6$$
$$W = 2.0 \qquad H = 1.8 \qquad \Delta x = 0.2 \qquad \Delta y = 0.2$$

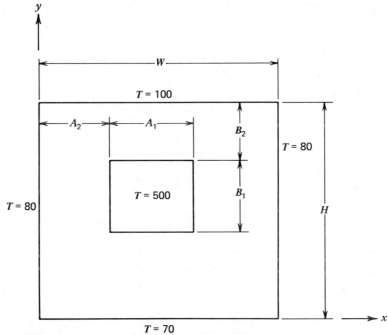

Figure P6.32L   Chimney cross section, Problem 6.32L.

6.33(L) Moisture concentrations, C, in a porous structural member, as shown in Figure P6.33(L) can be simulated from a Laplace equation model. The boundary conditions corresponding to the figure are as follows.

$i =$	1	2	3	4	5	6	7	8	9
$C_i =$	0.0	2.1	11.	20.	20.	20.	20.	20.	26.

Using a finite-difference grid approach, set up the appropriate simulation equations for the C values. Develop a FORTRAN or BASIC program, using a Gauss–Seidel solver (starters = 0.0, $\varepsilon = 0.05$), for simulating the

$C_{i,j}$ values. Assume consistent units and demonstrate your program using the following parameters. Present the results graphically if possible.

$$W=2.4 \qquad R=1.4 \qquad F=0.5 \qquad L=1.8$$
$$H=1.0 \qquad \Delta x=0.1 \qquad \Delta y=0.1$$

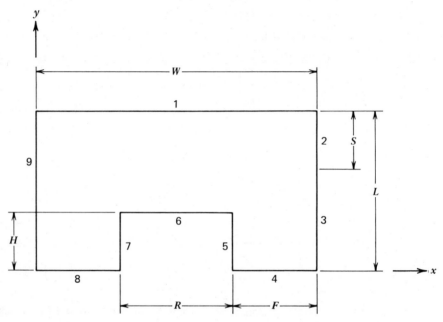

Figure P6.33L Porous Cross Section, Problem 6.33L.

6.34(L) Temperatures, T, within a long structural member, as shown in Figure P6.34(L), can be simulated from a Laplace equation model. Using a finite-difference grid scheme, set up the appropriate simulation equations. (*Hint:* Use an average of the adjacent nodal values for interior corner values.) Develop a FORTRAN or BASIC program, using a Gauss–Seidel solver (starters = 30., $\varepsilon=0.01$), for simulating the $T_{i,j}$ values. Assume consistent units and demonstrate your program using the following parameters. Present the results graphically if possible.

$$\Delta x=\Delta y=1.0$$

6.35(L) A rectangular $(L \times W)$ bar is used as a torque transmitter (subjected to pure twist). The local axial deformations, w, on a normal cross section can be simulated from a Laplace equation model.

$$\frac{\partial^2 w}{\partial x^2}+\frac{\partial^2 w}{\partial y^2}=0.0$$

$$\frac{\partial w}{\partial x}=\theta y \quad \text{at} \left(\pm\frac{W}{2},y\right)$$

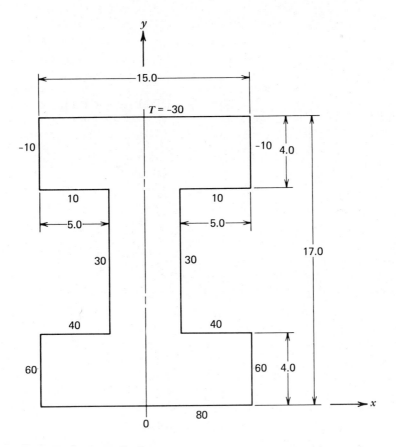

Figure P6.34L Structural-member cross section, Problem 6.34L.

$$\frac{\partial w}{\partial y} = -\theta x \quad \text{at}\left(x, \pm\frac{L}{2}\right)$$

$$w\left(\frac{-W}{2}, 0\right) = 0.0$$

Use a finite-difference grid approach and set up the appropriate simulation equations for w. Develop a FORTRAN or BASIC program, using a Gauss–Seidel solver (starters=0.0, $\varepsilon = 5 \times 10^{-6}$), for simulating the $w_{i,j}$ values. Assume consistent units and demonstrate your program using the following parameters. Present the results graphically if possible.

$$W = 1.4 \qquad L = 2.8 \qquad \theta = 0.005$$
$$\Delta x = 0.2 \qquad \Delta y = 0.2$$

6.36(L) Temperatures, T, on a guide-rail cross section, as shown in Figure P6.36(L), can be simulated from a Laplace equation model. Boundary conditions corresponding to the figure are

$$T_a = 130. \qquad T_b = 300. \qquad T_c = 85. \qquad T_d = 78.$$
$$T_e = 431. \qquad T_f = 280.$$

Use a finite-difference grid scheme and set up the required simulation equations for T. Develop a FORTRAN or BASIC program, using a Gauss–Seidel solver (starters $= 200.$, $\varepsilon = 0.01$), for simulating the $T_{i,j}$ values. Assume consistent units and demonstrate your program using the following parameters. Present the results graphically if possible.

$$\Delta x = \Delta y$$

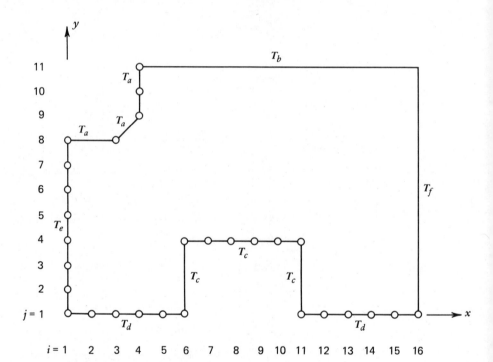

Figure P6.36L   Rail cross section, Problem 6.36L.

6.37(L) Repeat any problem from 6.25(L) through 6.36(L), except use an exact solver from your computer system library instead of the Gauss–Seidel solver.

6.38(L) The transient temperatures, T, in a large concrete floor can be simulated from the following model.

$$\frac{\partial^2 T}{\partial x^2} = \left(\frac{1}{\alpha}\right)\frac{\partial T}{\partial t}$$

$$T(x, 0) = -20.0$$

$$\frac{\partial T}{\partial t} = -\left(\frac{h}{k}\right)(T - T_a) \quad \text{at } (0, t)$$

$$\frac{\partial T}{\partial x} = 0.0 \quad \text{at } (D, t)$$

$$T_a = 20\left(\sin\frac{\pi t}{6} + 1.5\pi\right)$$

Use a fully explicit finite-difference approach and set up appropriate simulation equations. Develop a FORTRAN or BASIC program, based on the marching method in time, for simulating T_i values at various times, t. Assume consistent units and demonstrate your program using the following parameters. Present the results graphically at the times: $t = 2.5$ and 5.0.

$$\alpha = 0.025 \qquad \frac{h}{k} = 5.0 \qquad D = 0.75$$

$$\Delta x = 0.025 \qquad \Delta t = \frac{0.5(\Delta x)^2}{\alpha}$$

6.39(L) The transient pressures, P, in a porous underground containment system for a gas can be simulated from the following model.

$$\frac{\partial^2 W}{\partial x^2} = A \frac{\partial W}{\partial t}$$

$$W = P^2 \qquad \text{and} \qquad A = (2aP)^{-1}$$

$$W(x, 0) = 1.0$$

$$W(0, t) = 4.0$$

$$W(L, t) = 1.0$$

Use a fully explicit finite-difference approach and set up appropriate simulation equations. Develop a FORTRAN or BASIC program, based on the marching method in time, for simulating P_i values at various times, t. Assume consistent units and demonstrate your program using the following parameters. Present the results graphically at the times: $t = 600$, 1200, and 3600.

$$L = 78.0 \qquad a = 0.05 \qquad \Delta t = 20.0 \qquad \Delta x = 3.0$$

6.40(L) A well-known geotechnical heat transfer problem is how to predict near-surface underground temperatures, ϕ, at various times, t. One simple model from which ϕ simulations can be generated is

$$\frac{\partial^2 \phi}{\partial x^2} = \frac{1}{\alpha} \frac{\partial \phi}{\partial t}$$

$$\phi(0, t) = \phi_s$$

$$\phi(L, t) = 0.00$$

$$\phi_s = T_a \sin (0.0172t - 1.6)$$

x	$\phi(x, 0)$	x	$\phi(x, 0)$	x	$\phi(x, 0)$
0.0	−16.9	5.5	1.06	11.0	0.04
0.5	−13.8	6.0	1.07	11.5	0.01
1.0	−10.8	6.5	1.06	12.0	−0.03
1.5	−8.07	7.0	0.93	12.5	−0.03
2.0	−5.70	7.5	0.82	13.0	−0.04
2.5	−3.70	8.0	0.65	13.5	−0.03
3.0	−2.14	8.5	0.53	14.0	−0.03
3.5	−0.92	9.0	0.38	14.5	−0.01
4.0	−0.08	9.5	0.29	15.0	0.00
4.5	0.52	10.0	0.17	—	—
5.0	0.85	10.5	0.11	—	—

Use a fully explicit finite-difference approach and set up appropriate simulation equations. Develop a FORTRAN or BASIC program, based on the marching method in time, for simulating ϕ_i values at various times, t. Assume consistent units and demonstrate your program using the following parameters. Present the results graphically at the times: $t = 25, 200,$ and 300.

$$L = 15.0 \qquad \alpha = 0.05 \qquad T_a = 17.0$$

$$\Delta x = 0.5 \qquad \Delta t = \frac{0.5(\Delta x)^2}{\alpha}$$

6.41(L) The transverse oscillations, y, of an electrode wire in a flow-through gas-cleaning system can be simulated from the following model.

$$\frac{\partial^2 y}{\partial t^2} = \left[\frac{gF_o}{\gamma A}\right]\frac{\partial^2 y}{\partial x^2} - g$$

$$y(x, 0) = 0.1(L - x) \qquad \text{for } x > \frac{L}{2}, \qquad \text{otherwise } y(x, 0) = 0.1x$$

$$\frac{\partial y}{\partial t} = 0.0 \qquad \text{at } (x, 0)$$

$$y(0, t) = 0.0$$

$$y(L, t) = 0.0$$

Use a fully explicit finite-difference approach and set up appropriate simulation equations for y. Develop a FORTRAN or BASIC program, based on the marching method in time, for simulating y_i values at various times, t. Assume consistent units and demonstrate your program using the following parameters. Present the results graphically at the times: $t = 2 \times 10^{-4}$ and 98×10^{-4}

$$L = 4.6 \qquad A = 1.36 \times 10^{-5} \qquad \gamma = 480.$$

$$F_0 = 50.683 \qquad g = 32.2$$

$$\Delta x = 0.1 \qquad \Delta t = \frac{\Delta x}{\sqrt{F_0 g/(\gamma A)}}$$

6.42(L) An important cold-climate bioengineering problem is the cool-down behavior of exposed skin. One model that has been used to simulate transient near-surface skin temperatures, T, is

$$\frac{\partial^2 T}{\partial x^2} = \frac{1}{\alpha} \frac{\partial T}{\partial t}$$

$$T(x, 0) = T_m$$

$$T(0, t) = T_m$$

$$\frac{\partial T}{\partial x} = B(T_a - T) \qquad \text{at } (L, t)$$

Use a fully explicit finite-difference approach and set up the simulation equations. Develop a FORTRAN or BASIC program, based on the marching method in time, for simulating T_i values at various times, t. Assume consistent units and demonstrate your program using the following parameters. Present the results graphically at the times: $t = 12$, 30, and 60.

$$T_m = 34.7 \qquad \alpha = 1.14 \times 10^{-7} \qquad L = 0.0098$$
$$B = 0.05 \qquad T_a = -20.0 \qquad \Delta x = 2 \times 10^{-4}$$
$$\Delta t = 0.15$$

6.43(L) The transient diffusion of a pollutant, C, can be simulated from the following model.

$$D \frac{\partial^2 C}{\partial x^2} - kC = \frac{\partial C}{\partial t}$$

$$C(x, 0) = 0.0$$

$$C(L, t) = 0.0$$

$$\frac{\partial C}{\partial t} = -2.0 \times 10^{-6} \qquad \text{at } (0, t)$$

Use a fully explicit finite-difference approach and set up the simulation equations. Develop a FORTRAN or BASIC program, based on the marching method in time, for simulating C_i values at various times, t.

Assume consistent units and demonstrate your program using the following parameters. Present the results graphically at the times: $t=15$ and 400.

$$D=1.5 \times 10^{-5} \qquad k=0.03 \qquad L=0.2$$
$$\Delta t=0.5 \qquad \Delta x=0.005$$

6.44(L) The transient temperatures, T, of a road bed can be simulated from the following model.

$$\frac{\partial^2 T}{\partial x^2} = \frac{1}{\alpha}\frac{\partial T}{\partial t}$$

$$T(x, 0)=35.0$$

$$T(0, t)=12.5 \cos \frac{\pi t}{12}$$

$$T(L, t)=35.0$$

Use a fully explicit finite-difference approach and set up the appropriate simulation equations. Develop a FORTRAN or BASIC program, based on the marching method in time, for simulating T_i values at various times, t. Assume consistent units and demonstrate your program using the following parameters. Present the results graphically at the times: $t=24.5$, 34.5, and 44.5.

$$\alpha=0.01 \qquad L=4.0 \qquad \Delta t=0.5 \qquad \Delta x=0.1$$

6.45(L) The transient velocities, u, of certain fluids near a suddenly accelerated surface can be simulated from the following model.

$$\frac{\partial^2 u}{\partial x^2} = A\frac{\partial u}{\partial t}$$

$$u(x, 0)=0.0$$
$$u(0, t)=100.0$$
$$u(L, t)=0.0$$

Use a fully explicit finite-difference approach and set up the appropriate simulation equations. Develop a FORTRAN or BASIC program, based on the marching method in time, for simulating u_i values at various times, t. Assume consistent units and demonstrate your program using the following parameters. Present the results graphically at the times: $t=5$, 10, and 15.

$$A=5 \times 10^4 \qquad L=0.1 \qquad \Delta x=0.002 \qquad \Delta t=0.1$$

6.46(L) In a case-hardening process, the transient concentrations, C, of carbon can be simulated from the following model.

$$D\frac{\partial^2 C}{\partial x^2} = \frac{\partial C}{\partial t}$$

$$C(x, 0) = C_0$$
$$C(0, t) = C_s$$
$$C(L, t) = C_0$$

Use a fully explicit finite-difference approach and set up appropriate simulation equations. Develop a FORTRAN or BASIC program, based on the marching method in time, for simulating C_i values at various times, t. Assume consistent units and demonstrate your program using the following parameters. Present the results graphically at the times: $t = 400$, 2400, and 4800.

$$D = 1.0 \times 10^{-7} \qquad C_0 = 0.2 \qquad C_s = 0.8 \left[1.25 - \exp\left(\frac{-t}{1253}\right) \right]$$

$$L = 0.20 \qquad \Delta x = 0.004 \qquad \Delta t = 80.0$$

6.47(L) The transient vertical displacements, v, of a stretched cable can be simulated from the following model.

$$\frac{\partial^2 v}{\partial y^2} + 1.3 \times 10^{-4} = 4 \times 10^{-6} \frac{\partial^2 v}{\partial t^2}$$

$$\frac{\partial v}{\partial t} = 0.0 \qquad \text{at } (y, 0)$$

$$v(0, t) = 0.0$$

$$v(L, t) = 0.0$$

$$v(y, 0) = 0.4 \sin \frac{\pi y}{L}$$

Use a fully explicit finite-difference approach and set up the appropriate simulation equations. Develop a FORTRAN or BASIC program, based on the marching method in time, for simulating v_i values at various times, t. Assume consistent units and demonstrate your program using the following parameters. Present the results graphically at the times: $t = 6 \times 10^{-4}$, 30×10^{-4}, and 0.01.

$$L = 5.0 \qquad \Delta y = 0.1 \qquad \Delta t = 2 \times 10^{-4}$$

6.48(L) Repeat Problem 6.38(L), except use a fully implicit finite-difference approach with $\Delta t = 0.05$ and any good equation-set solver.

6.49(L) Repeat Problem 6.49(L), except use a Crank–Nicolson finite-difference approach.

Some Case-Study Projects

6.50(C) Set up a differential model and perform some simulations for transient temperatures in a two- or three-dimensional object.

6.51(C) Set up a differential model and perform some simulations for transient temperatures in a solid where the thermal flow resistance is a quadratic function of the temperature.

6.52(C) Set up a differential model and perform some simulations for transient pressures (and flows) in a two- or three-dimensional aquifer.

6.53(C) Repeat Problem 6.52(C), except use a flow resistance that is a complicated function of position in the aquifer and include several sinks (outflow wells).

6.54(C) Set up a differential model and perform some simulations for transient temperatures in a two- or three-dimensional object that has several sources (internal heating).

7
Numerical Simulations Using Finite Elements

AN INTRODUCTION TO FINITE-ELEMENT METHODS

The finite element method is a powerful numerical technique for obtaining approximate solutions (simulations) to boundary value problems. As pointed out previously, in boundary value problems, values for a dependent variable are sought in the interior of a given domain that has prescribed continuous, but not necessarily analytic, values on the boundary of the domain. Thus, typically, an elliptic-type differential model (or the elliptic portion of a parabolic model) is involved. Of course, if only one spatial coordinate is involved, a second-order ordinary differential equation model may be the starting point. However, it is interesting to note that, with at least one finite-element scheme, we can get directly to many reasonably good simulations without having to deal with the differential models. This has significant engineering appeal at times and is described in more detail later.

The finite-element method as applied to boundary value problems had its origin in the field of solid mechanics, but today it encompasses the whole area of continuum mechanics. Because of its diversity and flexibility when used in conjunction with the modern digital computer, the finite-element method currently is considered to be a general-purpose analysis tool applicable in almost all engineering-analysis situations.

The basic idea behind the method is that any domain can be considered to be a coupled group of subdomains, called finite elements. Uniformity of the subdomains is not a starting restriction, in contrast to the cell (or grid) restrictions in many finite-difference methods. The shapes of the finite elements are usually taken to be simple polygons, such as triangles and quadrilaterals for two-dimensional domains, and tetrahedrons and hexahedrons for three-dimensional domains, as illustrated in Figure 7.1. For one-dimensional domains, a simple line element may be used.

The process of dividing the domain into the finite-element subdomains is called discretization and, as with the discretization process for finite-difference methods, results in an approximation of the real domain. The discretization procedure

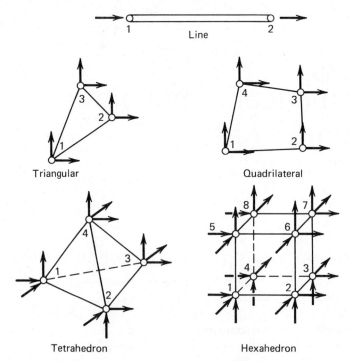

Figure 7.1 Typical finite elements for one, two, and three dimensions showing nodal degrees of freedom.

reduces the problem from one with an infinite number of unknowns (or field variables) to one that has a reasonably small number of unknowns, as shown in Figure 7.2.

Unique to the finite-element method is that the unknowns, or field variables, are now expressed in terms of *assumed* approximating functions within each element. Polynomials and trigonometric series are usually used for these approximating functions, since they are relatively simple in form and easy to manipulate. These approximating functions, which are also widely referred to as "interpolation functions," or "shape functions," are defined in terms of the unknowns at specified points called "nodes" or "nodal points." The reader is reminded here that these unknowns usually represent some type of physical quantity such as a local displacement, temperature, stress, fluid pressure, and so on, at each node. Generally, the nodal points are located at the intersections of the element sides, although additional nodes may be located within elements if necessary to accommodate more-complicated approximating functions. Obviously, the nodes will occur at the ends of one-dimensional line elements. For one and two-dimensional elements, the interfaces between adjacent elements are called interelement boundaries, as shown in Figure 7.3.

The local nodal values of the unknown variables and the approximating function for each element completely define the behavior of each element in terms of all the

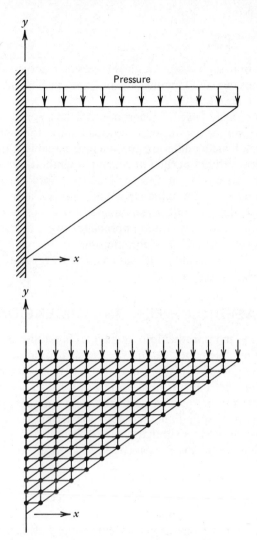

Figure 7.2 Discretization of thin triangular plate.

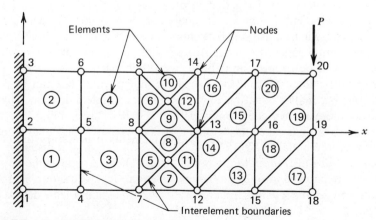

Figure 7.3 Typical finite-element model.

nodal values or in terms of the "nodal degrees of freedom." Of course, the remaining difficulty is how we should invoke the basic physical laws and principles governing element behavior.

Several approaches exist for incorporating basic governing principles and laws into the matrix-form relationship described above for each element. The fact that we have several widely differing approaches available is often viewed as evidence of the physical validity of the finite-element method. Basically, these approaches fall into four categories: an energy-balance approach, a variational (calculus) approach, a method of weighted residuals approach, or a direct approach. Since mathematical skills beyond those assumed for the users of this book are usually required to treat the first three approaches, we will concentrate on the direct approach. However, for the reader interested in a more comprehensive treatment, the references[1–11] should be consulted and/or a separate course in finite-element methods should be taken.

THE DIRECT APPROACH FOR ONE-DIMENSIONAL SYSTEMS

As described in the introduction, we have to break the system continuum up into a series of finite elements, calling the process discretization. Good engineering judgment is required in accomplishing this discretization; otherwise, sharp gradient zones may not get enough nodes and poor simulation values may result. The standard matrix techniques found in structural analyses [see references[2–6,8]] are chosen as the appropriate methodology for getting directly to the finite-element formulations required for simulations. We observe that this methodology will give us a matrix equation for each element of the form shown in Equation 7.1.

$$[k]\{x\} = \{f\} \tag{7.1}$$

Here $[k]$ is the element stiffness matrix, $\{f\}$ is a column vector representing loads at the ends of the element, and $\{x\}$ is a column vector representing the unknowns, which are one-dimensional displacements at the ends of the element. We are really most interested in the form of the equation, and not so much in the fact that the x's are displacements, realizing that in other boundary-value problems the x's will represent temperatures, pressures, and so on. Exactly how we get values for the k and f terms will be illustrated by example later. Obviously, since adjacent elements share common end nodes, the matrix equation of each element, as given by Equation 7.1, will be coupled directly to two other matrix equations. Consequently, it becomes evident that all these separate matrix equations can be put together into one big matrix equation representing the complete domain. The process of putting these matrix equations together is called "assembling," or "patching"; the overall result is called the "global matrix equation." The global set of equations relates all the nodal degrees of freedom (one-dimensional x displacements at the nodes in our illustration) to the nodal forces. Then, as soon as the boundary conditions are patched into the global matrix equation, an equation-set solver can be applied to get the x values. Examples 7.1, 7.2, and 7.3 illustrate details of the foregoing.

EXAMPLE 7.1

When a force f is applied to the end of a simple linear spring, as illustrated in Figure 7.4, a displacement Δ is produced; and there is a linear relationship between force and displacement as given by Equation 7.2, where k is the spring rate, or stiffness, of the spring. Since one end of the spring is fixed, the system can

$$f = k\Delta \tag{7.2}$$

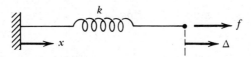

Figure 7.4   Linear spring of stiffness k, fixed at one end with an external force f applied at the other producing a displacement Δ.

have only one displacement, or degree of freedom, and a simple algebraic equation in one unknown can be solved to find the displacement. Next consider the linear spring shown in Figure 7.5, which has forces and displacements defined at each end. Let node 1 be given a displacement Δ_1, and node 2 a displacement Δ_2, the overall elongation of the spring becoming $\Delta_1 - \Delta_2$, assuming $\Delta_1 > \Delta_2$. Consider the spring as a single finite-element system and set up the appropriate element matrix equation.

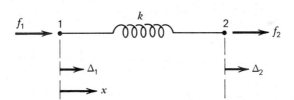

Figure 7.5   A linear spring of stiffness k, showing the displacements Δ_1, Δ_2, and forces f_1, f_2.

► Using Equation 7.2, the force-deflection relationship for node 1 can be written as:

$$f_1 = k(\Delta_1 - \Delta_2)$$

and since equilibrium of forces (static system) requires that $f_1 + f_2 = 0$, we have

$$f_2 = -k(\Delta_1 - \Delta_2)$$

These can be written in matrix form as:

$$\begin{Bmatrix} f_1 \\ f_2 \end{Bmatrix} = \begin{bmatrix} k & -k \\ -k & k \end{bmatrix} \begin{Bmatrix} \Delta_1 \\ \Delta_2 \end{Bmatrix}$$

or in standard matrix equation form as typified by Equation 7.1:

$$\{f\} = [k]\{\Delta\}$$

EXAMPLE 7.2

Consider a system as shown in Figure 7.6, consisting of two linear springs. Using the results of Example 7.1, set up a global matrix equation for simulating the nodal displacements, treating each spring as a finite element.

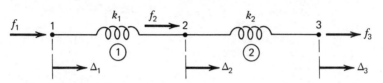

Figure 7.6 Two-element linear spring system of stiffness k_1 and k_2 with displacements Δ_1, Δ_2, Δ_3 and forces f_1, f_2, f_3.

▶ The element matrix equations can be written directly from Example 7.1.
   For element one:

$$\begin{Bmatrix} f_1 \\ f_2 \end{Bmatrix} = \begin{bmatrix} k_1 & -k_1 \\ -k_1 & k_1 \end{bmatrix} \begin{Bmatrix} \Delta_1 \\ \Delta_2 \end{Bmatrix}$$

For element two:

$$\begin{Bmatrix} f_2 \\ f_3 \end{Bmatrix} = \begin{bmatrix} k_2 & -k_2 \\ -k_2 & k_2 \end{bmatrix} \begin{Bmatrix} \Delta_2 \\ \Delta_3 \end{Bmatrix}$$

Noting that the matrix coefficients for the same node are added together in the assemblage process, we construct the following global matrix equation.

$$\begin{Bmatrix} f_1 \\ f_2 \\ f_3 \end{Bmatrix} = \begin{bmatrix} k_1 & -k_1 & 0 \\ -k_1 & k_1 + k_2 & -k_2 \\ 0 & -k_2 & k_2 \end{bmatrix} \begin{Bmatrix} \Delta_1 \\ \Delta_2 \\ \Delta_3 \end{Bmatrix}$$

We observe that the global stiffness matrix is of order 3×3; that is, the system in Figure 7.6 has three nodes and is a three-degree-of-freedom system. Also we have used a direct approach in simply adding the stiffness contributions from the elements joining each node. As shown, the global stiffness matrix is singular; however, we can remove the singularity by applying the boundary condition that $\Delta_1 = 0$. Furthermore, to make things even simpler, let us assume that $k_1 = k_2 = k$.

$$\begin{Bmatrix} f_1 \\ f_2 \\ f_3 \end{Bmatrix} = \begin{bmatrix} k & -k & 0 \\ -k & 2k & -k \\ 0 & -k & k \end{bmatrix} \begin{Bmatrix} 0 \\ \Delta_2 \\ \Delta_3 \end{Bmatrix}$$

thus, the final global matrix equation becomes

$$\left\{ \begin{matrix} f_2 \\ f_3 \end{matrix} \right\} = \left[\begin{matrix} 2k & -k \\ -k & k \end{matrix} \right] \left\{ \begin{matrix} \Delta_2 \\ \Delta_3 \end{matrix} \right\}$$

EXAMPLE 7.3

The system shown in Figure 7.7 consists of three rigid elements interconnected by linear springs with the spring rates (stiffnesses) k_i shown.

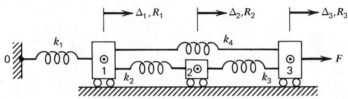

Figure 7.7 Three rigid elements interconnected by linear springs of stiffness k_1, k_2, k_3, k_4.

Determine the displacements of the rigid elements when subjected to the load F by using a finite-element method incorporating a direct approach.

▶ Each spring will be considered a finite element, making the discretization process elementary. An element matrix equation for each spring element can be taken directly from Example 7.1. We will now try to streamline our procedures a bit by assembling the global matrix equation right as we identify each element matrix equation. In fact, this is precisely the way most finite-element computer programs are arranged. We start with a skeleton global equation containing all zeros for the stiffness matrix coefficient terms and force vector terms. We will give the element equation first and then the resulting global skeleton equation after the element equation has been patched in.

For element (spring) one:

$$\left[\begin{matrix} k_1 & -k_1 \\ -k_1 & k_1 \end{matrix} \right] \left\{ \begin{matrix} \Delta_0 \\ \Delta_1 \end{matrix} \right\} = \left\{ \begin{matrix} f_0^1 \\ f_1^1 \end{matrix} \right\}$$

or

$$\left[\begin{matrix} k_1 & -k_1 & 0 & 0 \\ -k_1 & k_1 & 0 & 0 \\ 0 & 0 & 0 & 0 \\ 0 & 0 & 0 & 0 \end{matrix} \right] \left\{ \begin{matrix} \Delta_0 \\ \Delta_1 \\ \Delta_2 \\ \Delta_3 \end{matrix} \right\} = \left\{ \begin{matrix} f_0^1 \\ f_1^1 \\ 0 \\ 0 \end{matrix} \right\}$$

For element (spring) two:

$$\left[\begin{matrix} k_2 & -k_2 \\ -k_2 & k_2 \end{matrix} \right] \left\{ \begin{matrix} \Delta_1 \\ \Delta_2 \end{matrix} \right\} = \left\{ \begin{matrix} f_1^2 \\ f_2^2 \end{matrix} \right\}$$

or

$$\begin{bmatrix} 0 & 0 & 0 & 0 \\ 0 & k_2 & -k_2 & 0 \\ 0 & -k_2 & k_2 & 0 \\ 0 & 0 & 0 & 0 \end{bmatrix} \begin{Bmatrix} \Delta_0 \\ \Delta_1 \\ \Delta_2 \\ \Delta_3 \end{Bmatrix} = \begin{Bmatrix} 0 \\ f_1^2 \\ f_2^2 \\ 0 \end{Bmatrix}$$

For element (spring) three:

$$\begin{bmatrix} k_3 & -k_3 \\ -k_3 & k_3 \end{bmatrix} \begin{Bmatrix} \Delta_2 \\ \Delta_3 \end{Bmatrix} = \begin{Bmatrix} f_2^3 \\ f_3^3 \end{Bmatrix}$$

or

$$\begin{bmatrix} 0 & 0 & 0 & 0 \\ 0 & 0 & 0 & 0 \\ 0 & 0 & k_3 & -k_3 \\ 0 & 0 & -k_3 & k_3 \end{bmatrix} \begin{Bmatrix} \Delta_0 \\ \Delta_1 \\ \Delta_2 \\ \Delta_3 \end{Bmatrix} = \begin{Bmatrix} 0 \\ 0 \\ f_2^3 \\ f_3^3 \end{Bmatrix}$$

For element (spring) four:

$$\begin{bmatrix} k_4 & -k_4 \\ -k_4 & k_4 \end{bmatrix} \begin{Bmatrix} \Delta_1 \\ \Delta_3 \end{Bmatrix} = \begin{Bmatrix} f_1^4 \\ f_3^4 \end{Bmatrix}$$

or

$$\begin{bmatrix} 0 & 0 & 0 & 0 \\ 0 & k_4 & 0 & -k_4 \\ 0 & 0 & 0 & 0 \\ 0 & -k_4 & 0 & k_4 \end{bmatrix} \begin{Bmatrix} \Delta_0 \\ \Delta_1 \\ \Delta_2 \\ \Delta_3 \end{Bmatrix} = \begin{Bmatrix} 0 \\ f_1^4 \\ 0 \\ f_3^4 \end{Bmatrix}$$

In the foregoing the superscripts on the forces represent the element numbers; the subscripts represent the rigid elements, or the nodes.

The preceding global equations show only the element-by-element patching, where in fact all the terms have been superimposed as follows.

$$\begin{bmatrix} k_1 & -k_1 & 0 & 0 \\ -k_1 & k_1+k_2+k_4 & -k_2 & -k_4 \\ 0 & -k_2 & k_2+k_3 & -k_3 \\ 0 & -k_4 & -k_3 & k_3+k_4 \end{bmatrix} \begin{Bmatrix} \Delta_0 \\ \Delta_1 \\ \Delta_2 \\ \Delta_3 \end{Bmatrix} = \begin{Bmatrix} f_0^1 \\ f_1^1+f_1^2+f_1^4 \\ f_2^2+f_2^3 \\ f_3^3+f_3^4 \end{Bmatrix}$$

Under a given loading condition each individual element, as well as the entire system, must be in static equilibrium. Thus at each node i, the sum of the internal loads in the members must be equal to the external load at that node, which leads to the following node force relationships.

Node 1: $\qquad f_1^1+f_1^2+f_1^4=R_1=0$

Node 2: $\qquad f_2^2+f_2^3=R_2=0$

Node 3: $\qquad f_3^3+f_3^4=R_3=F$

Furthermore, the boundary condition on the left is $\Delta_0 = 0$. To make the situation even simpler, we will assume that $k_1 = k_2 = k_3 = k$. Incorporating all the preceding into the global matrix equation gives us the following.

$$(k) \begin{bmatrix} 3 & -1 & -1 \\ -1 & 2 & -1 \\ -1 & -1 & 2 \end{bmatrix} \begin{Bmatrix} \Delta_1 \\ \Delta_2 \\ \Delta_3 \end{Bmatrix} = \begin{Bmatrix} 0 \\ 0 \\ F \end{Bmatrix}$$

The preceding global matrix equation can be solved for the displacements (this is left as an exercise) to give us the following.

$$\begin{Bmatrix} \Delta_1 \\ \Delta_2 \\ \Delta_3 \end{Bmatrix} = \frac{F}{k} \begin{Bmatrix} 1.0000 \\ 1.3333 \\ 1.6667 \end{Bmatrix}$$

Once the displacements have been found, the forces in each spring element may be computed using the load-element stiffness matrix for each spring. The results are given here to complete the simulation.

$$f_1^1 = F$$
$$f_1^2 = -f_2^2 = -0.3333F$$
$$f_2^3 = -f_3^3 = -0.3333F$$
$$f_1^4 = -f_3^4 = -0.6667F$$

THE DIRECT APPROACH FOR SOME SIMPLE STRUCTURAL SYSTEMS

The finite-element method was first used in the aircraft industry to provide structural analyses for extremely complex shapes. As applied, the method was essentially a generalization of structural analysis procedures for framed structures.[10] The part to be analyzed was conceptually transformed into a system consisting of a relatively small number of elements interconnected at node points through which forces were transmitted. Thus an entire structure was visualized as made up of an assemblage of individual elements whose behavior could be treated with fundamental concepts involving basic stiffness and deflection equations. The process of assembling the overall, or global, stiffness relationship for the entire structure from the individual element stiffness equations became known as the "direct method" of structural analysis. As illustrated in very elementary form in the preceding section, the main thrust of this approach is the development of the local and global stiffness matrices. To enlarge the description of this approach, we will derive stiffness matrices for a pin-jointed bar, or truss, element; a beam element; and a plane-stress triangular plate element. Use in simple systems will be treated with examples. However, it must be pointed out that this direct approach is difficult, or impossible, to apply to the more complicated structural elements and certain special phenomena.[8]

Axial Bar-Element Systems

We can imagine a pin-jointed bar of length l, with a cross-sectional area A, and with a modulus of elasticity E, as shown in Figure 7.8. The end displacements u_1, v_1, u_2, and v_2 give us a four-degree-of-freedom system, where

$$\{q\} = [u_1 \quad v_1 \quad u_2 \quad v_2]^T \tag{7.3}$$

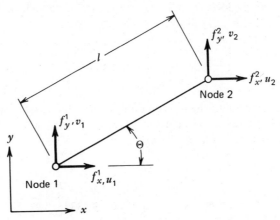

Figure 7.8   Pin-jointed bar element with forces and displacements.

From Figure 7.8 we find that

$$l = [(x_2 - x_1)^2 + (y_2 - y_1)^2]^{0.5} \tag{7.4}$$

and

$$S = \sin \theta = \frac{y_2 - y_1}{l}; \qquad C = \cos \theta = \frac{x_2 - x_1}{l} \tag{7.5}$$

The element stiffness matrix is formed by alternately displacing each degree of freedom while keeping all displacements of the other degrees of freedom zero. We assume that the nodal displacements are small in comparison to the length of the axial element so that linear theory applies. In Figure 7.9a node 1 is given a displacement $u_1 \neq 0$, $v_1 = u_2 = v_2 = 0$. An axial shortening $u_1 \cos \theta$ occurs and produces the following axial compressive force, as shown in Figure 7.10.

$$F = \left(\frac{AE}{l}\right) u_1 \cos \theta$$

The x and y components of the force (the superscripts shown here designate the node number) are given as:

$$f_x^1 = -f_x^2 = \frac{AE}{l} C^2 u_1$$

$$f_y^1 = -f_y^2 = \frac{AE}{l} CS u_1$$

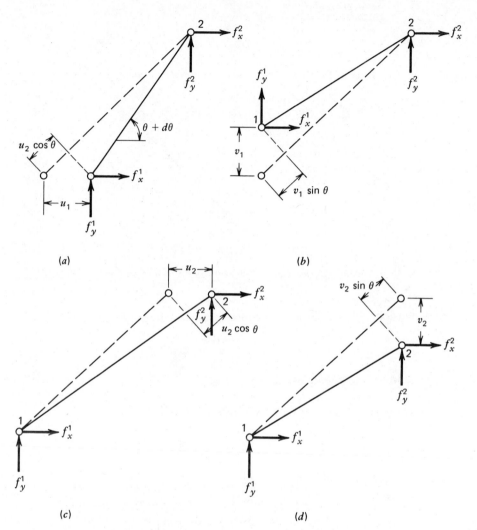

Figure 7.9 Node Displacements for pin-jointed bar element.

Since these force components provide static equilibrium, we have the following vector equation, Equation 7.6.

$$\frac{AE}{l} \begin{Bmatrix} C^2 \\ CS \\ -C^2 \\ -CS \end{Bmatrix} \begin{Bmatrix} u \\ v \end{Bmatrix} = \begin{Bmatrix} f_x^1 \\ f_y^1 \\ f_x^2 \\ f_y^2 \end{Bmatrix} \tag{7.6}$$

The remaining displacements are done in a similar manner. For example, node 2 is given a displacement $v_2 \neq 0$, $u_1 = u_2 = v_1 = 0$. An elongation $v_2 \sin \theta$ produces an

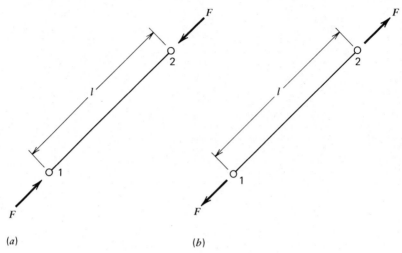

Figure 7.10 Axial compressive or tension load due to nodal displacements.

axial tension force as shown in Figure 10(b), and given as follows.

$$F = \frac{AE}{l} S v_2$$

The x and y components of this force are

$$f_x^2 = -f_x^1 = \frac{AE}{l} CS v_2$$

$$f_y^2 = -f_y^1 = \frac{AE}{l} S^2 v_2$$

Again, for static equilibrium, we can write a vector equation as shown by Equation 7.7. Performing the same calculations for $v_1 \neq 0$, $u_1 = u_2 = v_2 = 0$, and

$$\frac{AE}{l} \begin{Bmatrix} -CS \\ -S^2 \\ CS \\ S^2 \end{Bmatrix} \begin{Bmatrix} u \\ v \end{Bmatrix} = \begin{Bmatrix} f_x^1 \\ f_y^1 \\ f_x^2 \\ f_y^2 \end{Bmatrix} \tag{7.7}$$

For $u_2 \neq 0$, $u_1 = v_1 = v_2 = 0$, and using the superposition principle yields a complete matrix equation for the axial bar-element system.

$$\frac{AE}{l} \begin{bmatrix} C^2 & CS & -C^2 & -CS \\ CS & S^2 & -CS & -S^2 \\ -C^2 & -CS & C^2 & CS \\ -CS & -S^2 & CS & S^2 \end{bmatrix} \begin{bmatrix} u_1 \\ v_1 \\ u_2 \\ v_2 \end{bmatrix} = \begin{Bmatrix} f_x^1 \\ f_y^1 \\ f_x^2 \\ f_y^2 \end{Bmatrix} \tag{7.8}$$

We observe that this is of the same form as Equation 7.1, but with different symbols, which we acknowledge by showing Equation 7.9.

$$[k]\{q\} = \{f\} \tag{7.9}$$

If the axial pin-jointed bar element lies alone the x-axis and has only x-direction displacements, as shown in Figure 7.11, then Equation 7.8 reduces to Equation 7.10.

$$\frac{AE}{l} \begin{bmatrix} 1 & -1 \\ -1 & 1 \end{bmatrix} \begin{Bmatrix} u_1 \\ u_2 \end{Bmatrix} = \begin{Bmatrix} f_x^1 \\ f_x^2 \end{Bmatrix} \tag{7.10}$$

The use of the foregoing developments for a system made up of more than one bar element is illustrated in Example 7.4.

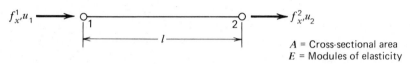

A = Cross-sectional area
E = Modules of elasticity

Figure 7.11 Pin-jointed bar element that lies on the x-axis with only x direction displacement.

EXAMPLE 7.4

Consider the pin-jointed truss shown in Figure 7.12. The numbering system for the nodes and members and other required information is indicated in the figure. Simulate the behavior of this truss under load by using the finite-element method and solving for the u and v displacements at node 2 and the forces in the bar elements.

▶ The equilibrium equation for a pin-jointed bar element is given by Equation 7.8. Since $S = \sin\theta = \frac{4}{5}$ and $C = \cos\theta = \frac{3}{5}$, the equilibrium equation for element (truss member) 1 becomes

$$\left[\frac{(AE/l)_1}{25}\right] \begin{bmatrix} 9 & 12 & -9 & -12 \\ 12 & 16 & -12 & -16 \\ -9 & -12 & 9 & 12 \\ -12 & -16 & 12 & 16 \end{bmatrix} \begin{Bmatrix} u_1 \\ v_1 \\ u_2 \\ v_2 \end{Bmatrix} = \begin{Bmatrix} P_{1x}^1 \\ P_{1y}^1 \\ P_{2x}^1 \\ P_{2y}^1 \end{Bmatrix}$$

In the same manner, since $S = \frac{4}{5}$ and $C = -\frac{3}{5}$, the equation for element 2 becomes

$$\left[\frac{(AE/l)_2}{25}\right] \begin{bmatrix} 9 & -12 & -9 & 12 \\ -12 & 16 & 12 & -16 \\ -9 & 12 & 9 & -12 \\ 12 & -16 & -12 & 16 \end{bmatrix} \begin{Bmatrix} u_2 \\ v_2 \\ u_3 \\ v_3 \end{Bmatrix} = \begin{Bmatrix} P_{1x}^2 \\ P_{1y}^2 \\ P_{2x}^2 \\ P_{2y}^2 \end{Bmatrix}$$

where P_{ix}^j and P_{iy}^j ($i, j = 1, 2$) are the x and y components of force on the jth element at the ith node.

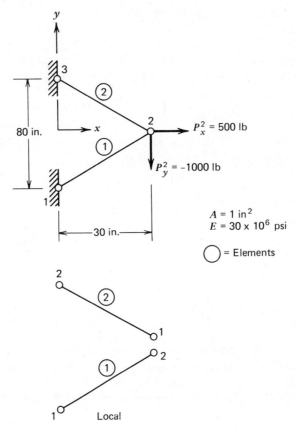

Figure 7.12 Two-element truss loaded with horizontal and vertical loads, global local configurations.

The element stiffness matrices can be combined in a direct fashion to form the overall "global," or assembled, stiffness matrix for the truss system using the patching techniques discussed earlier. Since u_2 and v_2 are common displacements, and $A_1 = A_2 = A$, $E_1 = E_2 = E$, and $l_1 = l_2 = 1$, the following global matrix equation evolves.

$$\left[\frac{AE/l}{25}\right]\begin{bmatrix} 9 & 12 & -9 & -12 & 0 & 0 \\ 12 & 16 & -12 & -16 & 0 & 0 \\ -9 & -12 & 18 & 0 & -9 & 12 \\ -12 & -16 & 0 & 32 & 12 & -16 \\ 0 & 0 & 9 & 12 & 9 & -12 \\ 0 & 0 & 12 & -16 & -12 & 16 \end{bmatrix}\begin{Bmatrix} u_1 \\ v_1 \\ u_2 \\ v_2 \\ u_3 \\ v_3 \end{Bmatrix} = \begin{Bmatrix} R_x^1 \\ R_y^1 \\ R_x^2 \\ R_y^2 \\ R_x^3 \\ R_y^3 \end{Bmatrix}$$

Here R_x^i, R_y^i $(i = 1, 2, 3)$ have been used to represent the external nodal forces. Nodes 1 and 3 are located at fixed reactions, giving $u_1 = u_3 = v_1 = v_3 = 0$, and the preceding global equation must be modified to incorporate these boundary conditions. We do this by eliminating the rows and columns associated with u_1, u_3, v_1, and v_3, which gives us a relatively small final matrix equation to solve.

$$\frac{AE/l}{25}\begin{bmatrix} 18 & 0 \\ 0 & 32 \end{bmatrix}\begin{Bmatrix} u_2 \\ v_2 \end{Bmatrix} = \begin{Bmatrix} R_x^2 \\ R_y^2 \end{Bmatrix} = \begin{Bmatrix} 500 \\ -1000 \end{Bmatrix}$$

The solution yields

$$\begin{Bmatrix} u_2 \\ v_2 \end{Bmatrix} = \begin{Bmatrix} 1.157 \times 10^{-3} \\ -1.302 \times 10^{-3} \end{Bmatrix}$$

The corresponding forces in member 1 of the truss system are given by

$$\begin{Bmatrix} P_{1x}^1 \\ P_{1y}^1 \\ P_{2x}^1 \\ P_{2y}^1 \end{Bmatrix} = \begin{bmatrix} \dfrac{AE/l}{25} \end{bmatrix} \begin{bmatrix} 9 & 12 & -9 & -12 \\ 12 & 16 & -12 & -16 \\ -9 & -12 & 9 & 12 \\ -12 & -16 & 12 & 16 \end{bmatrix} \begin{Bmatrix} 0 \\ 0 \\ 1.157 \times 10^{-3} \\ -1.302 \times 10^{-3} \end{Bmatrix}$$

$$= \begin{Bmatrix} 125 \\ 167 \\ -125 \\ -167 \end{Bmatrix}$$

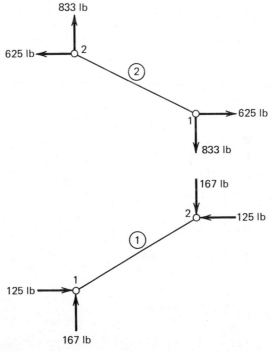

Figure 7.13 Members 1 and 2 with corresponding nodal forces.

It is left as an exercise to find the forces in member 2 of the system.

The forces are shown in Figure 7.13; we note that the sum of the nodal forces at the common node equals the applied external loading.

Beam-Element Systems

Again we use a direct approach to develop the appropriate stiffness matrix. The particular direct approach described herein follows Gallagher.[8] Consider a simple beam element such as shown in Figure 7.14. The unknowns (degrees of freedom) in this case are the nodal y displacements, w_1 and w_2, and the nodal rotations or angular displacements in the x-y plane, θ_1 and θ_2, giving us four degrees of freedom. The fact that we now have moments at the nodes, in contrast to the pin-ended bar elements, means that we have to account for local displacements between the nodes. We introduce for the first time an assumed polynomial approximating function as given by Equation 7.11.

$$w = a_0 + a_1 x + a_2 x^2 + a_3 x^3 \tag{7.11}$$

The θ_1 and θ_2 represent the slopes of the neutral axis at node 1 and 2.

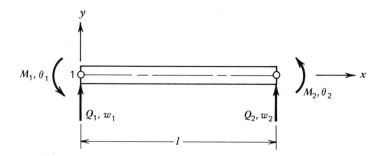

Figure 7.14 Beam element with shear forces, bending moments, and degrees of freedom.

A counterclockwise rotation (assumed positive) causes a positive transverse displacement, and we have the relationships given by Equation 7.12.

$$\theta_1 = \frac{dw}{dx}\bigg|_{x=0} \quad ; \quad \theta_2 = \frac{dw}{dx}\bigg|_{x=l} \tag{7.12}$$

also

$$w_1 = w|_{x=0}; \quad w_2 = w|_{x=l} \tag{7.13}$$

Using the latter two equations and recognizing that $dw/dx = a_1 + 2a_2 x + 3a_3 x^2$

gives us the matrix equation, Equation 7.14.

$$\{W\} = \begin{Bmatrix} w_1 \\ \theta_1 \\ w_2 \\ \theta_2 \end{Bmatrix} = \begin{bmatrix} 1 & 0 & 0 & 0 \\ 0 & 1 & 0 & 0 \\ 1 & l & l^2 & l^3 \\ 0 & 1 & 2l & 3l^2 \end{bmatrix} \begin{Bmatrix} a_0 \\ a_1 \\ a_2 \\ a_3 \end{Bmatrix} \quad (7.14)$$

Inverting the matrix, and solving for $[a_0 \quad a_1 \quad a_2 \quad a_3]^T$, gives Equation 7.15.

$$\begin{Bmatrix} a_0 \\ a_1 \\ a_2 \\ a_3 \end{Bmatrix} = l^{-3} \begin{bmatrix} l^3 & 0 & 0 & 0 \\ 0 & l^3 & 0 & 0 \\ -3l & -2l^2 & 3l & -l^2 \\ 2 & l & -2 & l \end{bmatrix} \begin{Bmatrix} w_1 \\ \theta_1 \\ w_2 \\ \theta_2 \end{Bmatrix} \quad (7.15)$$

We can rewrite Equation 7.11 as shown in Equation 7.16. Then we can substitute Equation 7.15 into Equation 7.16 and substitute $\xi = x/l$, $N_1 = 1 - 3\xi^2 + 2\xi^3$, $N_2 = x(1-\xi)^2$, $N_3 = 3\xi^2 - 2\xi^3$, and $N_4 = -x(\xi - \xi^2)$ to give us Equation 7.17.

$$w = \begin{bmatrix} 1 & x & x^2 & x^3 \end{bmatrix} \begin{Bmatrix} a_0 \\ a_1 \\ a_2 \\ a_3 \end{Bmatrix} \quad (7.16)$$

or

$$w = [N]\{W\} \quad (7.17)$$

where

$$[N] = [N_1 \quad N_2 \quad N_3 \quad N_4]$$

The N_i values from above are commonly known as the shape functions of the displacement field w.

To get at the stress–strain relationship through the internal moments, which we label m, and the physical characteristics of the beam element, we bring in the very basic linear beam formula (see Chapter 2), Equation 7.18.

$$m = EIw'' \quad (7.18)$$

Here we use the prime notation (indicating a differentiation with respect to x) to save space. Obviously, from the elementary strength of materials, the internal bending moments are related to the stresses, the EI to the material stiffness, and the w'' to the strains. From Equation 7.17 we can develop Equation 7.19.

$$w'' = [N'']\{W\} \quad (7.19)$$

The differentiated shape functions N_i'' are given by Equation 7.20.

$$N_1'' = \left(\frac{6}{l^2}\right)(2\xi - 1); \qquad N_2'' = \left(\frac{2}{l}\right)(3\xi - 2)$$

(7.20)

$$N_3'' = \left(\frac{6}{l^2}\right)(1 - 2\xi); \qquad N_4'' = \left(\frac{2}{l}\right)(3\xi - 1)$$

From Equations 7.19 and 7.20 we see that the second derivatives, or curvatures, vary linearly within the element. Thus the curvatures can be defined uniquely by the values of the second derivatives at nodes 1 and 2. Using Equation 7.19, we find that we can write a curvature matrix equation, Equation 7.21.

$$\begin{Bmatrix} w_1'' \\ w_2'' \end{Bmatrix} = l^{-2} \begin{bmatrix} -6 & -4l & 6 & -2l \\ 6 & 2l & -6 & 4l \end{bmatrix} \{W\}$$

(7.21)

We will follow the usual convention and assign a positive sign when the member is bent as shown in Figure 7.15, and a negative sign when the sense of m is reversed. Thus the internal moments at nodes 1 and 2 are given as $m_1 = -M_1$ and $m_2 = M_2$. Using the equilibrium relationship, the forces Q_1 and Q_2 can be determined in terms of m_1 and m_2, as follows.

$$\sum M_2 = 0$$

or

$$Q_1 l - M_1 - M_2 = 0$$

or

$$Q_1 = -\frac{m_1 - m_2}{l}$$

Similarly,

$$Q_2 = \frac{m_1 - m_2}{l}$$

These lead us to the following matrix equation, Equation 7.22.

$$\begin{Bmatrix} Q_1 \\ M_1 \\ Q_2 \\ M_2 \end{Bmatrix} = \frac{1}{l} \begin{bmatrix} -1 & 1 \\ -l & 0 \\ 1 & -1 \\ 0 & l \end{bmatrix} \begin{Bmatrix} m_1 \\ m_2 \end{Bmatrix}$$

(7.22)

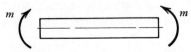

Figure 7.15 Beam sign convention.

Referring back to Equation 7.18, we observe that we can set up another matrix equation in terms of the element nodal values. We show this as Equation 7.23.

$$\begin{Bmatrix} m_1 \\ m_2 \end{Bmatrix} = EI \begin{bmatrix} 1 & 0 \\ 0 & 1 \end{bmatrix} \begin{Bmatrix} w_1'' \\ w_2'' \end{Bmatrix} \tag{7.23}$$

We can combine Equations 7.21, 7.22, and 7.23 to give a standard form matrix equation for the beam element. This is shown as Equation 7.24.

$$\begin{Bmatrix} Q_1 \\ M_1 \\ Q_2 \\ M_2 \end{Bmatrix} = [k]\{W\} \tag{7.24}$$

The stiffness matrix $[k]$ is given by Equation 7.25.

$$[k] = \frac{EI}{l^3} \begin{bmatrix} 12 & 6l & -12 & 6l \\ 6l & 4l^2 & -6l & 2l^2 \\ -12 & -6l & 12 & -6l \\ 6l & 2l^2 & -6l & 4l^2 \end{bmatrix} \tag{7.25}$$

The use of the above relationships for simulating the behavior of a simple beam system is demonstrated in Example 7.5.

EXAMPLE 7.5

Simulate the displacement behavior of the beam system shown in Figure 7.16 by the finite-element method, using the element relationships given previously. Assume consistent units within the equations and take the beam length to be 20, the two element lengths to be 10, the cross-sectional area to be 6, the modulus of elasticity to be 100, and the cross-sectional moment of inertia to be 4.5.

▶ The element equations are formed by using Equations 7.24 and 7.25.
   For element 1,

$$0.45 \begin{bmatrix} 12 & 60 & -12 & 60 \\ 60 & 400 & -60 & 200 \\ -12 & -60 & 12 & -60 \\ 60 & 200 & -60 & 400 \end{bmatrix} \begin{Bmatrix} w_1 \\ \theta_1 \\ w_2 \\ \theta_2 \end{Bmatrix} = \begin{Bmatrix} 0 \\ 5 \\ -0.5 \\ 0 \end{Bmatrix}$$

For element 2,

$$0.45 \begin{bmatrix} 12 & 60 & -12 & 60 \\ 60 & 400 & -60 & 200 \\ -12 & -60 & 12 & -60 \\ 60 & 200 & -60 & 400 \end{bmatrix} \begin{Bmatrix} w_2 \\ \theta_2 \\ w_3 \\ \theta_3 \end{Bmatrix} = \begin{Bmatrix} -0.5 \\ 0 \\ 0 \\ 0 \end{Bmatrix}$$

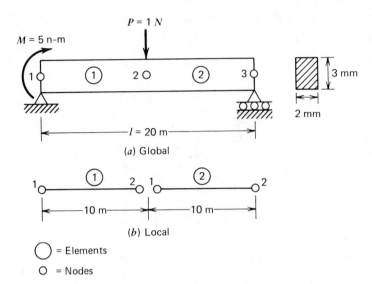

(a) Global

(b) Local

$\bigcirc$ = Elements

$\circ$ = Nodes

Figure 7.16 Beam example showing the definition of nodes, degrees of freedom, and global and local configurations.

The element stiffness matrices shown above can be combined (patched) together in the direct fashion illustrated earlier to give a global matrix equation for the system.

$$0.45 \begin{bmatrix} 12 & 60 & -12 & 60 & 0 & 0 \\ 60 & 400 & -60 & 200 & 0 & 0 \\ -12 & -60 & 24 & 0 & -12 & 60 \\ 0 & 0 & 0 & 800 & -60 & 200 \\ 0 & 0 & -12 & -60 & 12 & -60 \\ 0 & 0 & 60 & 200 & -60 & 400 \end{bmatrix} \begin{Bmatrix} w_1 \\ \theta_1 \\ w_2 \\ \theta_2 \\ w_3 \\ \theta_3 \end{Bmatrix} = \begin{Bmatrix} 0 \\ 5 \\ -1 \\ 0 \\ 0 \\ 0 \end{Bmatrix}$$

Since the beam system is supported at the two ends, the boundary conditions become $w_1 = w_3 = 0$. Introducing these conditions into the above equation, which as shown before allows us to eliminate the matrix rows and columns belonging to these conditions, gives us the final global matrix equation.

$$0.45 \begin{bmatrix} 400 & -60 & 200 & 0 \\ -60 & 24 & 0 & 60 \\ 200 & 0 & 800 & 200 \\ 0 & 60 & 200 & 400 \end{bmatrix} \begin{Bmatrix} \theta_1 \\ w_2 \\ \theta_2 \\ \theta_3 \end{Bmatrix} = \begin{Bmatrix} 5 \\ -1 \\ 0 \\ 0 \end{Bmatrix}$$

Solution of this set of equations gives us the following simulation values.

$w_1 = 0.000 \text{ (given)} \qquad w_2 = -9.259 \times 10^{-2} \qquad w_3 = 0.000 \text{ (given)}$

$\theta_1 = 1.851 \times 10^{-2} \qquad \theta_2 = -0.925 \times 10^{-2} \qquad \theta_3 = 1.851 \times 10^{-2}$

The comparison of these results with an analytical solution shows an exact match.

Plane Triangular-Element Systems

The simplest element for two-dimensional continuum problems is the triangular element shown in Figure 7.17. It can be used effectively to approximate practically any system configuration. Such elements were introduced by early investigators to study the torsion problem and to derive the stiffness matrix for plane stress problems.

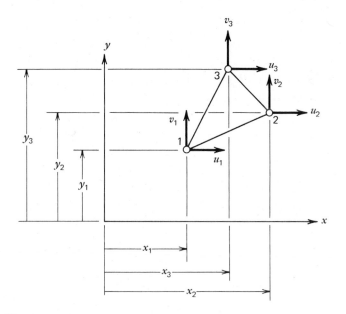

Figure 7.17 Triangular element of thickness t.

Consider the triangular element of relatively small thickness t shown in Figure 7.17. The nodes are located at the vertices and are numbered 1, 2, and 3. These are the local node numbers for the element. The corresponding global coordinates for these nodes are (x_1, y_1), (x_2, y_2), and (x_3, y_3). The displacement behavior of the element will be described with u and v values, corresponding to x and y movements respectively, at each node. Thus the displacement simulations with this element will require six degrees-of-freedom as represented by the vector relationship shown in Equation 7.26.

$$\{q\} = [u_1 \quad v_1 \quad u_2 \quad v_2 \quad u_3 \quad v_3]^T \qquad (7.26)$$

We make the direct assumption of linear displacements in the x-y plane so that we can introduce the following approximating functions.

$$u_i = a_1 + a_2 x_i + a_3 y_i$$
$$v_i = b_1 + b_2 x_i + b_3 y_i \tag{7.27}$$

The a's and b's can be evaluated using the matrix equations given by Equations 7.28, 7.29, and 7.30.

$$\begin{Bmatrix} u_1 \\ u_2 \\ u_3 \end{Bmatrix} = [A] \begin{Bmatrix} a_1 \\ a_2 \\ a_3 \end{Bmatrix}; \qquad \begin{Bmatrix} v_1 \\ v_2 \\ v_3 \end{Bmatrix} = [A] \begin{Bmatrix} b_1 \\ b_2 \\ b_3 \end{Bmatrix} \tag{7.28}$$

The A matrix is given by Equation 7.29.

$$[A] = \begin{bmatrix} 1 & x_1 & y_1 \\ 1 & x_2 & y_2 \\ 1 & x_3 & y_3 \end{bmatrix} \tag{7.29}$$

Solving for the a and b vectors requires the inverse of $[A]$, as shown in Equation 7.30.

$$\begin{Bmatrix} a_1 \\ a_2 \\ a_3 \end{Bmatrix} = [A]^{-1} \begin{Bmatrix} u_1 \\ u_2 \\ u_3 \end{Bmatrix} \qquad \begin{Bmatrix} b_1 \\ b_2 \\ b_3 \end{Bmatrix} = [A]^{-1} \begin{Bmatrix} v_1 \\ v_2 \\ v_2 \end{Bmatrix} \tag{7.30}$$

The inverse of $[A]$ is given by Equation 7.31.

$$[A]^{-1} = \frac{0.5}{A} \begin{bmatrix} x_2 y_3 - x_3 y_2 & x_3 y_1 - x_1 y_3 & x_1 y_2 - x_2 y_1 \\ y_2 - y_3 & y_3 - y_1 & y_1 - y_2 \\ x_3 - x_2 & x_1 - x_3 & x_2 - x_1 \end{bmatrix} \tag{7.31}$$

In Equation 7.31, $A = 0.5 \det [A]$ and is the area of the triangle, as given by Equation 7.32.

$$A = 0.5 [x_2 y_3 - x_3 y_2 + x_3 y_1 - x_1 y_3 + x_1 y_2 - x_2 y_1] \tag{7.32}$$

It is left as an exercise to show that we can get Equations 7.33 and 7.34 from Equations 7.28 through 7.31.

$$u = [N_1 \quad N_2 \quad N_3] \begin{Bmatrix} u_1 \\ u_2 \\ u_3 \end{Bmatrix} \tag{7.33}$$

$$v = [N_1 \quad N_2 \quad N_3] \begin{Bmatrix} v_1 \\ v_2 \\ v_3 \end{Bmatrix} \tag{7.34}$$

In Equations 7.33 and 7.34, N_1, N_2, and N_3 represent the shape functions for the triangular element. A generalized form can be used to express the shape functions, as shown in Equations 7.35 and 7.36.

$$N_i = \frac{0.5}{A} [\alpha_i + \beta_i x + \gamma_i y] \qquad i = 1, 2, 3 \tag{7.35}$$

The α_i, β_i, and γ_i are determined from Equation 7.36.

$$
\begin{aligned}
\alpha_1 &= x_2 y_3 - x_3 y_2 & \beta_1 &= y_2 - y_3 & \gamma_1 &= x_3 - x_2 \\
\alpha_2 &= x_3 y_1 - x_1 y_3 & \beta_2 &= y_3 - y_1 & \gamma_2 &= x_1 - x_3 \\
\alpha_3 &= x_1 y_2 - x_2 y_1 & \beta_3 &= y_1 - y_2 & \gamma_3 &= x_2 - x_1
\end{aligned}
\tag{7.36}
$$

Equations 7.33 and 7.34 can be written as a single matrix equation, as shown in Equation 7.37.

$$
\left\{ \begin{array}{c} u \\ v \end{array} \right\} =
\begin{bmatrix}
N_1 & 0 & N_2 & 0 & N_3 & 0 \\
0 & N_1 & 0 & N_2 & 0 & N_3
\end{bmatrix}
\left\{ \begin{array}{c} u_1 \\ v_1 \\ u_2 \\ v_2 \\ u_3 \\ v_3 \end{array} \right\}
\tag{7.37}
$$

The two-dimensional strain displacement relations are

$$\varepsilon_x = \frac{\partial u}{\partial x} ; \qquad \varepsilon_y = \frac{\partial v}{\partial y} ; \qquad \gamma_{xy} = \frac{\partial v}{\partial x} + \frac{\partial u}{\partial y}$$

and the corresponding stress–strain relations for a homogeneous, isotropic plane stress element are given by

$$\sigma_x = \frac{E}{1 - v^2} (\varepsilon_x + v\varepsilon_y)$$

$$\sigma_y = \frac{E}{1 - v^2} (\varepsilon_y + v\varepsilon_x)$$

$$\tau_{xy} = \frac{0.5 E \gamma_{xy}}{1 + v} = 0.5 E \gamma_{xy} \frac{1 - v}{1 - v^2}$$

The latter can be written in matrix form as shown in Equation 7.38.

$$\{\sigma\} = [C]\{\varepsilon\} \tag{7.38}$$

where

$$\{\sigma\} = \left\{ \begin{array}{c} \sigma_x \\ \sigma_y \\ \tau_{xy} \end{array} \right\} ; \qquad \{\varepsilon\} = \left\{ \begin{array}{c} \varepsilon_x \\ \varepsilon_y \\ \gamma_{xy} \end{array} \right\}$$

and

$$[C] = \left[\frac{E}{1 - v^2} \right]
\begin{bmatrix}
1 & v & 0 \\
v & 1 & 0 \\
0 & 0 & \dfrac{1 - v}{2}
\end{bmatrix}$$

Substituting the foregoing derivatives of the shape functions (given by Equation 7.35) gives us a general matrix equation for the strains, as shown in Equation 7.39.

$$\begin{Bmatrix} \varepsilon_x \\ \varepsilon_y \\ \gamma_{xy} \end{Bmatrix} = \frac{0.5}{A} \begin{bmatrix} \beta_1 & 0 & \beta_2 & 0 & \beta_3 & 0 \\ 0 & \gamma_1 & 0 & \gamma_2 & 0 & \gamma_3 \\ \gamma_1 & \beta_1 & \gamma_2 & \beta_2 & \gamma_3 & \beta_3 \end{bmatrix} \begin{Bmatrix} u_1 \\ v_1 \\ u_2 \\ v_2 \\ u_3 \\ v_3 \end{Bmatrix} \tag{7.39}$$

It is noted that Equation 7.39 can be written in shortened form as

$$\{\varepsilon\} = [B]\{q\}$$

To complete the derivation of the stiffness matrix, the joint force stress matrix needs to be constructed. The constant stresses relative to the triangular element are illustrated in Figure 7.18. Since the stresses are just the stress distributions, they can be converted into equivalent nodal forces. For example, the σ_x stress on the right-hand face [see Figure 7.19] yields a total force:

$$F_x^{\text{total}} = \sigma_x t(y_3 - y_1) = \sigma_x t \beta_2$$

This force produces the equivalent nodal forces f_1^x, f_2^x, and f_3^x. Part of the total load is taken by nodes 1 and 2, and part by nodes 2 and 3. That part of the force taken by nodes 1 and 2 is now assumed to be divided equally between the two nodes and is given by Equation 7.40.

$$f_1^x = f_2^x = -0.5\sigma_x t \beta_3 \tag{7.40}$$

That part of the force taken by nodes 2 and 3, which again is assumed to be divided equally, is given by Equation 7.41.

$$f_2^x = f_3^x = -0.5\sigma_x t \beta_1 \tag{7.41}$$

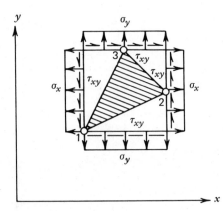

Figure 7.18   Plane stress triangular element.

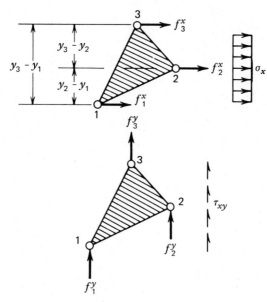

Figure 7.19 Direct stress σ_x and shear stress τ_{xy} on right-hand face with equivalent nodal forces.

Hence, for the right-hand face under the influence of σ_x, we have the equivalent nodal forces grouped under Equation 7.42. Note that $-(\beta_3+\beta_1)=-(y_1-y_3)=\beta_2$.

$$f^x_{1r}=-0.5\sigma_x t\beta_3$$
$$f^x_{2r}=-0.5\sigma_x t\beta_3-0.5\sigma_x t\beta_1=0.5\sigma_x t\beta_2 \qquad (7.42)$$
$$f^x_{3r}=-0.5\sigma_x t\beta_1$$

Consulting Figure 7.19 again, we see that the shear force on the right-hand face gives us a total force of

$$F_y^{total}=\tau_{xy}t(y_3-y_1)=\tau_{xy}t\beta_2$$

This force produces the equivalent nodal forces f^y_1, f^y_2, and f^y_3. Again part of the load is taken by nodes 1 and 2 and part by nodes 2 and 3. Assuming that it is equally divided, as before, we can express the part for nodes 1 and 2 as Equation 7.43.

$$f^y_1=f^y_2=0.5\tau_{xy}t(y_2-y_1)=-0.5\tau_{xy}t\beta_3 \qquad (7.43)$$

Equation 7.44 gives us the part taken by nodes 2 and 3.

$$f^y_2=f^y_3=0.5\tau_{xy}t(y_3-y_2)=-0.5\tau_{xy}t\beta_1 \qquad (7.44)$$

Thus nodal forces owing to shear forces on the right-hand face can be given by the grouping under Equation 7.45.

$$f^y_{1r} = -0.5\tau_{xy}t\beta_3$$

$$f^y_{2r} = -0.5\tau_{xy}t\beta_3 - 0.5\tau_{xy}t\beta_1 = 0.5\tau_{xy}t\beta_2 \qquad (7.45)$$

$$f^y_{3r} = -0.5\tau_{xy}t\beta_1$$

Using the same reasoning, we can develop nodal forces owing to total stress forces on the left-hand face and the top and bottom faces. From the left-hand face, we get the nodal forces shown grouped under Equation 7.46.

$$f^x_{1l} = -0.5\sigma_x t\beta_2$$

$$f^x_{2l} = 0$$

$$f^x_{3l} = -0.5\sigma_x t\beta_2$$

$$f^y_{1l} = -0.5\tau_{xy}t\beta_2 \qquad (7.46)$$

$$f^y_{2l} = 0$$

$$f^y_{3l} = -0.5\tau_{xy}t\beta_2$$

From the top face, we get the nodal forces shown grouped under Equation 7.47.

$$f^y_{1t} = -0.5\sigma_y t\gamma_2$$

$$f^y_{2t} = -0.5\sigma_y t\gamma_1$$

$$f^y_{3t} = 0.5\sigma_y t\gamma_3$$

$$f^x_{1t} = -0.5\tau_{xy}t\gamma_2 \qquad (7.47)$$

$$f^x_{2t} = -0.5\tau_{xy}t\gamma_1$$

$$f^x_{3t} = 0.5\tau_{xy}t\gamma_3$$

From the bottom face, we get the nodal forces shown grouped under Equation 7.48.

$$f^y_{1b} = -0.5\sigma_y t\gamma_3$$

$$f^y_{2b} = -0.5\sigma_y t\gamma_3$$

$$f^y_{3b} = 0$$

$$f^x_{1b} = -0.5\tau_{xy}t\gamma_3 \qquad (7.48)$$

$$f^x_{2b} = -0.5\tau_{xy}t\gamma_3$$

$$f^x_{3b} = 0$$

The total force on any node is the sum of the nodal forces given for that node above. For example,

$$f^x_1 = f^x_{1t} + f^x_{1b} + f^x_{1r} + f^x_{1l}$$

$$f^x_1 = -0.5\tau_{xy}t(\gamma_2 + \gamma_3) - 0.5\sigma_x t(\beta_3 + \beta_2)$$

This gives us Equation 7.49.

$$f_1^x = 0.5\sigma_x t\beta_1 + 0.5\tau_{xy}t\gamma_1 \tag{7.49}$$

In the same manner, we can get five more similar equations. However, we can express all six of these nodal force equations easier by using one matrix equation, as shown in Equation 7.50.

$$\{f\} = [\Lambda]\{\sigma\} \tag{7.50}$$

where

$$\{\sigma\} = \begin{Bmatrix} \sigma_x \\ \sigma_y \\ \tau_{xy} \end{Bmatrix}$$

$$[\Lambda] = 0.5t \begin{bmatrix} \beta_1 & 0 & \gamma_1 \\ 0 & \gamma_1 & \beta_1 \\ \beta_2 & 0 & \gamma_2 \\ 0 & \gamma_2 & \beta_2 \\ \beta_3 & 0 & \gamma_3 \\ 0 & \gamma_3 & \beta_3 \end{bmatrix}$$

$$\{f\} = \begin{Bmatrix} f_1^x \\ f_1^y \\ f_2^x \\ f_2^y \\ f_3^x \\ f_3^y \end{Bmatrix}$$

Substituting Equation 7.38 and the shortened form of Equation 7.39 into Equation 7.50 gives us our final element equation, Equation 7.51, relating nodal forces and displacements.

$$\{f\} = [k]\{q\} \tag{7.51}$$

The element stiffness matrix, $[k]$, consists of the product of three other matrices, as indicated in Equation 7.52.

$$[k] = [\Lambda][C][B] \tag{7.52}$$

Performing the matrix multiplication implied in Equation 7.52 gives the required stiffness matrix for the plane triangular element. Details are given as Equation 7.53.

$$[k] = \left[\frac{Et}{4A(1-v^2)}\right] \begin{bmatrix} k_{11} & k_{12} & k_{13} & k_{14} & k_{15} & k_{16} \\ & k_{22} & k_{23} & k_{24} & k_{25} & k_{26} \\ & & k_{33} & k_{34} & k_{35} & k_{36} \\ & & & k_{44} & k_{45} & k_{46} \\ & \text{symmetric} & & & k_{55} & k_{56} \\ & & & & & k_{66} \end{bmatrix} \tag{7.53}$$

where

$$k_{11} = \beta_1^2 + v_m\gamma_1^2$$
$$k_{12} = v_p\gamma_1\beta_1$$
$$k_{13} = \beta_1\beta_2 + v_m\gamma_1\gamma_2$$
$$k_{14} = v\beta_1\gamma_2 + v_m\beta_2\gamma_1$$
$$k_{15} = \beta_1\beta_3 + v_m\gamma_1\gamma_3$$
$$k_{16} = v\beta_1\gamma_3 + v_m\gamma_1\beta_3$$
$$k_{22} = \gamma_1^2 + v_m\beta_1^2$$
$$k_{23} = v\beta_2\gamma_1 + v_m\gamma_2\beta_1$$
$$k_{24} = \gamma_1\gamma_2 + v_m\beta_1\beta_2$$
$$k_{25} = v\gamma_1\beta_3 + v_m\beta_1\gamma_3$$
$$k_{26} = \gamma_1\gamma_3 + v_m\beta_1\beta_3$$
$$k_{33} = \beta_2^2 + v_m\gamma_2^2$$

$$k_{34} = v_p\gamma_2\beta_2$$
$$k_{35} = \beta_2\beta_3 + v_m\gamma_2\gamma_3$$
$$k_{36} = v\beta_2\gamma_3 + v_m\gamma_2\beta_3$$
$$k_{44} = \gamma_2^2 + v_m\beta_2^2$$
$$k_{45} = v\gamma_2\beta_3 + v_m\beta_2\gamma_3$$
$$k_{46} = \gamma_2\gamma_3 + v_m\beta_2\beta_3$$
$$k_{55} = \beta_3^2 + v_m\gamma_3^2$$
$$k_{56} = v\beta_3\gamma_3 + v_m\beta_3\gamma_3$$
$$k_{66} = \gamma_3^2 + v_m\beta_3^2$$

and

$$v_m = \frac{1-v}{2}$$

$$v_p = \frac{1+v}{2}$$

The stiffness matrix for a triangular element in plane strain may be obtained in the same manner as described in the foregoing except that the stress–strain relationships (Equation 7.38) must be modified. The modified stress–strain equation is given as Equation 7.54.

$$\{\sigma\} = [C']\{\varepsilon\} \tag{7.54}$$

where

$$[C'] = E[(1+v)(1-2v)]^{-1} \begin{bmatrix} 1-v & v & 0 \\ v & 1-v & 0 \\ 0 & 0 & \dfrac{1-2v}{2} \end{bmatrix}$$

and $\{\sigma\}$ and $\{\varepsilon\}$ are as defined before.

The required matrix multiplications are as shown in Equation 7.52, except that $[C]$ must be replaced by $[C']$. The resulting matrix is given as Equation 7.55.

$$[k'] = CK \begin{bmatrix} K_{11} & K_{12} & K_{13} & K_{14} & K_{15} & K_{16} \\ & K_{22} & K_{23} & K_{24} & K_{25} & K_{26} \\ & & K_{33} & K_{34} & K_{35} & K_{36} \\ & & & K_{44} & K_{45} & K_{46} \\ & \text{symmetric} & & & K_{55} & K_{56} \\ & & & & & K_{66} \end{bmatrix} \tag{7.55}$$

where

$$(CK) = Et[2(1+v)(1-2v)]^{-1} \qquad K_{25} = v\beta_3\gamma_1 + \lambda_2\beta_1\gamma_3$$
$$\lambda_1 = (1-v) \qquad\qquad K_{26} = \lambda_1\gamma_1\gamma_3 + \lambda_2\beta_1\beta_3$$

$$\lambda_2 = \frac{1-2v}{v}$$

$$K_{33} = \lambda_1 \beta_2^2 + \lambda_2 \gamma_2^2$$

and

$$K_{34} = v\beta_2 \gamma_2 + \lambda_2 \beta_2 \gamma_2$$

$$K_{11} = \lambda_1 \beta_1^2 + \lambda_2 \gamma_1^2$$

$$K_{35} = \lambda_1 \beta_2 \beta_3 + \lambda_2 \gamma_2 \gamma_3$$

$$K_{12} = v\beta_1 \gamma_1 + \lambda_2 \beta_1 \gamma_1$$

$$K_{36} = v\beta_2 \gamma_3 + \lambda_2 \gamma_2 \beta_3$$

$$K_{13} = \lambda_1 \beta_1 \beta_2 + \lambda_2 \gamma_1 \gamma_2$$

$$K_{44} = \lambda_1 \gamma_2^2 + \lambda_2 \beta_2^2$$

$$K_{14} = v\beta_1 \gamma_2 + \lambda_2 \gamma_1 \beta_2$$

$$K_{45} = v\gamma_2 \beta_3 + \lambda_2 \beta_2 \gamma_3$$

$$K_{15} = \lambda_1 \beta_1 \beta_3 + \lambda_2 \gamma_1 \gamma_3$$

$$K_{46} = \lambda_1 \gamma_2 \gamma_3 + \lambda_2 \beta_2 \beta_3$$

$$K_{16} = v\beta_1 \gamma_3 + \lambda_2 \gamma_1 \beta_3$$

$$K_{55} = \lambda_1 \beta_3^2 + \lambda_2 \gamma_3^2$$

$$K_{22} = \lambda_1 \gamma_1^2 + \lambda_2 \beta_1^2$$

$$K_{56} = v\beta_3 \gamma_3 + \lambda_2 \gamma_3 \beta_3$$

$$K_{23} = v\gamma_1 \beta_2 + \lambda_2 \beta_1 \gamma_2$$

$$K_{66} = \lambda_1 \gamma_3^2 + \lambda_2 \beta_3^2$$

$$K_{24} = \lambda_1 \gamma_1 \gamma_2 + \lambda_2 \beta_1 \beta_2$$

Examples 7.6 and 7.7 illustrate use of the foregoing element formulations in structural plate systems.

EXAMPLE 7.6

Consider the thin flat plate system shown in Figure 7.20. Assume consistent units and take a thickness of 0.25, an elastic modulus of 30×10^6, and a Poisson ratio of 0.3. The right edge of the plate is loaded with a distributed load of 400 force units per unit area. Using a two triangular-element approximation, simulate the two-dimensional deflections and stresses.

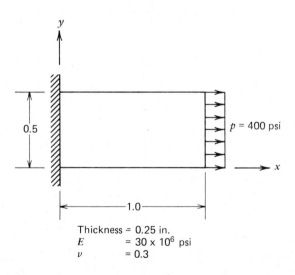

Thickness = 0.25 in.
E = 30 × 10⁶ psi
v = 0.3

Figure 7.20 Flat plate, Example 7.6.

▶ We choose the plane stress case and visualize the finite elements and nodes as shown in Figure 7.21.

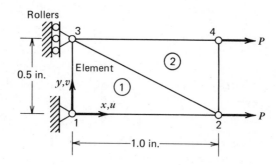

Figure 7.21 Finite-element idealization of flat plate (plane stress condition). Two elements.

For element 1 (nodes 1, 2, and 3), we have $A=0.25$, $\beta_1=y_2-y_3=-0.5$, $\beta_2=y_3-y_1=0.5$, $\beta_3=y_1-y_2=0$, $\gamma_1=x_3-x_2=-1.0$, $\gamma_2=x_1-x_3=0$, and $\gamma_3=x_2-x_1=1.0$. The stiffness matrix, using Equation 7.53, becomes

$$[k^1]=Et[4A(1-v^2)]^{-1}\begin{bmatrix} 0.6000 & 0.3250 & -0.2500 & -0.1750 & -0.3500 & -0.1500 \\ & 1.0875 & -0.1500 & -0.0875 & -0.0175 & -1.000 \\ & & 0.2500 & 0 & 0 & 0.1500 \\ & & & 0.0875 & 0.1750 & 0 \\ & \text{symmetric} & & & 0.3500 & 0 \\ & & & & & 1.000 \end{bmatrix}$$

For element 2 (nodes 2, 4, and 3), we have $A=0.25$, $\beta_1=0$, $\beta_2=0.5$, $\beta_3=-0.5$, $\gamma_1=-1.0$, $\gamma_2=1.0$, and $\gamma_3=0$. The stiffness matrix becomes

$$[k^2]=Et[4A(1-v^2)]^{-1}$$
$$\times\begin{bmatrix} 0.3500 & 0 & -0.3500 & -0.1750 & 0 & 0.1750 \\ & 1.000 & -0.1500 & -1.000 & 0.1500 & 0 \\ & & 0.6000 & 0.3250 & -0.2500 & -0.1750 \\ & & & 1.0875 & -0.1500 & -0.0875 \\ & \text{symmetric} & & & 0.2500 & 0 \\ & & & & & 0.0875 \end{bmatrix}$$

The element stiffness matrices can be combined in the direct fashion previously illustrated to form the overall global stiffness matrix. The resulting matrix equation follows.

$$Et[4A(1-v^2)]^{-1}\,[KG]\begin{Bmatrix} u_1 \\ v_1 \\ u_2 \\ v_2 \\ u_3 \\ v_3 \\ u_4 \\ v_4 \end{Bmatrix}=\begin{Bmatrix} F_1^x \\ F_1^y \\ F_2^x \\ F_2^y \\ F_3^x \\ F_3^y \\ F_4^x \\ F_4^y \end{Bmatrix}$$

Since the system is fixed at node 1, and restrained in v only at node 3, we get $u_1=v_1=u_3=0$. Introducing these constraints into the above equation yields the final global matrix equation to be solved. The $[KG]$ matrix is shown first for reference.

$[KG]=$

$$\begin{bmatrix} 0.6000 & 0.3250 & -0.2500 & -0.1750 & -0.3500 & -0.1500 & 0 & 0 \\ & 1.0875 & -0.1500 & -0.0875 & -0.0175 & -1.000 & 0 & 0 \\ & & 0.6000 & 0 & 0 & 0.3250 & -0.3500 & -0.1750 \\ & & & 1.0875 & 0.3250 & 0 & -0.1500 & -01.000 \\ & & & & 0.6000 & 0 & -0.2500 & -0.1500 \\ & & & & & 1.0875 & -0.1750 & -0.0875 \\ & \text{symmetric} & & & & & 0.6000 & 0.3250 \\ & & & & & & & 1.0875 \end{bmatrix}$$

$$Et[4A(1-v^2)]^{-1}\begin{bmatrix} 0.6000 & 0 & 0.3250 & -0.3500 & -0.1750 \\ & 1.0875 & 0 & -0.1500 & -1.000 \\ & & 1.0875 & -0.1750 & -0.0875 \\ & & & 0.6000 & 0.3250 \\ & \text{symmetric} & & & 1.0875 \end{bmatrix}\begin{Bmatrix} u_2 \\ v_2 \\ v_3 \\ u_4 \\ v_4 \end{Bmatrix}=\begin{Bmatrix} F_2^x \\ F_2^y \\ F_3^y \\ F_4^x \\ F_4^y \end{Bmatrix}$$

The external loading is applied only to the right-hand face as shown in Figure 7.20. Consequently, the loading is divided equally between nodes 2 and 4.

$$F_2^x=F_4^x=0.5P(y_4-y_2)t$$

or

$$F_2^x=f_4^x=0.5(400)(0.5)(0.25)=25$$

also

$$F_2^y=F_3^y=F_4^y=0$$

With the nodal loads specified, the global matrix equation shown above can be solved to give the displacements.

$$
\begin{Bmatrix} u_2 \\ v_2 \\ v_3 \\ u_4 \\ v_4 \end{Bmatrix} = \left[\frac{4A(1-v^2)}{Et} \right] \begin{Bmatrix} 109.8901 \\ 0 \\ -16.4835 \\ 109.8901 \\ -16.4835 \end{Bmatrix} = \begin{Bmatrix} 1.3333 \\ 0 \\ -0.2000 \\ 1.3333 \\ -0.2000 \end{Bmatrix} \times 10^{-5}
$$

Once the displacements have been determined, the reaction forces at the support nodes 1 and 3 can be computed. Using the appropriate equations from the original global set (e.g., for R_1^x, we use the first equation), we get the following.

$$
R_1^x = (Et)[4A(1-v^2)]^{-1}(-0.25u_2 - 0.175v_2 - 0.15v_3)
$$
$$
R_1^y = (Et)[4A(1-v^2)]^{-1}(-0.15u_2 - 0.0875v_2 - v_3)
$$
$$
R_3^x = (Et)[4A(1-v^2)]^{-1}(0.325v_2 - 0.25u_4 - 0.15v_4)
$$

The result is $R_1^x = -25$, $R_1^y = 0$, and $R_3^x = -25$, all in consistent force units.

The stresses in the elements can be obtained by using Equations 7.38 and 7.39. Thus, for plane stress, we have the following.

$$
\{\sigma\} = [C]\{\varepsilon\} \qquad \text{or} \qquad \{\sigma\} = [C][B]\{q\}
$$

Performing the matrix multiplications indicated in the foregoing gives the following matrix equation.

$$
\{\sigma\} = E[2A(1-v^2)]^{-1} \begin{bmatrix} \beta_1 & v\gamma_1 & \beta_2 & v\gamma_2 & \beta_3 & v\gamma_3 \\ v\beta_1 & \gamma_1 & v\beta_2 & \gamma_2 & v\beta_3 & \gamma_3 \\ v_1\gamma_1 & v_1\beta_1 & v_1\gamma_2 & v_1\beta_2 & v_1\gamma_3 & v_1\beta_3 \end{bmatrix} \{q\}
$$

where $v_1 = (1-v)/2$.

For element 1 ($E = 30 \times 10^6$, $v = 0.3$, $A = 0.25$, $v_1 = 0.30$) we have the following.

$$
\{\sigma\}^1 = 6.5934 \times 10^7 \begin{bmatrix} -0.5 & -0.3 & 0.5 & 0 & 0 & 0.3 \\ -0.15 & -1.0 & 0.15 & 0 & 0 & 1.0 \\ -0.35 & -0.175 & 0 & 0.175 & 0.35 & 0 \end{bmatrix} \begin{Bmatrix} u_1 \\ v_1 \\ u_2 \\ v_2 \\ u_3 \\ v_3 \end{Bmatrix}
$$

As expected, we get the following element stress values from the multiplication.

$$
\{\sigma\}^1 = \begin{Bmatrix} 400 \\ 0 \\ 0 \end{Bmatrix}
$$

In a similar manner we find the following stress values for element 2.

$$\{\sigma\}^2 = \begin{Bmatrix} 400 \\ 0 \\ 0 \end{Bmatrix}$$

We observe that, in this example, large stress gradients are not present. Consequently, we would not expect that improved accuracy could be obtained in the simulated values simply by increasing the number of elements. A BASIC computer program that can be used to simulate the displacements in this example is given as Figure 7.22.

```
100 REM EXAMPLE 7.6 AND 7.7 BASIC PROGRAM
110 CLS! CLEAR50! DEFINT I,J,L,M,N
120 PRINT"2D TRIANGULAR-ELEMENT FINITE ELEMENT EXAMPLE PROGRAM"
130 INPUT"GIVE PLATE THICKNESS";TH
140 INPUT"GIVE ELASTIC MODULUS";EM
150 INPUT"GIVE POISSON RATIO";PR
160 INPUT"GIVE NUMBER OF TRIANGULAR ELEMENTS";NT
170 INPUT"GIVE NUMBER OF NODES";NN
180 DIM N1(NT),N2(NT),N3(NT),A1(NT),A2(NT),A3(NT),B1(NT),B2(NT),B3(NT),G1(NT),G2(NT),G3(NT),X(NN),Y(NN),A(NT)
190 M=2*NN! DIM K(M,M),F(M),D(10,M),ND(10),U(NN),V(NN)
200 REM INPUT ELEMENT INFORMATION
210 FOR I=1TONT
220    PRINT"GIVE GLOBAL NODE NUMBERS (LOCAL 1,2,&3 CCW) FOR ELEMENT";I
230    INPUT N1(I),N2(I),N3(I)
240 NEXTI
250 FORI=1TONN
260    PRINT"GIVE GLOBAL COORDINATES, X , Y , FOR NODE";I
270    INPUT X(I),Y(I)
280 NEXTI
290 PRINT!PRINT">>>>> ELEMENT NODE LIST <<<<<"! L=0
300 FORI=1TONT
310    PRINT"NODES 1, 2, & 3 FOR ELEMENT";I;" ARE!";N1(I);N2(I);N3(I)
320    L=L+1! IF L<14 GOTO350
330    INPUT"FOR MORE, PRESS 'ENTER'";ZZ! L=0
340    PRINT"ELEMENT NODES CONTINUED"
350 NEXTI
360 INPUT"TO CONTINUE, PRESS 'ENTER'";ZZ
370 PRINT!PRINT"////// NODE COORDINATES ///////"! L=0
380 FORI=1TONN
390    PRINT" X  AND  Y   FOR NODE ";I;" ARE!"; X(I);Y(I)
400    L=L+1! IF L<14 GOTO430
410    INPUT"FOR MORE, PRESS 'ENTER'";ZZ! L=0
420    PRINT"NODE COORDINATES CONTINUED"
430 NEXTI
440 PRINT!INPUT"DO YOU DESIRE TO MAKE A CHANGE IN A NODE NO.?--Y/N";Q$
450    IF Q$<>"Y" GOTO490
460    INPUT"ENTER THE ELEMENT NUMBER";J
470    INPUT"GIVE THE NEW GLOBAL NODE NUMBERS ( 1, 2, 3 CCW LOCAL)";N1(J),N2(J),N3(J)
480    GOTO440
490 PRINT!INPUT"DO YOU DESIRE TO MAKE A CHANGE IN A NODE'S COORDINATES?--Y/N";Q1$
500    IF Q1$<>"Y" GOTO540
510    INPUT"GIVE THE NODE NUMBER";J
520    INPUT"GIVE THE NEW COORDINATES, X , Y";X(J),Y(J)
530    GOTO490
540 PRINT"DO YOU WISH TO SEE NODE ASSIGNMENTS & COORDINATES AGAIN?--Y/N"
550 INPUT AA$! IF AA$="Y" GOTO290
560 REM COMPUTE ELEMENT AREAS AND SHAPE FUNCTION PARAMETERS
570 FOR I=1TONT
580    A(I)=0.5*(X(N2(I))*Y(N3(I))-X(N3(I))*Y(N2(I))+X(N3(I))*Y(N1(I))-X(N1(I))*Y(N3(I))+X(N1(I))*Y(N2(I))-X(N2(I))*Y(N1(I)))
```

Figure 7.22*a*   BASIC program, examples 7.6 and 7.7.

```
590    A1(I)=X(N2(I))*Y(N3(I))-X(N3(I))*Y(N2(I))
600    A2(I)=X(N3(I))*Y(N1(I))-X(N1(I))*Y(N3(I))
610    A3(I)=X(N1(I))*Y(N2(I))-X(N2(I))*Y(N1(I))
620    B1(I)=Y(N2(I))-Y(N3(I))
630    B2(I)=Y(N3(I))-Y(N1(I))
640    B3(I)=Y(N1(I))-Y(N2(I))
650    G1(I)=X(N3(I))-X(N2(I))
660    G2(I)=X(N1(I))-X(N3(I))
670    G3(I)=X(N2(I))-X(N1(I))
680 NEXTI
690 REM CALL SUB1500 TO ASSEMBLE STIFFNESS MATRICES
700 GOSUB1500
710 REM CALL SUB2000 TO INSERT DISPLACEMENT CONSTRAINTS
720 GOSUB2000
730 REM CALL SUB2500 TO INSERT BOUNDARY CONDITIONS
740 GOSUB2500
750 REM SOLVE GLOBAL MATRIX EQUATION FOR DISPLACEMENTS
760 REM USE SHORT VERSION OF GAUSSIAN EQUATION-SET SOLVER
770 REM--THIS SOLVER DESTROYS THE K(M,M) AND F(M) ARRAYS
780 REM--THIS SOLVER DOES NOT HAVE A MAXIMUM PIVOT STRATEGY
790 FORII=1TOM: I1=II+1: F(II)=F(II)/K(II,II)
800    IF II=M GOTO880
810    FORJ=I1TOM: K(II,J)=K(II,J)/K(II,II): NEXTJ
820    FORI=I1TOM: F(I)=F(I)-K(I,II)*F(II)
830      FORJ=I1TOM
840        K(I,J)=K(I,J)-K(I,II)*K(II,J)
850      NEXTJ
860    NEXTI
870 NEXTII
880 I1=II: II=II-1: IFII<=0 GOTO910
890 FORJ=I1TOM: F(II)=F(II)-K(II,J)*F(J): NEXTJ
900 GOTO880
910 REM PRINT DISPLACEMENT VALUES
920 FORI=1TONN: NU=2*I-1: NV=2*I
930    U(I)=F(NU): V(I)=F(NV)
940 NEXTI
950 CLS: L=0: PRINT">>>>>> DISPLACEMENT VALUES <<<<<<"
960 FORI=1TONN
970    PRINT"U=";U(I);" V=";V(I);" FOR NODE ";I: L=L+1
980    IF L<14 GOTO1000
990    INPUT"FOR MORE PRESS 'ENTER'";ZZ: L=0
1000 NEXTI
1010 INPUT"DO YOU WISH TO SEE VALUES AGAIN? -Y/N";A$
1020 IF A$="Y" GOTO950
1030 END
1500 REM ASSEMBLES GLOBAL STIFFNESS MATRIX ELEMENT BY ELEMENT
```

Figure 7.22*b*   BASIC program, Examples 7.6 and 7.7 *(continued)*.

```
1510 P1=(1-PR)/2! P2=(1+PR)/2
1520 FORI=1TOM! FORJ=1TOM! K(I,J)=0! NEXTJ,I
1530 FORI=1TONT
1540   CK=EM*TH/(4*A(I)*(1-PR[2))
1550   J1=2*N1(I)! J2=2*N2(I)! J3=2*N3(I)! I1=J1-1! I2=J2-1! I3=J3-1
1560   K(I1,I1)=K(I1,I1)+CK*(B1(I)[2+P1*G1(I)[2)
1570   K(I1,J1)=K(I1,J1)+CK*(P2*G1(I)*B1(I))
1580     K(J1,I1)=K(I1,J1)
1590   K(I1,I2)=K(I1,I2)+CK*(B1(I)*B2(I)+P1*G1(I)*G2(I))
1600     K(I2,I1)=K(I1,I2)
1610   K(I1,J2)=K(I1,J2)+CK*(PR*B1(I)*G2(I)+P1*B2(I)*G1(I))
1620     K(J2,I1)=K(I1,J2)
1630   K(I1,I3)=K(I1,I3)+CK*(B1(I)*B3(I)+P1*G1(I)*G3(I))
1640     K(I3,I1)=K(I1,I3)
1650   K(I1,J3)=K(I1,J3)+CK*(PR*B1(I)*G3(I)+P1*G1(I)*B3(I))
1660     K(J3,I1)=K(I1,J3)
1670   K(J1,J1)=K(J1,J1)+CK*(G1(I)[2+P1*B1(I)[2)
1680   K(J1,I2)=K(J1,I2)+CK*(PR*B2(I)*G1(I)+P1*G2(I)*B1(I))
1690     K(I2,J1)=K(J1,I2)
1700   K(J1,J2)=K(J1,J2)+CK*(G1(I)*G2(I)+P1*B1(I)*B2(I))
1710     K(J2,J1)=K(J1,J2)
1720   K(J1,I3)=K(J1,I3)+CK*(PR*G1(I)*B3(I)+P1*B1(I)*G3(I))
1730     K(I3,J1)=K(J1,I3)
1740   K(J1,J3)=K(J1,J3)+CK*(G1(I)*G3(I)+P1*B1(I)*B3(I))
1750     K(J3,J1)=K(J1,J3)
1760   K(I2,I2)=K(I2,I2)+CK*(B2(I)[2+P1*G2(I)[2)
1770   K(I2,J2)=K(I2,J2)+CK*(P2*G2(I)*B2(I))
1780     K(J2,I2)=K(I2,J2)
1790   K(I2,I3)=K(I2,I3)+CK*(B2(I)*B3(I)+P1*G2(I)*G3(I))
1800     K(I3,I2)=K(I2,I3)
1810   K(I2,J3)=K(I2,J3)+CK*(PR*B2(I)*G3(I)+P1*G2(I)*B3(I))
1820     K(J3,I2)=K(I2,J3)
1830   K(J2,J2)=K(J2,J2)+CK*(G2(I)[2+P1*B2(I)[2)
1840   K(J2,I3)=K(J2,I3)+CK*(PR*G2(I)*B3(I)+P1*B2(I)*G3(I))
1850     K(I3,J2)=K(J2,I3)
1860   K(J2,J3)=K(J2,J3)+CK*(G2(I)*G3(I)+P1*B2(I)*B3(I))
1870     K(J3,J2)=K(J2,J3)
1880   K(I3,I3)=K(I3,I3)+CK*(B3(I)[2+P1*G3(I)[2)
1890   K(I3,J3)=K(I3,J3)+CK*(PR*B3(I)*G3(I)+P1*B3(I)*G3(I))
1900     K(J3,I3)=K(I3,J3)
1910   K(J3,J3)=K(J3,J3)+CK*(G3(I)[2+P1*B3(I)[2)
1920 NEXTI
1930 RETURN
1940 END
2000 REM--MODIFICATION OF GLOBAL MATRIX EQUATION
2010 REM--FOR DISPLACEMENT CONSTRAINTS
2020 FORI=1TOM! F(I)=0! NEXTI! J=0
2030 INPUT"DO YOU WISH TO ENTER A DISPLACEMENT CONSTRAINT?-Y/N"!D$
```

Figure 7.22c BASIC program, Examples 7.6 and 7.7 *(continued)*.

```
2040   IFD$="N" GOTO2120
2050 J=J+1
2060 INPUT"GIVE THE NODE NUMBER";ND(J)
2070 INPUT"IS THIS A 'U' , OR 'V', CONSTRAINT?--U/V";UV$
2080   NC=2*ND(J); IF UV$="U" THEN NC=NC-1
2090 FORI=1TOM; D(J,NC)=K(NC,I); K(NC,I)=0; K(I,NC)=0; NEXTI
2100 K(NC,NC)=1.0
2110 GOTO2030
2120 RETURN
2130 END
2500 REM--MODIFICATION OF GLOBAL MATRIX EQUATION
2510 REM--FOR BOUNDARY CONDITIONS
2520 INPUT"DO YOU WISH TO ENTER A BOUNDARY CONDITION?--Y/N";B$
2530   IFB$="N" GOTO2700
2540 INPUT"IS THIS A 'FORCE',OR 'DISPLACEMENT', CONDITION?--F/D";C$
2550   IFC$="D" GOTO2620
2560 INPUT"GIVE NODE NUMBER";ND
2570 INPUT"IS THIS AN 'X', OR 'Y', DIRECTION FORCE?--X/Y";XY$
2580   NC=2*ND; IFXY$="X" THEN NC=NC-1
2590 PRINT"GIVE VALUE FOR THE  ";XY$;" FORCE,"
2600 INPUT F(NC)
2610 GOTO2520
2620 INPUT"GIVE NODE NUMBER";ND
2630 INPUT"IS THIS A 'U', OR 'V', DISPLACEMENT VALUE?--U/V";BC$
2640   NC=2*ND; IFBC$="U" THEN NC=NC-1
2650 PRINT"GIVE VALUE OF THE";BC$;" DISPLACEMENT,"
2660 INPUT DV
2670 FORI=1TOM; F(I)=F(I)-K(I,NC)*DV; K(NC,I)=0; K(I,NC)=0; NEXTI
2680 K(NC,NC)=1.0; F(NC)=DV
2690 GOTO2510
2700 RETURN
2710 END
```

Figure 7.22*d*   BASIC program, Examples 7.6 and 7.7 *(continued)*.

EXAMPLE 7.7

Consider the uniform cross section cantilever beam shown in Figure 7.23. The beam has a thickness of 0.5, a length of 10.0, and a depth of 1.0. The material parameters include an elastic modulus of $E = 30 \times 10^6$, and a Poisson ratio of 0.3.

Assume consistent units throughout and perform a finite-element simulation of the displacements at the midpoint and free end if a transverse load of $P = 300$ is applied at the free end as shown in the figure.

► Again we choose the plane stress case and visualize the finite elements and nodes as shown in Figure 7.24.

However, we recognize that now, in contrast to the last example, extremely large stress gradients will exist, and our simplistic elements may not give us the best results. We proceed anyway, but quickly discover that it becomes much too involved to try to write down each element's stiffness matrix and show all the manipulations necessary, even for the relatively small number of elements used here. We retreat to our computer program shown in Figure 7.22. This

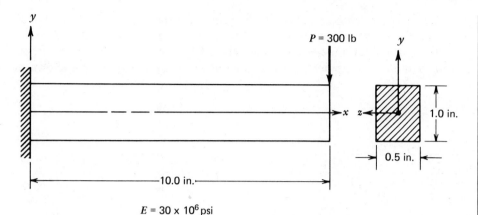

$E = 30 \times 10^6$ psi

Figure 7.23   Cantilever beam example.

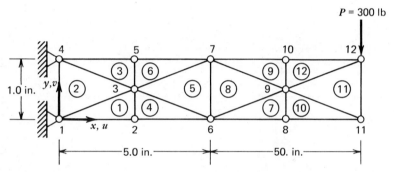

Figure 7.24   Finite-element idealization of flat plate cantilever beam (plane stress condition). 12 elements.

experience is typical of most finite-element analyses. The results (rounded to two digits) of running the program are given here.

Node	$u \times 10^4$	$v \times 10^3$
1	0	0
2	−3.1	−0.91
3	−0	−0.88
4	0	0
5	3.1	−0.91
6	−5.3	−3.1
7	5.3	−3.1
8	−6.6	−6.2
9	−0.0021	−6.2
10	6.6	−6.2
11	−7.0	−9.8
12	7.1	−9.8

As we suspected before starting, the results are not very good; the analytical exact v displacement at the free end is -0.08. Since the use of more elements often improves a finite-element simulation where gradients are prevalent, we go to another computer program (not shown) and the refinements shown in Figure 7.25.

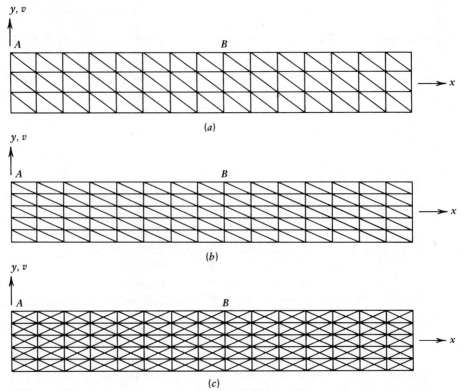

Figure 7.25 Refined finite models of cantilever beam using plane stress triangular elements. *(a)* 64 nodes, 90 elements. *(b)* 96 nodes, 150 elements. *(c)* 171 nodes, 300 elements.

Case 1, Figure 7.25(*a*), shows 64 nodes; Case 2, Figure 7.25(*b*), shows 96 nodes; and Case 3, Figure 7.25(*c*), exhibits 171 nodes. The simulation results are given in Table 7.1.

Table 7.1 Tip Deflections and Stresses in a Plane-Stress Element Cantilever Beam as Illustrated in Figure 7.25

Case	Tip Deflection	Axial Stress at A	Axial Stress at B
1	0.046	21,000	7,548
2	0.050	23,100	9,800
3	0.064	25,240	12,000
Exact	0.080	35,000	16,800

Obviously, the more elements, the better the simulation. However, it is apparent that improvement comes only slowly with the mesh refinement. We conclude that the constant strain elements are not very satisfactory for this kind of problem. Further, at locations where large stresses and gradients are expected, a fine mesh (many small elements) should be used; away from these areas a coarse mesh can be employed. An approach often preferred in professional settings to get around the accuracy problems shown in this example is to use higher-order elements. For example, for our two-dimensional problem we could use quadratic or cubic approximations for the displacements within each element; but, of course, this would require additional nodes to be specified for each element and significant complications in evaluating the stiffness matrices. For the interested reader, aspects of the foregoing are discussed in references.[1-11]

THE DIRECT APPROACH FOR SOME SIMPLE, STEADY FLOW SYSTEMS

Many of the flow systems discussed in Chapters 2 and 3 can be simulated using finite-element techniques. The two- and three-dimensional steady flow of thermal energy through a solid (conduction heat transfer) and the steady fluid flow through a porous medium are typical of many engineering flow problems now treated with finite-element methods. As with the structural systems, we can avoid working from the full differential models by using a direct approach, wherein we invoke basic principles directly within finite elements. Again the reader is reminded that this approach, although illustrating most features of finite-element methodology, cannot show us some of the most powerful features of the method. The reader interested in a fuller treatment of the subject should consult the references,[1-11] and take a separate course in finite-element methods in engineering. In this section, we look at how some of the previous material can be extended to flow systems, and then apply the techniques to some simple one- and two-dimensional systems.

One-Dimensional Conduction Heat Transfer

To develop the method for one dimension, we consider the three-layer system shown in Figure 7.26. Each layer consists of a different material, and the heat flow takes place only in the x direction, implying that the left-hand face is kept at a uniform temperature that is higher than the uniform temperature on the right-hand face. The basic law governing this type of flow is Fourier's law of conduction, which can be stated as the algebraic formula given in Equation 7.56.

$$q = kA \frac{\Delta T}{L} \tag{7.56}$$

In Equation 7.56, ΔT represents the temperature decrease across a homogeneous layer of thickness l, of cross sectional area A, and with a thermal conductivity of k.

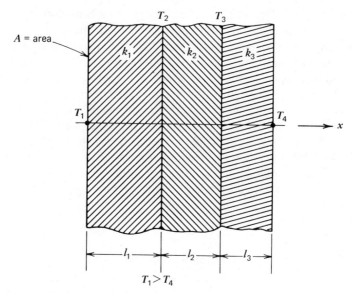

Figure 7.26 One-dimensional heat flow through a three-layered wall.

This law could have been expressed in universal form as discussed in Chapter 2, where the flow resistance is $R = l/(kA)$, but we find it convenient to deal with the separate terms in this special situation and thus use Equation 7.56. We now analyze each layer as a separate finite element in the system and observe that, in fact, each finite element is a control volume. A typical element, its nodes, and the heat flow symbols most convenient for this development are shown in Figure 7.27.

First we consider the heat flow through node 1, approximated by Equation 7.57.

$$q_1 = \left(\frac{kA}{l}\right)(T_1 - T_2) \tag{7.57}$$

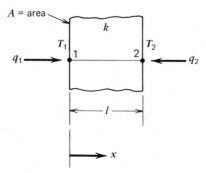

Figure 7.27 Single homogeneous layer with thermal conductivity k showing Temperatures T_1 and T_2 and heat transfers q_1 and q_2.

$$q_2 = -\left(\frac{kA}{l}\right)(T_1 - T_2) \tag{7.58}$$

Here it is assumed that $T_1 > T_2$, and that the basic conservation principle for the steady flow control volume (finite element) means that $q_1 + q_2 = 0$. Thus we get $q_2 = -q_1$, and the flow through node 2 becomes the approximation shown in Equation 7.58. The combined Equations 7.57 and 7.58 give us the matrix equation, Equation 7.59, for a single element.

$$\{Q\} = [k]\{T\} \tag{7.59}$$

where

$$\{Q\} = \begin{Bmatrix} q_1 \\ q_2 \end{Bmatrix}$$

$$[k] = \frac{kA}{l}\begin{bmatrix} 1 & -1 \\ -1 & 1 \end{bmatrix}$$

$$\{T\} = \begin{Bmatrix} T_1 \\ T_2 \end{Bmatrix}$$

As before, we can combine (patch together) the element matrix equations to give a global matrix equation for the whole system. Boundary conditions would be imposed on this global equation and an equation set solver applied to give a solution for the unknown nodal temperatures. This is best illustrated by a two-element example; we will leave further development of a global equation for the three-element system shown in Figure 7.26 as an exercise.

EXAMPLE 7.8

A fire wall consists of two layers of masonry, one layer 10 units thick and the other 6 units thick, as shown in Figure 7.28.

The environment temperature on the inside (left-hand side) is 2500, and on the outside, 75. Thermal conductivities are $k_1 = 9.6$ and $k_2 = 1.2$. The heat transfer coefficient on the inside is 10, and on the outside, 1.5. Assume consistent units throughout, neglect joint effects and demonstrate a finite-element method simulation of the nodal temperatures.

▶ Applying Equation 7.59, we get for element 1 the following matrix equation.

$$0.96\begin{bmatrix} 1 & -1 \\ -1 & 1 \end{bmatrix}\begin{Bmatrix} T_1 \\ T_2 \end{Bmatrix} = \begin{Bmatrix} q_1 \\ q_2 \end{Bmatrix}$$

And for element 2,

$$0.2\begin{bmatrix} 1 & -1 \\ -1 & 1 \end{bmatrix}\begin{Bmatrix} T_2 \\ T_3 \end{Bmatrix} = \begin{Bmatrix} q_2 \\ q_3 \end{Bmatrix}$$

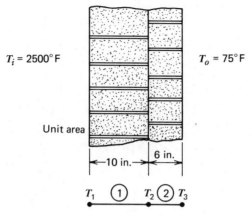

Figure 7.28 Typical finite-element idealization, Example 7.8.

Patching these together, we get a global matrix equation.

$$\begin{bmatrix} 0.96 & -0.96 & 0 \\ -0.96 & 1.16 & -0.2 \\ 0 & -0.2 & 0.2 \end{bmatrix} \begin{Bmatrix} T_1 \\ T_2 \\ T_3 \end{Bmatrix} \begin{Bmatrix} q_1^1 \\ q_2^1 + q_2^2 \\ q_3^2 \end{Bmatrix}$$

where the superscripts shown on the nodal heat transfers represent the element numbers. For the steady flow system we are working with here, the conservation principle applied at the boundary between element 1 and 2 requires that

$$q_2^2 = -q_2^1$$

Furthermore, we observe that the heat transfer terms can be interpreted as heat flux terms if we so desire because a unit area $(A=1.)$ is being used. Thus the second term in the heat flux vector of the global matrix equation becomes zero. We can make an important generalization from this; that is, the heat flux terms between elements will always be zero in the global equation and can be written this way initially. On the other hand, the heat flux terms representing physical boundaries have to be retained and usually require special modification of the global equation to account for the boundary conditions. Although we cannot demonstrate it with this one-dimensional example, the easiest boundary condition to incorporate is the insulated (heat flux =0) case; this is often referred to in finite-element work as the natural boundary condition. Consequently, the global equation is often deliberately written with a zero heat flux vector initially. The actual heat flux vector is then developed as the boundary conditions are incorporated.

In this example, we have two, convective, heat transfer boundary conditions, and we will have to apply Newton's basic law of cooling at each boundary.

$$q_s = hA(T_\infty - T_s)$$

Here the s subscript refers to the boundary surface; the h is the heat transfer coefficient; A is the area normal to the q_s heat transfer vector; and T_∞ is the convective environment temperature. In Chapter 2 we treated some convective boundaries using a universal flow formula, and it is evident that the flow resistance for such an approach has to be $R = 1/(hA)$. For our example, the left-hand boundary condition becomes

$$q_1^1 = (10)(1)(2500 - T_1)$$

and the right-hand one becomes

$$q_3^2 = (1.5)(1)(75 - T_3)$$

These can be incorporated into our previous global equation to give the final global matrix equation.

$$\begin{bmatrix} 10.96 & -0.96 & 0 \\ -0.96 & 1.16 & -0.20 \\ 0 & -0.20 & 1.70 \end{bmatrix} \begin{Bmatrix} T_1 \\ T_2 \\ T_3 \end{Bmatrix} = \begin{Bmatrix} 25,000. \\ 0 \\ 112.5 \end{Bmatrix}$$

Applying an equation-set solver, we get our simulated nodal temperatures.

$$\begin{Bmatrix} T_1 \\ T_2 \\ T_3 \end{Bmatrix} = \begin{Bmatrix} 2464 \\ 2093 \\ 312 \end{Bmatrix}$$

These temperatures can in turn be used to determine the heat flux by substituting back into one of the basic laws, either Fourier's or Newton's cooling.

One-Dimensional Fluid Flow

Many steady fluid flow systems can be considered a collection of one-dimensional finite-element control volumes, as illustrated in Figure 7.29. Each element is taken as the flow passage between two designated junctions, the junctions in this case being labeled with subscripted pressure, P, symbols. Obviously the junctions become the end nodes for our finite elements. We will consider only volumetric flow rates, designated by VV; however, these can be changed to mass flows easily by multiplying by a fluid density (see Chapter 2). But to avoid some thermodynamic complications, we also limit our development to so-called incompressible flows, that is, those with a constant density fluid.

To develop our element matrix equation, we isolate one flow element as shown in Figure 7.30. The most general flow relationship within the element can be expressed using the universal flow principle from Equation 2.6 of Chapter 2. Thus, using our symbols, we can write Equation 7.60 to give us the flow through node 1.

$$VV_1 = \frac{P_1 - P_2}{R} \tag{7.60}$$

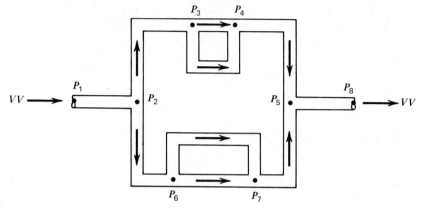

Figure 7.29 Typical one-dimensional fluid flow system.

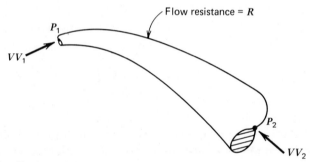

Figure 7.30 Typical one-dimensional fluid flow finite element.

Applying the conservation principle (Equation 2.3) to the element and noting that both flow arrows point into the element, $VV_2 = -VV_1$, we can write Equation 7.61 to represent the flow through node 2.

$$VV_2 = -\frac{P_1 - P_2}{R} \qquad (7.61)$$

Thus our element matrix equation becomes, upon combining Equations 7.60 and 7.61, Equation 7.62.

$$\{VV\} = [k_f]\{P\} \qquad (7.62)$$

where

$$\{VV\} = \begin{Bmatrix} VV_1 \\ VV_2 \end{Bmatrix}$$

$$[k_f] = R^{-1} \begin{bmatrix} 1. & -1. \\ -1. & 1. \end{bmatrix}$$

$$\{P\} = \begin{Bmatrix} P_1 \\ P_2 \end{Bmatrix}$$

We can combine (patch together) the element matrix equations to give a global matrix equation for the whole system, which can be solved after application of boundary conditions to complete our simulation of nodal pressures. However, it should be noted that for most flow problems, the flow resistances, the R's, depend on the simulated pressures; and so astute guesses have to be made to get good values for these. In other words, local linearizations may have to be done on a trial-and-error basis to get a good global equation to solve. Nevertheless, one class of flow problems exhibits relatively constant R values. These problems are known as the laminar flow problems and typically include flows through porous media or very slow flows through tubes. Development of a global equation for one of the latter cases is left as an exercise.

Two-Dimensional Conduction Heat Transfer

We will consider only plane triangular elements, with only linear temperature variations within any element. The development of the element matrix equations is very similar to that described in the preceding sections for the plane stress, plane strain structural systems, and follows Jennings.[13]

Consider the simple triangular element shown in Figure 7.31. The rectangular Cartesian coordinates for the three vertex nodes are (x_1, y_1), (x_2, y_2), and (x_3, y_3). We now assume that the temperature varies linearly within the element as shown in Equation 7.63.

$$T = a_1 + a_2 x + a_3 y \tag{7.63}$$

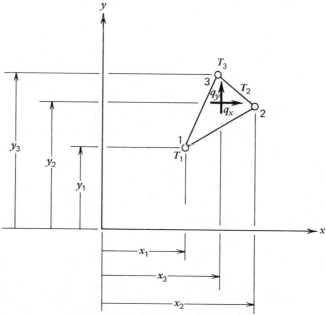

Figure 7.31 Triangular element with linear temperature variation.

From this a matrix equation can be written for the temperatures at the three nodes, Equation 7.64.

$$\begin{Bmatrix} T_1 \\ T_2 \\ T_3 \end{Bmatrix} = \begin{bmatrix} 1 & x_1 & y_1 \\ 1 & x_2 & y_2 \\ 1 & x_3 & y_3 \end{bmatrix} \begin{Bmatrix} a_1 \\ a_2 \\ a_3 \end{Bmatrix} \tag{7.64}$$

or

$$\{a\} = [A]^{-1}\{T\}$$

where the $[A]$ matrix is the same as the $[A]$ matrix given in Equation 7.29. Then we introduce shape functions, N_i, to describe any interior temperature, as shown in Equation 7.65.

$$T = [N_1 \quad N_2 \quad N_3] \begin{Bmatrix} T_1 \\ T_2 \\ T_3 \end{Bmatrix} \tag{7.65}$$

The shape functions have already been developed from the $[A]^{-1}$ matrix and are given by Equations 7.32, 7.35, and 7.36.

The differential form of Fourier's basic conduction law, $q_x = -k_x \, dT/dx$, is applied directly to the element to establish the relationship between the local heat flux vector component, q_x, and the local temperatures. Performing the appropriate differentiations on Equation 7.65 leads us to Equation 7.66.

$$\begin{Bmatrix} q_x \\ q_y \end{Bmatrix} = \begin{bmatrix} \dfrac{-0.5k}{A} \end{bmatrix} \begin{bmatrix} \beta_1 & \beta_2 & \beta_3 \\ \gamma_1 & \gamma_2 & \gamma_3 \end{bmatrix} \begin{Bmatrix} T_1 \\ T_2 \\ T_3 \end{Bmatrix} \tag{7.66}$$

The heat transfer from the edge of one element must equal the heat transfer to the edge of the adjacent element. However, we do not have enough temperature variables to describe this conservation principle fully. We proceed anyway by approximating the conservation only at the nodes. For example, the heat transfer across the edge between nodes 1 and 2 is divided equally between nodes 1 and 2, and so on. For our element with nodes 1, 2, and 3, the heat transfers across the edges (1-2), (2-3), and (3-1) are given by the formulations shown in Equation 7.67.

$$\begin{aligned} q_{1-2} &= (y_2 - y_1)q_x - (x_2 - x_1)q_y \\ q_{2-3} &= (y_3 - y_2)q_x - (x_3 - x_2)q_y \\ q_{3-1} &= (y_1 - y_3)q_x - (x_1 - x_3)q_y \end{aligned} \tag{7.67}$$

Distributing these edge heat transfers to the nodes is done as follows. For node 1,

$$\begin{aligned} q_1 &= \frac{q_{3-1} + q_{1-2}}{2} \\ &= 0.5(y_2 - y_3)q_x + 0.5(x_3 - x_2)q_y \\ &= 0.5\beta_1 q_x + 0.5\gamma_1 q_y \end{aligned}$$

Similarly for node 2, $q_2 = 0.5\beta_2 q_x + 0.5\gamma_2 q_y$

Similarly for node 3, $q_3 = 0.5\beta_3 q_x + 0.5\gamma_3 q_x$

Combining the foregoing relationships for the nodal heat transfers with Equation 7.66 gives us the element matrix equation, Equation 7.68.

$$[k]\{T\} = \{q\} \tag{7.68}$$

where the conductance matrix, $[k]$, is given by

$$[k] = \left[\frac{-0.25k}{A}\right]\begin{bmatrix} \beta_1^2 + \gamma_1^2 & \beta_1\beta_2 + \gamma_1\gamma_2 & \beta_1\beta_3 + \gamma_1\gamma_3 \\ \beta_1\beta_2 + \gamma_1\gamma_2 & \beta_2^2 + \gamma_2^2 & \beta_2\beta_3 + \gamma_2\gamma_3 \\ \beta_1\beta_3 + \gamma_1\gamma_3 & \beta_2\beta_3 + \gamma_2\gamma_3 & \beta_3^2 + \gamma_3^2 \end{bmatrix}$$

the nodal heat transfer vector, $\{q\}$, is

$$\{q\} = \begin{Bmatrix} q_1 \\ q_2 \\ q_3 \end{Bmatrix}$$

and the nodal temperature vector, $\{T\}$, is

$$\{T\} = \begin{Bmatrix} T_1 \\ T_2 \\ T_3 \end{Bmatrix}$$

In assembling the global matrix equation for a system, we observe that the net heat transfer for any interior node will always be zero (provided we do not have any internal heat generation). Consequently, it is usually convenient to simply express the complete heat transfer vector as zero initially, thus immediately giving the so-called natural conditions. Then, as boundary conditions are incorporated, the appropriate values can be added to the q driving vector. Otherwise, the combining of the element equations (the patching process) to give the global matrix equation proceeds exactly as described before. This is illustrated in Example 7.9.

EXAMPLE 7.9

Consider the cross section shown in Figure 7.32, where we have already decided upon using the finite-element method and have chosen 12 triangular elements and 12 nodes. Both the left-hand (slanted) and right-hand edges are insulated, that is, show zero heat transfer boundaries. In fact, these are symmetry planes for a larger cross section. The upper edge has a surface temperature of $T = 600$; the lower

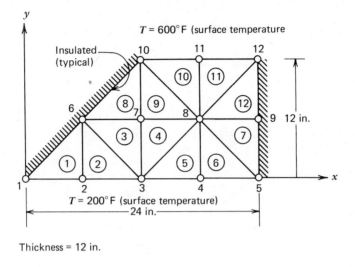

Thickness = 12 in.

◯ = Elements

O = Nodes

Figure 7.32 Cross section, Example 7.9.

edge has a surface temperature of $T = 200$. The system is long in the axial (z direction); since boundary temperatures vary so little in the z direction, we take a uniform axial thickness of 12 and ignore any z direction variations, giving us a two-dimensional steady heat transfer system with a uniform thermal conductivity of $k = 1.0$. Assume consistent units throughout and perform a finite element simulation of the interior node temperatures.

▶ Using Equation 7.68, we get the following element matrix equation for element 1.

$$\begin{bmatrix} 0.5 & -0.5 & 0 \\ -0.5 & 1.0 & -0.5 \\ 0 & -0.5 & 0.5 \end{bmatrix} \begin{Bmatrix} T_1 \\ T_2 \\ T_6 \end{Bmatrix} = - \begin{Bmatrix} q_1^1 \\ q_2^1 \\ q_6^1 \end{Bmatrix}$$

The superscripts on the heat transfer terms simply designate that the contributions come from element 1. Similarly, we get the following element matrix equation for element 2.

$$\begin{bmatrix} 1.0 & -0.5 & -0.5 \\ -0.5 & 0.5 & 0 \\ -0.5 & 0 & 0.5 \end{bmatrix} \begin{Bmatrix} T_2 \\ T_3 \\ T_6 \end{Bmatrix} = - \begin{Bmatrix} q_2^2 \\ q_3^2 \\ q_6^2 \end{Bmatrix}$$

Continuing in a similar manner, we generate a total of 12 element equations. We then patch (assemble) these together into a preliminary global matrix equation, as follows.

$$
\begin{bmatrix}
0.5 & -0.5 & 0 & 0 & 0 & 0 & 0 & 0 & 0 & 0 & 0 & 0 \\
-0.5 & 2. & -0.5 & 0 & 0 & -1. & 0 & 0 & 0 & 0 & 0 & 0 \\
0 & -0.5 & 2. & -0.5 & 0 & 0 & -1. & 0 & 0 & 0 & 0 & 0 \\
0 & 0 & -0.5 & 2. & -0.5 & 0 & 0 & -1. & 0 & 0 & 0 & 0 \\
0 & 0 & 0 & -0.5 & 1. & 0 & 0 & 0 & -0.5 & 0 & 0 & 0 \\
0 & -1. & 0 & 0 & 0 & 2. & -1. & 0 & 0 & 0 & 0 & 0 \\
0 & 0 & -1. & 0 & 0 & -1. & 4. & -1. & 0 & -1. & 0 & 0 \\
0 & 0 & 0 & -1. & 0 & 0 & -1. & 4. & -1. & 0 & -1. & 0 \\
0 & 0 & 0 & 0 & -0.5 & 0 & 0 & -1. & 2. & 0 & 0 & -0.5 \\
0 & 0 & 0 & 0 & 0 & 0 & -1. & 0 & 0 & 1.5 & -0.5 & 0 \\
0 & 0 & 0 & 0 & 0 & 0 & 0 & -1. & 0 & -0.5 & 2. & -0.5 \\
0 & 0 & 0 & 0 & 0 & 0 & 0 & 0 & -0.5 & 0 & -0.5 & 1.
\end{bmatrix}
\begin{Bmatrix}
T_1 \\ T_2 \\ T_3 \\ T_4 \\ T_5 \\ T_6 \\ T_7 \\ T_8 \\ T_9 \\ T_{10} \\ T_{11} \\ T_{12}
\end{Bmatrix}
=
\begin{Bmatrix}
0 \\ 0 \\ 0 \\ 0 \\ 0 \\ 0 \\ 0 \\ 0 \\ 0 \\ 0 \\ 0 \\ 0
\end{Bmatrix}
$$

We now modify the global equation to account for the boundary conditions, noting that the natural conditions (setting the q vector to zero) have already accounted for the insulated boundaries. But to modify for node 1, which has a specified temperature of 200, we first transfer all the off-diagonal terms in the conductance matrix that multiply T_1 over to the right-hand side of the equation. All these off-diagonal terms are then set to zero in the conductance matrix. Then, in the first row, all off-diagonal terms are set to zero, and the diagonal term is set equal to 1. The resulting global equation follows.

$$
\begin{bmatrix}
1. & 0 & 0 & 0 & 0 & 0 & 0 & 0 & 0 & 0 & 0 & 0 \\
0 & 2. & -0.5 & 0 & 0 & -1. & 0 & 0 & 0 & 0 & 0 & 0 \\
0 & -0.5 & & & & & & & & & & \\
0 & 0 & & & & & & & & & & \\
0 & 0 & & & & & & & & & & \\
0 & -1. & \text{(This part of matrix the same as before)} & & & & & & & & & \\
0 & 0 & & & & & & & & & & \\
0 & 0 & & & & & & & & & & \\
0 & 0 & & & & & & & & & & \\
0 & 0 & & & & & & & & & & \\
0 & 0 & & & & & & & & & & \\
0 & 0 & & & & & & & & & &
\end{bmatrix}
\begin{Bmatrix}
T_1 \\ T_2 \\ T_3 \\ T_4 \\ T_5 \\ T_6 \\ T_7 \\ T_8 \\ T_9 \\ T_{10} \\ T_{11} \\ T_{12}
\end{Bmatrix}
=
\begin{Bmatrix}
200 \\ 100 \\ 0 \\ 0 \\ 0 \\ 0 \\ 0 \\ 0 \\ 0 \\ 0 \\ 0 \\ 0
\end{Bmatrix}
$$

We can repeat this modification process for all nodes that have temperatures specified as boundary conditions. The final global equation follows.

$$
\begin{bmatrix}
1. & 0 & 0 & 0 & 0 & 0 & 0 & 0 & 0 & 0 & 0 & 0 \\
0 & 1. & 0 & 0 & 0 & 0 & 0 & 0 & 0 & 0 & 0 & 0 \\
0 & 0 & 1. & 0 & 0 & 0 & 0 & 0 & 0 & 0 & 0 & 0 \\
0 & 0 & 0 & 1. & 0 & 0 & 0 & 0 & 0 & 0 & 0 & 0 \\
0 & 0 & 0 & 0 & 1. & 0 & 0 & 0 & 0 & 0 & 0 & 0 \\
0 & 0 & 0 & 0 & 0 & 2. & -1. & 0 & 0 & 0 & 0 & 0 \\
0 & 0 & 0 & 0 & 0 & -1. & 4. & -1. & 0 & 0 & 0 & 0 \\
0 & 0 & 0 & 0 & 0 & 0 & -1. & 4. & -1. & 0 & 0 & 0 \\
0 & 0 & 0 & 0 & 0 & 0 & 0 & -1. & 2. & 0 & 0 & 0 \\
0 & 0 & 0 & 0 & 0 & 0 & 0 & 0 & 0 & 1. & 0 & 0 \\
0 & 0 & 0 & 0 & 0 & 0 & 0 & 0 & 0 & 0 & 1. & 0 \\
0 & 0 & 0 & 0 & 0 & 0 & 0 & 0 & 0 & 0 & 0 & 1. \\
\end{bmatrix}
\begin{Bmatrix}
T_1 \\ T_2 \\ T_3 \\ T_4 \\ T_5 \\ T_6 \\ T_7 \\ T_8 \\ T_9 \\ T_{10} \\ T_{11} \\ T_{12}
\end{Bmatrix}
=
\begin{Bmatrix}
200 \\ 200 \\ 200 \\ 200 \\ 200 \\ 200 \\ 800 \\ 800 \\ 400 \\ 600 \\ 600 \\ 600
\end{Bmatrix}
$$

In this example the global matrix can be partitioned so that only the central part of the complete global equation has to be solved to complete the simulation.

$$
\begin{bmatrix}
2. & -1. & 0 & 0 \\
-1. & 4. & -1. & 0 \\
0 & -1. & 4. & -1. \\
0 & 0 & -1. & 2.
\end{bmatrix}
\begin{Bmatrix}
T_6 \\ T_7 \\ T_8 \\ T_9
\end{Bmatrix}
=
\begin{Bmatrix}
200. \\ 800. \\ 800. \\ 400.
\end{Bmatrix}
$$

Applying an equation-set solver gives us the simulated interior temperatures. $T_6 = 284.$, $T_7 = 369.$, $T_8 = 391.$, and $T_9 = 396.$

Most two-dimensional heat transfer modeling involves some convection boundaries governed by Newton's basic law of cooling, but in the preceding material we have deliberately avoided discussing this to minimize the complexity of the presentation. However, now we will show that only a few additional features have to be added to our methodology to allow it to handle such boundaries.

Recall that the heat transfers across the edges of an element were divided up and assigned to the adjacent nodes. A unity depth (thickness) of the element was assumed in the formulas. For all interior nodes in a system, the heat transfer terms canceled each other in the global matrix equation. Consequently, as a matter of convenience we set the initial global driving vector to zero, which coincidently gave us our so-called natural conditions automatically. Later the temperature boundary conditions showed up in this vector as part of a boundary condition modification process. Now we shall use this same modification process to get the convection boundary conditions incorporated into the global matrix also.

Consider the element in Figure 7.31 again. Suppose that any edge i-j (the i and j represent node numbers) is exposed to a convection environment at a temperature of T_∞, and with a heat transfer coefficient of h. The heat transfer rate out of this edge owing to convection will be given (using Newton's basic cooling principle) by

Equation 7.69.

$$q_{i-j}=0.5htS_{i-j}[(T_i-T_\infty)+(T_j-T_\infty)] \qquad (7.69)$$

In this equation the hyphenated subscripts indicate the edge spanned by the nodes i and j; t is the thickness of the element; and S_{i-j} is the length of the edge. We have simply divided the overall convection transfer from the edge equally between nodes i and j. Obviously, S_{i-j} can be found from Equation 7.70.

$$S_{i-j}=\sqrt{(x_i-x_j)^2+(y_i-y_j)^2} \qquad (7.70)$$

Thus, for a node i, we have the convection heat transfer term given by Equation 7.71.

$$q_i=\frac{htS_{i-j}(T_i-T_\infty)}{2} \qquad (7.71)$$

A like term will always exist for node j. Consequently the convection condition at a given node can always be accounted for, if we take a unity element thickness, by adding $-hS_{i-j}T_\infty/2$ to the driving vector at the node i and j positions, and $-hS_{i-j}/2$ to the $K_{i,i}$ and $K_{j,j}$ terms in the conductance matrix. Application of the foregoing is illustrated in Example 7.10, where an interactive computer program coded in BASIC (Figure 7.34) provides the simulation results. This computer program can also be used for Example 7.9 and other two-dimensional heat transfer problems.

EXAMPLE 7.10

Consider the two-dimensional conduction heat transfer system shown in Figure 7.33. The system has a uniform thermal conductivity of $k=1.0$ and a thickness of $t=12$. The upper and lower boundaries are insulated (zero heat transfer), while the left-hand boundary is exposed to a convective environment at a temperature of $T_\infty=600$ with a heat transfer coefficient of $h=0.5$. The right-hand boundary has a surface temperature held at $T=200$. The desired elements and nodes are indicated by number. Assume consistent units throughout and perform a finite-element simulation of the nodal temperatures, using a computer program if desired.

▶ We will develop the global matrix equation by hand first so that we can observe how the convection boundaries affect the numbers during the modification step. Using Equation 7.68 and patching the terms into the global matrix element by element gives us the following preliminary global equation.

$$
\begin{bmatrix}
-1.08 & 0.333 & 0 & 0 & 0 & 0 & 0 & 0.750 \\
0.333 & -1.75 & 0.333 & 0.333 & 0 & 0.417 & 0 & 0.333 \\
0 & 0.333 & -1.08 & 0.750 & 0 & 0 & 0 & 0 \\
0 & 0.333 & 0.750 & -2.17 & 0.750 & 0.333 & 0 & 0 \\
0 & 0 & 0 & 0.750 & -1.08 & 0.333 & 0 & 0 \\
0 & 0.417 & 0 & 0.333 & 0.333 & -1.75 & 0.333 & 0.333 \\
0 & 0 & 0 & 0 & 0 & 0.333 & -1.08 & 0.750 \\
0.750 & 0.333 & 0 & 0 & 0 & 0.333 & 0.750 & -2.17
\end{bmatrix}
\begin{Bmatrix}
T_1 \\ T_2 \\ T_3 \\ T_4 \\ T_5 \\ T_6 \\ T_7 \\ T_8
\end{Bmatrix}
=
\begin{Bmatrix}
0 \\ 0 \\ 0 \\ 0 \\ 0 \\ 0 \\ 0 \\ 0
\end{Bmatrix}
$$

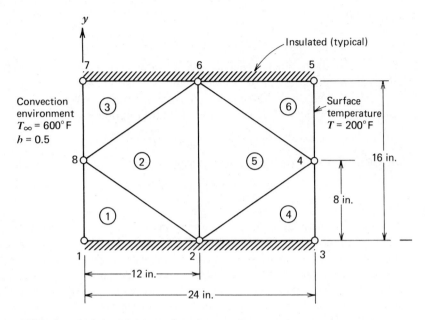

Figure 7.33 Cross section, Example 7.10.

Then, after the above preliminary global matrix has been modified for the convective boundary between nodes 1 and 8, we get the following equation.

$$
\begin{bmatrix}
-3.08 & 0.333 & 0 & 0 & 0 & 0 & 0 & 0.750 \\
0.333 & & & & & & & \\
0 & & & & & & & \\
0 & & \text{(This part of matrix the same as before)} & & & & \\
0 & & & & & & & \\
0 & & & & & & & \\
0 & & & & & & & \\
0.750 & 0.333 & 0 & 0 & 0 & 0.333 & 0.750 & -4.17
\end{bmatrix}
\begin{Bmatrix}
T_1 \\ T_2 \\ T_3 \\ T_4 \\ T_5 \\ T_6 \\ T_7 \\ T_8
\end{Bmatrix}
=
\begin{Bmatrix}
-1200 \\ 0 \\ 0 \\ 0 \\ 0 \\ 0 \\ 0 \\ -1200
\end{Bmatrix}
$$

After the modification for the convective condition between nodes 8 and 7, we get the following global matrix equation.

$$
\begin{bmatrix}
-3.08 & 0.333 & 0 & 0 & 0 & 0 & 0 & 0.750 \\
0.333 & & & & & & & \\
0 & & & & & & & \\
0 & & \multicolumn{4}{c}{\text{(This part of matrix the same as before)}} & & \\
0 & & & & & & & \\
0 & & & & & & & \\
0 & 0 & 0 & 0 & 0 & 0.333 & -3.08 & 0.750 \\
0.750 & 0.333 & 0 & 0 & 0 & 0.333 & 0.750 & -6.17
\end{bmatrix}
\begin{Bmatrix}
T_1 \\ T_2 \\ T_3 \\ T_4 \\ T_5 \\ T_6 \\ T_7 \\ T_8
\end{Bmatrix}
=
\begin{Bmatrix}
-1200 \\ 0 \\ 0 \\ 0 \\ 0 \\ 0 \\ -1200 \\ -2400
\end{Bmatrix}
$$

Of course, the modification has to be continued to account for the surface temperature conditions on the right-hand edge of the system; but since we

```
100 REM EXAMPLE 7.10 BASIC PROGRAM
110 CLS! CLEAR50! DEFINT I,J,L,M,N
120 PRINT"2D FINITE-ELEMENT CONDUCTION HEAT TRANSFER"
130 INPUT"GIVE ELEMENT UNIFORM THICKNESS";TH
140 INPUT"GIVE ELEMENT UNIFORM THERMAL CONDUCTIVITY";TC
150 INPUT"GIVE NUMBER OF TRIANGULAR ELEMENTS USED";NT
160 INPUT"GIVE NUMBER OF NODES";NN! M=NN
170 DIM N1(NT),N2(NT),N3(NT),A1(NT),A2(NT),A3(NT),B1(NT),B2(NT),B3(NT),G1(NT),G2(NT),G3(NT),A(NT)
180 DIM X(M),Y(M),F(M),T(M),K(M,M)
190 REM INPUT ELEMENT INFO.
200 FOR I=1TONT
210   PRINT"GIVE GLOBAL NODE NUMBERS (LOCAL 1,2,&3 CCW) FOR ELEMENT";I
220   INPUT N1(I),N2(I),N3(I)
230 NEXTI
240 FORI=1TONN
250   PRINT"GIVE GLOBAL COORDINATES, X , Y , FOR NODE";I
260   INPUT X(I),Y(I)
270 NEXTI
280 PRINT!PRINT">>>>> ELEMENT NODE LIST <<<<<"! L=0
290 FORI=1TONT
300   PRINT"NODES 1, 2, & 3 FOR ELEMENT";I;" ARE!";N1(I);N2(I);N3(I)
310   L=L+1! IF L<14 GOTO340
320   INPUT"FOR MORE, PRESS 'ENTER'";ZZ! L=0
330   PRINT"ELEMENT NODES CONTINUED"
340 NEXTI
350 INPUT"TO CONTINUE, PRESS 'ENTER'";ZZ
360 PRINT!PRINT"////// NODE COORDINATES //////"! L=0
370 FORI=1TONN
380   PRINT" X AND Y  FOR NODE ";I;" ARE!"; X(I);Y(I)
390   L=L+1! IF L<14 GOTO420
400   INPUT"FOR MORE, PRESS 'ENTER'";ZZ! L=0
410   PRINT"NODE COORDINATES CONTINUED"
420 NEXTI
430 PRINT!INPUT"DO YOU DESIRE TO MAKE A CHANGE IN A NODE NO.?--Y/N";Q$
440   IF Q$<>"Y" GOTO480
450   INPUT"ENTER THE ELEMENT NUMBER";J
460   INPUT"GIVE THE NEW GLOBAL NODE NUMBERS ( 1, 2, 3 CCW LOCAL)";N1(J),N2(J),N3(J)
470   GOTO430
480 PRINT!INPUT"DO YOU DESIRE TO MAKE A CHANGE IN A NODE'S COORDINATES?--Y/N";Q1$
490   IF Q1$<>"Y" GOTO530
500   INPUT"GIVE THE NODE NUMBER";J
510   INPUT"GIVE THE NEW COORDINATES, X , Y";X(J),Y(J)
520   GOTO480
530 PRINT"DO YOU WISH TO SEE NODE ASSIGNMENTS & COORDINATES AGAIN?--Y/N"
540 INPUT AA$! IF AA$="Y" GOTO280
550 REM COMPUTE ELEMENT AREAS AND SHAPE FUNCTION PARAMETERS
560 FOR I=1TONT
570   A(I)=0.5*(X(N2(I))*Y(N3(I))-X(N3(I))*Y(N2(I))+X(N3(I))*Y(N1(I))-X(N1(I))*Y(N3(I))+X(N1(I))*Y(N2(I))-X(N2(I))*Y(N1(I)))
580   A1(I)=X(N2(I))*Y(N3(I))-X(N3(I))*Y(N2(I))
```

Figure 7.34*a* BASIC program, Example 7.10.

```
590   A2(I)=X(N3(I))*Y(N1(I))-X(N1(I))*Y(N3(I))
600   A3(I)=X(N1(I))*Y(N2(I))-X(N2(I))*Y(N1(I))
610   B1(I)=Y(N2(I))-Y(N3(I))
620   B2(I)=Y(N3(I))-Y(N1(I))
630   B3(I)=Y(N1(I))-Y(N2(I))
640   G1(I)=X(N3(I))-X(N2(I))
650   G2(I)=X(N1(I))-X(N3(I))
660   G3(I)=X(N2(I))-X(N1(I))
670 NEXTI
680 REM CALL SUB1000 TO ASSEMBLE CONDUCTANCE MATRICES
690 GOSUB1000
700 REM CALL SUB2000 TO INSERT BOUNDARY CONDITIONS
710 GOSUB2000
720 REM SOLVE GLOBAL MATRIX EQUATION FOR TEMPERATURES
730 REM USE SHORT VERSION OF GAUSSIAN EQUATION-SET SOLVER
740 REM--THIS SOLVER DESTROYS THE K(M,M) AND F(M) ARRAYS
750 FORII=1TOM: I1=II+1: F(II)=F(II)/K(II,II)
760   IF II=M GOTO840
770   FORJ=I1TOM: K(II,J)=K(II,J)/K(II,II): NEXTJ
780   FORI=I1TOM: F(I)=F(I)-K(I,II)*F(II)
790     FORJ=I1TOM
800       K(I,J)=K(I,J)-K(I,II)*K(II,J)
810     NEXTJ
820   NEXTI
830 NEXTII
840 I1=II: II=II-1: IFII<=0 GOTO870
850 FORJ=I1TOM: F(II)=F(II)-K(II,J)*F(J): NEXTJ
860 GOTO840
870 REM PRINT TEMPERATURE VALUES
880 CLS: L=0: PRINT">>>>>> TEMPERATURE VALUES <<<<<<"
890 FORI=1TONN
900   PRINT"T=";F(I);" FOR NODE ";I: L=L+1
910   IF L<14 GOTO930
920   INPUT"FOR MORE PRESS 'ENTER'";ZZ: L=0
930 NEXTI
940 INPUT"DO YOU WISH TO SEE VALUES AGAIN? -Y/N";A$
950 IF A$="Y" GOTO880
960 END
1000 REM ASSEMBLES GLOBAL CONDUCTANCE MATRIX, ELEMENT BY ELEMENT
1010 FORI=1TOM: F(I)=0: FORJ=1TOM: K(I,J)=0: NEXTJ,I
1020 FORI=1TONT
1030   CK=-TC*TH/(4*A(I))
1040   I1=N1(I): I2=N2(I): I3=N3(I)
1050   K(I1,I1)=K(I1,I1)+CK*(B1(I)[2+G1(I)[2)
1060   K(I1,I2)=K(I1,I2)+CK*(B1(I)*B2(I)+G1(I)*G2(I))
1070     K(I2,I1)=K(I1,I2)
1080   K(I1,I3)=K(I1,I3)+CK*(B1(I)*B3(I)+G1(I)*G3(I))
1090     K(I3,I1)=K(I1,I3)
1100   K(I2,I2)=K(I2,I2)+CK*(B2(I)[2+G2(I)[2)
```

Figure 7.34*b*   BASIC program, Example 7.10 *(continued)*.

```
1110   K(I2,I3)=K(I2,I3)+CK*(B2(I)*B3(I)+G2(I)*G3(I))
1120     K(I3,I2)=K(I2,I3)
1130   K(I3,I3)=K(I3,I3)+CK*(B3(I)[2+G3(I)[2)
1140 NEXTI
1150 RETURN
1160 END
2000 REM--MODIFICATION OF GLOBAL MATRIX EQUATION
2010 REM--FOR BOUNDARY CONDITIONS
2020 INPUT"DO YOU WISH TO ENTER A BOUNDARY CONDITION?--Y/N";B$
2030   IFB$="N" THEN RETURN
2040 INPUT"IS THIS A 'TEMP,',OR 'Q', CONDITION?--T/Q";C$
2050   IFC$="Q" GOTO2110
2060 INPUT"GIVE THE NUMBER OF THE AFFECTED NODE";ND
2070 INPUT"GIVE THE VALUE OF THE NODAL TEMPERATURE";BV
2080 FORI=1TOM: F(I)=F(I)-K(I,ND)*BV: K(ND,I)=0: K(I,ND)=0: NEXTI
2090 K(ND,ND)=1,0: F(ND)=BV
2100 GOTO 2020
2110 INPUT"IS THIS A CONVECTION CONDITION?--Y/N";CC$
2120   IFCC$="Y" GOTO2170
2130 INPUT"GIVE THE NUMBER OF THE AFFECTED NODE";ND
2140 INPUT"GIVE THE HEAT TRANSFER RATE, OUT=+, IN=-";Q
2150 F(ND)=F(ND)+Q
2160 GOTO2020
2170 INPUT"GIVE THE 2 NODE NUMBERS OF THE AFFECTED EDGE";N8,N9
2180 INPUT"GIVE THE HEAT TRANSFER COEFFICIENT";H
2190 INPUT"GIVE THE ENVIRONMENT TEMPERATURE";TE
2200   AH=SQR((X(N8)-X(N9))[2+(Y(N8)-Y(N9))[2)*TH
2210   K(N8,N8)=K(N8,N8)-H*AH/2: K(N9,N9)=K(N9,N9)-H*AH/2
2220   F(N8)=F(N8)-H*AH*TE/2: F(N9)=F(N9)-H*AH*TE/2
2230 GOTO2020
2240 RETURN
2250 END
```

Figure 7.34c   BASIC program, Example 7.10 (continued).

have been through this in the previous example, we will now resort to our computer program, shown in Figure 7.34.

Results of the simulation using the computer program follow. We observe that the values are the same as the exact values that can be generated analytically.

$$
\begin{Bmatrix} T_1 \\ T_2 \\ T_3 \\ T_4 \\ T_5 \\ T_6 \\ T_7 \\ T_8 \end{Bmatrix} = \begin{Bmatrix} 569.23 \\ 384.62 \\ 200.00 \\ 200.00 \\ 200.00 \\ 384.62 \\ 569.23 \\ 569.23 \end{Bmatrix}
$$

Two-Dimensional Fluid Flow

Simulations involving general two-dimensional fluid flow by finite elements are beyond the scope of this book. However, slow (low velocity) flows through porous media, driven by pressure differences only, can be handled by extensions of the heat transfer methods described in the preceding sections. The basic flow law applied here is the differential form of Darcy's law, which can be written in the same form as Fourier's basic law of heat conduction, as shown in Equation 7.72.

$$\left(\frac{VV}{a}\right)_x = -\frac{K}{\mu}\frac{dP}{dx} \tag{7.72}$$

Here K represents a physical characteristic of the porous media called permeability (which has length squared units); μ is the dynamic viscosity of the flowing fluid (with units of force-time/length squared); and a is the system area normal to the volumetric flow rate, VV. The derivative represents the fluid pressure, P, gradient with respect to the flow direction, x.

Following the same triangular-element development as used for the two-dimensional conduction heat transfer, the element matrix equation shown by Equation 7.73 emerges.

$$[k]\{P\} = \{VV\} \tag{7.73}$$

where $[k]$, the conductance matrix is given by

$$[k] = \left[\frac{Kt}{4A\mu}\right]\begin{bmatrix} \beta_1^2 + \gamma_1^2 & \beta_1\beta_2 + \gamma_1\gamma_2 & \beta_1\beta_3 + \gamma_1\gamma_3 \\ \beta_1\beta_2 + \gamma_1\gamma_2 & \beta_2^2 + \gamma_2^2 & \beta_2\beta_3 + \gamma_2\gamma_3 \\ \beta_1\beta_3 + \gamma_1\gamma_3 & \beta_2\beta_3 + \gamma_2\gamma_3 & \beta_3^2 + \gamma_2^2 \end{bmatrix}$$

the nodal flow rate vector, $\{VV\}$, is

$$\{VV\} = \begin{Bmatrix} VV_1 \\ VV_2 \\ VV_3 \end{Bmatrix}$$

and the nodal pressure vector, $\{P\}$, is

$$\{P\} = \begin{Bmatrix} P_1 \\ P_2 \\ P_3 \end{Bmatrix}$$

For a system of uniform thickness, the thickness term, t, can be removed; but then the nodal volumetric flows, VV_i, must be given on a unit thickness basis.

Assembly of the global matrix equation proceeds in exactly the same manner as discussed for the heat transfer case. However, the modification process to incorporate boundary conditions can be simpler because the counterpart of the convective boundary is not present. The modified global equation is solved by the usual techniques to give the simulated pressures at all nodes in the system.

In many of these simulation problems, the local flows and/or flow velocities are sought. Equation 7.74 can be used to get the local x-y velocity components within any element, while Equation 7.75 gives local volumetric flow rates through the edges of an element (out of the element is positive).

$$
\begin{Bmatrix} \left(\dfrac{VV}{a}\right)_x \\ \left(\dfrac{VV}{a}\right)_y \end{Bmatrix} = \begin{bmatrix} \dfrac{K}{2\mu A} \end{bmatrix} \begin{bmatrix} \beta_1 & \beta_2 & \beta_3 \\ \gamma_1 & \gamma_2 & \gamma_3 \end{bmatrix} \begin{Bmatrix} P_1 \\ P_2 \\ P_3 \end{Bmatrix}
\tag{7.74}
$$

$$
\begin{Bmatrix} VV_{1-2} \\ VV_{2-3} \\ VV_{3-1} \end{Bmatrix} = t \begin{bmatrix} (y_2 - y_1) & -(x_2 - x_1) \\ (y_3 - y_2) & -(x_3 - x_2) \\ (y_1 - y_3) & -(x_1 - x_3) \end{bmatrix} \begin{Bmatrix} \left(\dfrac{VV}{a}\right)_x \\ \left(\dfrac{VV}{a}\right)_y \end{Bmatrix}
\tag{7.75}
$$

Application of the foregoing in simulating pressures in a porous-media flow system is illustrated in Example 7.11. It should be noted that the computer program used in the Example 7.10 heat transfer simulation can also be used here, with only minor changes required.

EXAMPLE 7.11

Consider the flow system shown in Figure 7.35. Flow occurs from left to right through a uniformly porous block. No flow occurs through the upper and lower surfaces labeled as impermeable boundaries; it is also assumed that no flow is found directed into or out of the page. The thickness of the block is $t = 12$; the permeability to viscosity ratio is $K/\mu = 3.5$; pressure on the left-hand face is $P = 50$; and pressure on the right-hand face is $P = 10$. Assume consistent units throughout; use the elements and nodes designated and perform a finite element simulation of the system pressures and flow rates.

▶ Applying Equation 7.73, we get the following element matrix equation for element number one.

$$
\begin{bmatrix} -22.07 & 15.27 & 6.801 \\ 15.27 & -30.55 & 15.27 \\ 6.801 & 15.27 & -22.07 \end{bmatrix} \begin{Bmatrix} P_1 \\ P_4 \\ P_7 \end{Bmatrix} = \begin{Bmatrix} VV_1 \\ VV_4 \\ VV_7 \end{Bmatrix}
$$

Similarly, we generate the other five element matrix equations. These are then patched together into a preliminary global equation. After modifications to account for boundary conditions at nodes 1, 7, 3, and 5, our final global matrix

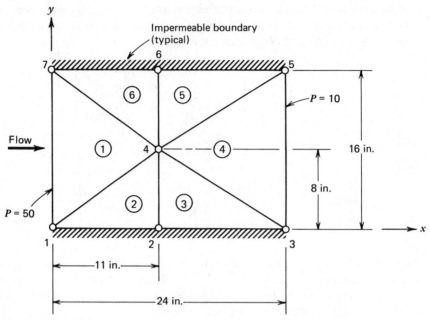

Figure 7.35 Two-dimensional porous-media flow system, Example 7.11.

equation becomes

$$\begin{bmatrix} 1. & 0 & 0 & 0 & 0 & 0 & 0 \\ 0 & -91.2 & 0 & 63. & 0 & 0 & 0 \\ 0 & 0 & 1. & 0 & 0 & 0 & 0 \\ 0 & 63. & 0 & -182. & 0 & 63. & 0 \\ 0 & 0 & 0 & 0 & 1. & 0 & 0 \\ 0 & 0 & 0 & 63. & 0 & -91.2 & 0 \\ 0 & 0 & 0 & 0 & 0 & 0 & 1. \end{bmatrix} \begin{Bmatrix} P_1 \\ P_2 \\ P_3 \\ P_4 \\ P_5 \\ P_6 \\ P_7 \end{Bmatrix} = \begin{Bmatrix} 50 \\ -893 \\ 10 \\ -1786 \\ 10 \\ -893 \\ 50 \end{Bmatrix}$$

Applying an equation-set solver, we get the following pressure values at the internal nodes.

$$P_2 = 31.7$$
$$P_4 = 31.7$$
$$P_6 = 31.7$$

Since we are also interested in simulating flows in this case, we turn around and use the simulated pressures in Equation 7.74 to get the element 1 internal velocity components.

$$\left(\frac{VV}{a}\right)_x = -\frac{3.5}{176}\left[(-8)(50)+(16)(31.7)+(-8)(50)\right]=5.82$$

$$\left(\frac{VV}{a}\right)_y = -\frac{3.5}{176}\left[(-11)(50)+(0)(31.7)+(11)(50)\right]=0.$$

The volumetric flows through the edges of element 1 are obtained from Equation 7.75.

$$VV_{1-4}=12[(8)(5.82)-(11)(0)]=559$$
$$VV_{4-7}=12[(8)(5.82)-(-11)(0)]=559$$
$$VV_{7-1}=12[(-16)(5.82)-(0)(0)]=-1120$$

Slight inaccuracies exist in the simulated values because of roundoff in the hand computations. Nevertheless, the flow values compare favorably to the exact values obtained analytically.

$$\left(\frac{VV}{a}\right)_x = 5.83333$$

and

$$VV_{7-1}=-1119.97$$

REFERENCES

1. John Robinson, *Integrated Theory of Finite Element Methods*, John Wiley & Sons, London, 1973.

2. Robert D. Cook, *Concepts and Applications of Finite Element Analysis*, 2nd ed., John Wiley & Sons, New York, 1981.

3. Kenneth H. Huebner and Earl A. Thornton, *The Finite Element Method for Engineers*, 2nd ed., John Wiley & Sons, New York, 1982.

4. C. S. Desai, *Elementary Finite Element Method*, Prentice-Hall, Inc., Englewood Cliffs, New Jersey, 1979.

5. C. S. Desai and John F. Abel, *Introduction to the Finite Element Method*, Van Nostrand, New York, 1972.

6. Klaus-Jürgen Bathe, *Finite Element Procedures in Engineering Analysis*, Prentice-Hall, Inc., Englewood Cliffs, New Jersey, 1982.

7. O. C. Zienkiewicz and K. Morgan, *Finite Elements and Approximation*, John Wiley & Sons, New York, 1983.

8. R. H. Gallagher, *Finite Element Analysis Fundamentals*, Prentice-Hall, Inc., New Jersey, 1975.

9. L. J. Segerlind, *Applied Finite Element Analysis*, John Wiley & Sons, New York, 1976.

10. S. S. Przemieniecki, *Theory of Matrix Structural Analysis*, McGraw-Hill Book Company, New York, 1968.

11. D. H. Norrie and G. DeVries, *An Introduction to Finite Element Analysis*, Academic Press, Inc., New York, 1978.

12. M. J. Turner, R. W. Clough, H. C. Martin, and L. J. Topp, "Stiffness and Deflection Analysis of Complex Structures," *J. Aerospace Science*, Vol. 23, No. 9, pp. 805–823, 1956.

13. A. Jennings, *Matrix Computation for Engineers and Scientists*, John Wiley & Sons, New York, 1977.

14. A. Mironer, *Engineering Fluid Mechanics*, McGraw-Hill Book Company, New York, 1979.

15. Glen E. Myers, *Analytical Methods in Conduction Heat Transfer*, McGraw-Hill Book Company, New York, 1971.

PROBLEMS

7.1 Show that the inverse of the stiffness matrix, $[k]^{-1}$, derived in Example 7.1 is singular; that is, its determinant is zero. How would the force-displacement equations have to be posed to give a stiffness matrix whose inverse would not be singular? (*Hints:* What kind of boundary conditions might be specified? Why is $f_2 = -f_1$?)

7.2 Referring to Example 7.2, demonstrate how the global stiffness matrix can be reduced from a 3×3 matrix to a 2×2 matrix by applying boundary condition $\Delta_1 = 0$.

7.3 Using the final global matrix equation given in Example 7.2, solve for the displacements Δ_2 and Δ_3. Compute the f_1 (which represents a reaction force in this case).

7.4 Referring to Example 7.3, use the final global matrix equation and solve for the displacements Δ_1, Δ_2, and Δ_3.

7.5 Using the element matrix equations, derive the initial (before conditions are applied) global matrix equation for the system u and v displacements described in Example 7.4. Note that the desired global equation is given.

7.6 Referring to Example 7.4, compute the forces P_{1x}^2, P_{1y}^2, and so on, in the member 2 of the truss system, using the appropriate matrix relationships.

7.7 Using Equations 7.15 and 7.16, and the following definitions,

$$\xi = \frac{x}{l}$$

$$N_1 = 1 - 3\xi^2 + 2\xi^3$$

$$N_2 = x(1 - \xi)^2$$

$$N_3 = 3\xi^2 - 2\xi^3$$
$$N_4 = -x(\xi - \xi^2)$$

show that the displacement, w, can always be expressed as Equation 7.17.

7.8 Using Equations 7.19 and 7.20, demonstrate that a curvature matrix equation can be written as shown in Equation 7.21.

7.9 Using Equations 7.21, 7.22, and 7.23, demonstrate that the beam system element stiffness matrix can be written as shown in Equation 7.25.

7.10 Show that Equations 7.28 through 7.31 can be used to get Equations 7.33 and 7.34.

7.11 Demonstrate that the matrix multiplications shown in Equation 7.52 yield the stiffness matrix given as Equation 7.53.

7.12 Demonstrate that the matrix multiplications shown in Equation 7.52, but with $[C]$ replaced by $[C']$ (see Equation 7.54), yield the stiffness matrix given as Equation 7.55.

7.13 Consider the four-element structural system shown in Figure P7.13. Using a finite-element method, simulate the displacements and reaction forces. Compare results with Example 7.6. (*Hint:* It may be desirable to use a computer program for this problem.)

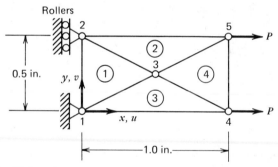

Figure P7.13 Finite-element idealization of flat plate (plane stress condition). Four elements.

7.14 A simple two-dimensional pin-jointed truss is shown in Figure P7.14. It is loaded by a single vertical force of 1000, as shown. The cross-sectional areas of the bars making up the system are 0.5 and the elastic modulus is 30×10^6. Using 6 elements and four nodes (the joints), simulate the joint displacements and forces in the members by a finite-element method. Assume consistent units throughout.

7.15 A stepped cross-sectional bar is loaded by two concentrated axial forces, as shown in Figure P7.15. Using three elements and four nodes, simulate the nodal displacements, and element forces and stresses with a finite-element method. Assume consistent units throughout.

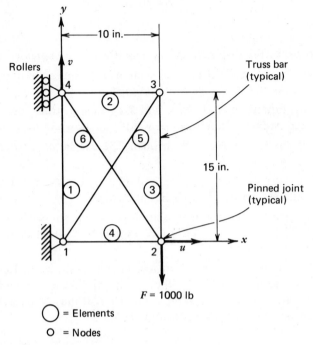

Figure P7.14 Truss system, Problem 7.14.

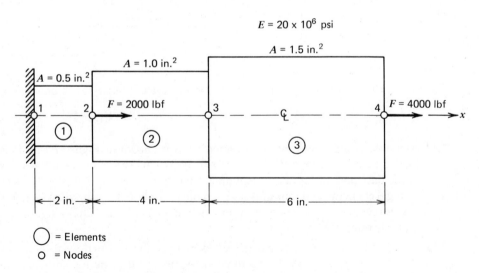

Figure P7.15 Stepped-bar system, Problem 7.15.

7.16 A simple supported beam with a distributed uniform load over part of the span is shown in Figure P7.16. Using the two elements shown, determine the displacement of the center of the beam, end rotations, and stresses in the members by a finite-element method. Assume consistent units throughout and leave simulated values in terms of EI (elastic modulus × area moment of inertia) units.

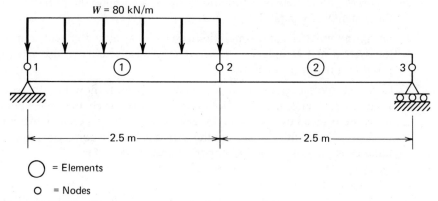

Figure P7.16 Simple beam system, Problem 7.16.

7.17 Simulate the joint displacements and member forces in the two-dimensional pin-jointed truss shown in Figure P7.17, by use of a finite-element method. Assume consistent units throughout and leave the results in terms of P, L, and member properties.

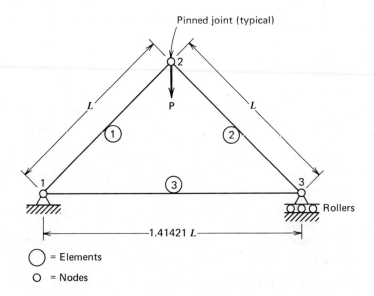

7.18 Set up a finite-element global matrix equation for simulation of nodal temperatures in the system shown in Figure 7.26.

7.19 Set up the element and global matrix equations using thermal resistances (see the universal flow formula, Equation 2.6) for the system in Example 7.8. Let the thermal resistance be $R = l/kA$ for conduction in the solid, and $R = 1/ha$ for the convection transfers at a boundary surface.

7.20 Using the finite element simulation results, determine the heat transfers for the system in Example 7.8.

7.21 Using the finite-element simulation results, determine the heat transfers for the system in Example 7.9. (*Hint:* It may be desirable to use a computer program for this problem.)

7.22 Consider the single element of a one-dimensional flow system shown in Figure P7.22. Assume very slow (laminar) flow such that the volumetric flow rate is given by $VV_1 = [\pi D^4/(128\mu)](P_1 - P_2)/L$, where μ is the dynamic viscosity of the fluid. Develop the element matrix equation required for finite-element simulations. Compare with Equation 7.62.

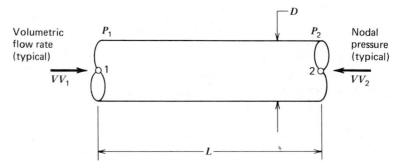

Figure P7.22 One-dimensional fluid-flow element, Problem 7.22.

7.23 The one-dimensional flow network system shown in Figure P7.23 is to be analyzed by the finite-element method. If the element flow resistances are $R_1 = 0.5$, $R_2 = 1.5$, $R_3 = 0.3$, $R_4 = 3.5$, and $R_5 = 2.6$, and the boundary pressures are $P_1 = 100$, and $P_4 = 0.0$, simulate the internal pressures P_2 and P_3. Assume consistent units throughout and compute the flow rate through element 4.

7.24 A two-dimensional porous-media flow system is shown in Figure P7.24. Assume consistent units and simulate the internal nodal pressures P_2, P_4, and P_6, and the volumetric flows through side 1–7 of element 3, and side 3–5 of element 5, by the finite-element method.

The following problems require significant use of a computer.

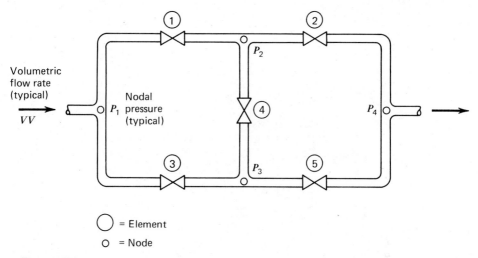

Figure P7.23 One-dimensional flow system, Problem 7.23.

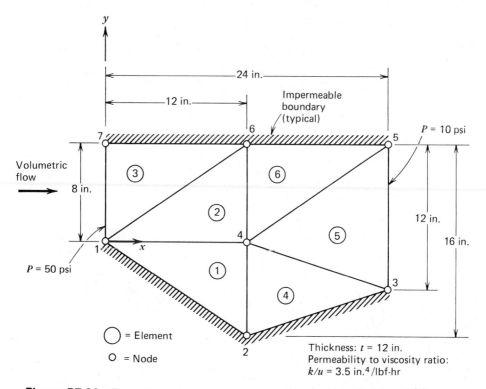

Figure P7.24 Two-dimensional porous-media flow system, Problem 7.24.

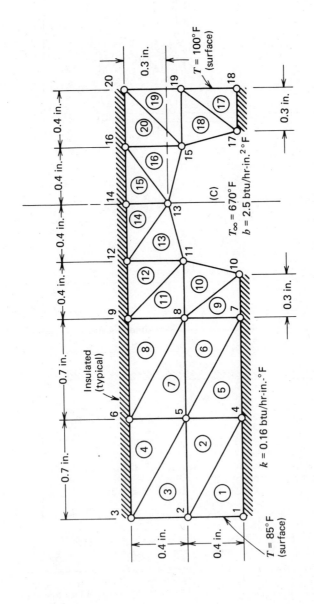

Figure P7.25L Two-dimensional heat transfer system. Problem 7.25L.

◯ = Elements

○ = Nodes

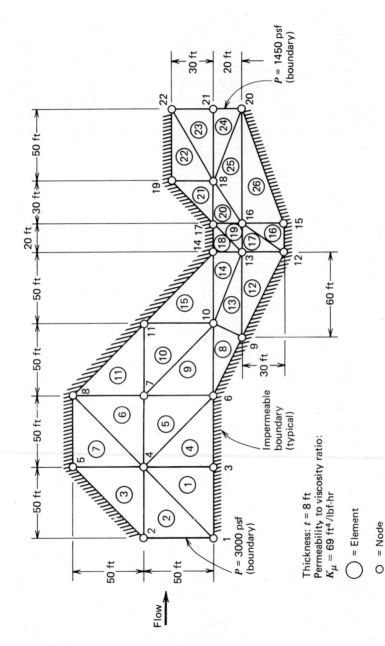

Figure P7.26L Two-dimensional porous-media flow system. Problem 7.26L.

Thickness: t = 8 ft
Permeability to viscosity ratio:
K_μ = 69 ft⁴/lbf-hr

◯ = Element

○ = Node

P = 3000 psf
(boundary)

P = 1450 psf
(boundary)

Impermeable
boundary
(typical)

Flow

7.25(L) Use the finite-element method to simulate the temperatures, and the heat transfers out of the region labeled C, of the system shown in Figure P7.25(L). Assume consistent units.

7.26(L) Use the finite element method to simulate the pressures and system flow for the system shown in Figure P7.26(L). Assume consistent units.

7.27(L) A simply supported beam has the parameters shown in Figure P7.27(L). Develop and demonstrate an appropriate FORTRAN or BASIC computer program for simulating displacements and stresses by a finite-element method, using 50 uniform beam elements. Display the displacements with a trend print-plot. Compare the results with an exact solution and/or with a finite-difference simulation, as specified by the instructor.

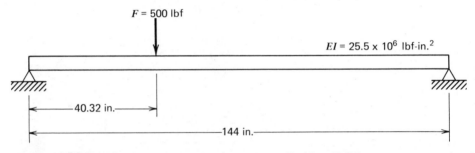

Figure P7.27L Simply supported beam system, Problem 7.27L.

7.28(L) Repeat Example 7.7, except use graded-size elements, with the smallest elements placed near the fixed boundary. Develop a FORTRAN or BASIC program to simulate the stresses in conjunction with the displacements.

8
Closure

THE PROBLEM OF STIFF MODELS

Inevitably, in certain engineering applications of finite-difference simulations, the simulation method will appear to be extremely sensitive to choice of the independent-variable step size and/or to the number of steps. The usual unhappy outcome is a blowup of the simulation numbers. Although the blowup often seems to indicate an instability similar to that which can occur when the explicit method is applied incorrectly to parabolic models, it is, in fact, more likely the result of working with a "stiff" model. If this occurs, it is not worthwhile, as a rule, to try to force the simulation method to work by trial-and-error modification of step sizes. Instead, it is probably more productive to consider use of an intrinsically stable predictor–corrector method, or to start over with a new model.[1] It is beyond the scope of this book to discuss this topic in depth, but it is possible to point out the types of models to watch for in regard to stiffness. If a model is identified as likely to be stiff, a reevaluation of the problem and model setup may be the best approach.

The model displaying stiffness probably has at least two widely different characteristic values for one of the independent variables. Physical situations exhibiting boundary layer behaviors are prime candidates for stiff models. Also transient systems having widely different time constants will always be stiff to some degree. Models with more than one dispersion, or transfer-rate parameter, can behave stiffly if the magnitudes of the parameters differ greatly. Almost all finite-difference simulation methods applied to parabolic models will exhibit some stiffness. Another evaluation of the original system will usually show that astute use of lumped analysis will eliminate the stiffness problem and give a simpler model.

EIGENVALUE PROBLEMS

Many differential-modeling situations in engineering lead to the so-called matrix eigenvalue problem, where a matrix equation of the following form occurs.

$$[[A] - \lambda[I]]\{x\} = 0$$

where $[A]$ is a square matrix of coefficients, $[I]$ is the identity matrix, $\{x\}$ represents the vector of dependent variables, called in this case the eigenvector, and λ is a constant called the eigenvalue. In general, there will be a different eigenvector for each different eigenvalue; but only certain eigenvalues are allowed. How these are determined is a distinct subject in itself. The interested reader is directed to the references[2–4] for additional information and computational details.

VERIFICATION TASKS

A simulation is an artificial description of a system's real behavior. Unfortunately, because simulations are often so cheap to generate, and because a few have been so spectacularly successful in replicating real observed behavior, it has become an accepted practice in many settings to accept a simulation as the real thing, without questions. Generally speaking, such a practice is very dangerous for the engineer analyst. Just as with any kind of computation, a verification should be part of any modeling simulation project.

Two separate efforts always exist in verification. One effort has to be directed at the model. The other must be directed at the simulation method. The former has to be associated at the end with comparisons with prototype systems. Typically, test engineering (laboratory and/or field testing) is involved. It is not uncommon to have to set up a new model or make major revisions to the model at this stage. The importance of this effort should not be underestimated, but within the scope of this book it cannot be addressed properly; so we go on and look at the second effort. Here we assume that the model is perfect and try to verify that particular algorithms and computer programs are doing what is expected of them.

Two approaches are generally followed to verify algorithms and programs. First, a consistency test can be done by choosing a different algorithm to perform the simulation while using the same model. This approach is often used to provide relative-error estimates and is recommended as a relatively easy first step in all verifications. However, it can be pursued to a relatively sophisticated level if so desired. Such a pursuit makes up the "error analysis" part of formal numerical analysis topics. We forego getting into this aspect here and refer the interested reader to the Chapter 6 references and courses in numerical analysis. In the second approach we compare results from a computer simulation directly with results from a classical analytical solution of a carefully selected (benchmark) model. Generally, an attempt is made to select the benchmark model to be of the same order, and with the same number of terms, but locally linearized, as the original model. Obviously, in most cases, accuracy behaviors have to be inferred when going back to the real model. The analyst must also be able to generate the classical solutions quickly. More about this is discussed in the following. Overall to date, this approach seems to have been more accepted in engineering applications than the error-analysis approach.

THE ROLE OF EXACT SOLUTIONS AND TRANSFORM METHODS

A comparison with simulation numbers generated from a classical mathematics solution is an accepted means of verifying a simulation algorithm and/or method.

For this reason some techniques for getting classical solutions are reviewed in Appendix A1. The basic approach here is to get a locally linear version of the actual model identified as a benchmark model. This in itself may be a relatively difficult task. However, using standard-form models and making everything dimensionless usually helps. A classical solution is then generated over a reasonable testing span. The numerical method is then applied to the same model over the same span. Thus the magnitudes of the inaccuracies can be designated clearly, point by point.

Aside from the accuracy testing described above, having an exact solution often gives the engineer-analyst some other subtle advantages. For example, the overall (global) nature of the system's behavior is usually evident from a single formulation. The role of each parameter is probably a clear effect. Singular points and asymptotic behaviors are easy to spot. Optimization schemes, which usually depend on parameter changes, can be applied directly. Much sophisticated analysis material from the area of optimal control theory can be applied if a global solution is available.

For many models with step, or impulse, conditions and/or driving vectors, the Laplace transform method may provide the easiest way to get an exact solution. Further, the transfer functions, which come out of application of the method, make up an intrinsic part of much of so-called input-output methodology from the area of controls. The transform methods are beyond the scope of this book; the interested reader is directed to optimal control literature for more information.

ADDITIONAL SOURCES OF MODELING INFORMATION

Although many sources have been listed already as references in the preceding chapters, additional good references are available. For more on numerical simulation topics, references (5–10) are recommended. A wide variety of model setup procedures, and model descriptions from alternate perspectives can be found in references.[11–18] Much of the finite-element application in engineering is oriented around large POL type packages.[19] Also, it should be apparent to any engineer involved in modeling that many of the new engineering periodicals are a rich source of numerical simulation topics.

REFERENCES

1. J. H. Ferziger, *Numerical Methods for Engineering Application*, John Wiley & Sons, Inc., New York, 1981.

2. R. W. Hornbeck, *Numerical Methods*, Quantum Publishers, Inc., New York, 1975.

3. M. L. James, G. M. Smith, and J. C. Wolford, *Applied Numerical Methods for Digital Computation with Fortran and CSMP*, 2nd ed., IEP, Thomas Y. Crowell Co., Inc., New York, 1977.

4. B. Carnahan, H. A. Luther, and J. O. Wilkes, *Applied Numerical Methods*, John Wiley & Sons, Inc., New York, 1969.

5. S. A. Hovanessian and L. A. Pipes, *Digital Computer Methods in Engineering*, McGraw-Hill Book Company, New York, 1969.

6. R. W. Southworth and S. L. DeLeeuw, *Digital Computation and Numerical Methods*, McGraw-Hill Book Company, New York, 1965.

7. W. Cheney and D. Kincaid, *Numerical Mathematics and Computing*, Brooks/Cole Publishing Co., Monterey, Calif., 1980.

8. S. S. Kuo, *Computer Applications of Numerical Methods*, Addison-Wesley Publishing Co., Reading, Mass., 1972.

9. R. Beckett and J. Hurt, *Numerical Calculations and Algorithms*, McGraw-Hill Book Company, New York, 1967.

10. E. Isaacson and H. B. Keller, *Analysis of Numerical Methods*, John Wiley & Sons, Inc., New York, 1966.

11. W. R. Bennett, *Scientific and Engineering Problem Solving with the Computer*, Prentice-Hall, Inc., Englewood Cliffs, N.J., 1976.

12. C. M. Close and D. K. Frederick, *Modeling and Analysis of Dynamic Systems*, Houghton Mifflin Company, Boston, Mass., 1979.

13. C. L. Dym and E. S. Ivey, *Principles of Mathematical Modeling*, Academic Press, Inc., New York, 1980.

14. W. J. Gajda and W. E. Biles, *Engineering Modeling and Computation*, Houghton Mifflin Company, Boston, Mass., 1978.

15. R. Haberman, *Mathematical Models-Mechanical Vibrations, Population Dynamics, and Traffic Flow*, Prentice-Hall, Inc., Englewood Cliffs, N.J., 1977.

16. C. R. Mischke, *Mathematical Model Building (An Introduction to Engineering)*, Iowa State University Press, Ames, Iowa, 1980.

17. J. A. Adams and D. F. Rogers, *Computer-Aided Heat Transfer Analysis*, McGraw-Hill Book Company, New York, 1973.

18. N. Roberts, D. Andersen, R. Deal, M. Garet, and W. Shaffer, *Introduction to Computer Simulation, A Systems Dynamics Modeling Approach*, Addison-Wesley Publishing Co., Reading, Mass., 1983.

19. K. J. Bathe, E. L. Wilson, and F. E. Peterson, *SAP IV—A Structural Analysis Program for Static and Dynamic Response of Linear Systems*, Report EERC 73-11, College of Engineering, University of California, Berkeley, Calif., 1974.

Appendix A1
Some Classical Solution Methods

A1.1 SIMPLE VARIABLES-SEPARABLE EQUATIONS

1. If the differential model is of the form given by

$$\frac{dy}{dx} = f(x); \qquad y(x_0) = y_0$$

then the solution can be generated by separating all independent variable terms to the right-hand side and integrating:

$$\int dy = \int f(x)\, dx$$

or

$$y = \int f(x)\, dx + C_i$$

where C_i is evaluated by substituting the y_0 condition. If $f(x)$ is nonlinear in x, then tables might have to be used to integrate. The use of tables is illustrated in Example A1.1.

2. If the differential model is of the form given by

$$\frac{dy}{dx} - g(y) = 0; \qquad y(x_0) = y_0$$

then the solution can be generated by separating all the dependent-variable terms to the left-hand side and the independent-variable terms to the right-hand side, and integrating.

$$\int \frac{dy}{g(y)} = \int dx$$

or

$$\int \frac{dy}{g(y)} = x + C_i$$

where C_i is evaluated by substituting the y_0 condition. If $g(y)$ is nonlinear in y, then tables might have to be used to integrate. The use of tables is illustrated in Example A1.2.

3. If the differential model is of the form given by

$$\frac{d^n y}{dx^n} = f(x); \qquad y(x_0) = y_0$$

$$\vdots \qquad \vdots$$

$$n \text{ conditions}$$

then the solution can be generated by reposing the model as

$$\frac{d}{dx}(Z) = f(x)$$

where

$$Z = \frac{d^{n-1} y}{dx^{n-1}}$$

and repetitively applying the method of item 1. Constants of integration are evaluated by substituting the original y conditions. Use is illustrated in Example A1.3.

EXAMPLE A1.1

Solution of Type-1 Variables-Separable Model

Model

$$\frac{du}{dt} = 6 \sin 0.5t; \quad u(0) = u_0$$

Separate Variables

$$du = (6 \sin 0.5t)\, dt$$

Integrate

$$u = \int (6 \sin 0.5t)\, dt + C_i$$

$$u = -12 \cos 0.5t + C_i$$

Substitute Condition to Evaluate C_i

$$u_0 = -12(1) + C_i \qquad \text{and} \qquad C_i = u_0 + 12$$

Solution Becomes

$$u = 12(1 - \cos 0.5t) + u_0$$

EXAMPLE A1.2

Solution of Type-2 Variables-Separable Model

Model

$$\frac{du}{dt} - 6e^{2u} = 0; \qquad u(0) = u_0$$

Separate variables

$$\frac{du}{e^{2u}} = 6\,dt$$

Integrate

$$\int e^{-2u}\,du = 6 \int dt + C_i$$

$$e^{-2u} = -12t + C_i$$

$$u = -0.5 \ln(-12t + C_i)$$

Substitute condition to evaluate C_i

$$C_i = e^{-2u_0}$$

Solution becomes

$$u = \ln\left(\frac{1}{e^{-2u_0} - 12t}\right)^{0.5}$$

EXAMPLE A1.3

Solution of Type-3 Variables-Separable Model

Model

$$\frac{d^3u}{dt^3} = 3t^2 + 1; \qquad \begin{aligned} u(0) &= 0 \\ u'(0) &= 0 \\ u''(0) &= 1 \end{aligned}$$

Substitute Z **and apply Type-1 procedure to get**

$$\frac{d^2u}{dt^2} = Z = t^3 + t + C_1$$

Substitute Z and apply Type-1 procedure again to get

$$\frac{du}{dt} = Z = \frac{1}{4}t^4 + \frac{1}{2}t^2 + C_1 t + C_2$$

The model is now a Type-1 and can be solved directly to give

$$u = (\tfrac{1}{20})t^5 + (\tfrac{1}{6})t^3 + (\tfrac{1}{2})C_1 t^2 + C_2 t + C_3$$

Substituting conditions to evaluate the constants gives

$$C_1 = 1; \qquad C_2 = 0; \qquad C_3 = 0$$

Solution becomes

$$u = (\tfrac{1}{20})t^5 + (\tfrac{1}{6})t^3 + (\tfrac{1}{2})t^2$$

A1.2 AN OPERATOR METHOD FOR SELECTED HOMOGENEOUS EQUATIONS

1. This method is good for linear, ordinary, homogeneous models with constant coefficients. The differential model is assumed to be of the following form, where the a's are constants.

$$\frac{d^n y}{dx^n} + a_1 \frac{d^{n-1} y}{dx^{n-1}} + \cdots a_n y = 0; \qquad n \text{ conditions}$$

The symbol D is used in place of the derivatives as a first step, which gives

$$D^n y + a_1 D^{n-1} y + \cdots a_n y = 0$$

Algebraic properties of the D equation allow us to simplify it to

$$D^n + a_1 D^{n-1} + \cdots a_n = 0$$

This equation is known as the "characteristic equation" (we abbreviate it CE). Replacing the D operators by r's in the CE gives us

$$r^n + a_1 r^{n-1} + \cdots a_n = 0$$

which is simply a general polynomial in r, having n roots. Solving for the roots, which may be either real or complex, we list them as $r_1, r_2, \ldots, r_n$.

2. If none of the $r_1 \ldots r_n$ roots is the same, then we can write the solution immediately, as follows.

$$y = C_1 e^{r_1 x} + C_2 e^{r_2 x} + \cdots C_n e^{r_n x}$$

The $C_1, C_2, \ldots, C_n$ constants must be evaluated by substituting the original conditions into the above equation (and its derivative equations). This procedure is illustrated in Example A1.4.

3. If any of the $r_1 \ldots r_n$ roots are the same, we must suppress duplication in the solution by multiplying each replicate term by the independent variable, raised to an integer power just large enough to avoid duplication. This is illustrated as follows and in Example A1.5.

$$y = C_1 e^{r_1 x} + C_2 x e^{r_2 x} + C_3 x^2 e^{r_3 x} + \cdots C_n e^{r_n x}$$

where $r_1 = r_2 = r_3$.

4. An alternate form of solution, popular for some engineering work, can be used if two of the r roots are real, equal in magnitude, but opposite in sign.

$$y = C_1 \cosh r_1 x + C_2 \sinh r_2 x + \cdots C_n e^{r_n x}$$

where: $r_1 = -r_2$. Such a solution can be shown to be equivalent to the exponential form solution by using the following identities.

$$e^{bx} = \sinh bx + \cosh bx$$

$$e^{-bx} = \cosh bx - \sinh bx$$

5. Complex roots occur in conjugate pairs, such as $r_1 \pm i r_2$, where $i = \sqrt{-1}$. Typically, the real part of the solution can be separated from the imaginary part as follows.

$$y = C_1 e^{(r_1 + i r_2)x} + C_2 e^{(r_1 - i r_2)x} + \cdots C_n e^{r_n x}$$
$$y = e^{r_1 x}(C_1 e^{i r_2 x} + C_2 e^{-i r_2 x}) + \cdots C_n e^{r_n x}$$

An alternate form of this solution is found in much of engineering.

$$y = e^{r_1 x}(A_1 \cos r_2 x + A_2 \sin r_2 x) + \cdots C_n e^{r_n x}$$

Here the following identities have been used.

$$e^{ibx} = \cos bx + i \sin bx$$

$$e^{-ibx} = \cos bx - i \sin bx$$

Other trigometric identities can be used to give still another equivalent solution form popular in some engineering disciplines.

$$y = e^{r_1 x} B_1 \sin (r_2 x + B_2) + \cdots C_n e^{r_n x}$$

It is observed that the constant B_2 can now be viewed as a phase shift in cyclic behaviors.

6. Occasionally, only imaginary roots will occur. The foregoing procedures are still valid except that r_1 is set to zero. Aspects of this are illustrated in Example A1.6.

EXAMPLE A1.4

Solution for a Homogeneous Model Having Real, Nonrepeated, Characteristic Roots

Model

$$\frac{d^2y}{dx^2} - 4y = 0; \qquad \begin{array}{l} y(0) = 0 \\ y'(0) = 1 \end{array}$$

Characteristic equation

$$D^2 - 4 = 0$$
$$r^2 - 4 = 0$$

Roots (by inspection)

$$r_1 = 2; \qquad r_2 = -2$$

General solution

$$y = C_1 e^{2x} + C_2 e^{-2x}$$

Substitution of conditions into y and y'

$$0 = C_1(1) + C_2(1)$$
$$1 = 2C_1(1) - 2C_2(1)$$

Integration constants

$$C_1 = 0.25; \qquad C_2 = -0.25$$

Solution

$$y = 0.25(e^{2x} - e^{-2x})$$

EXAMPLE A1.5

Solution for a Homogeneous Model Having Real, Repeated, Characteristic Roots

Model

$$\frac{d^2y}{dx^2} - 4\frac{dy}{dx} + 4y = 0; \qquad \begin{array}{l} y(0) = 0 \\ y'(0) = 1 \end{array}$$

Characteristic equation

$$D^2 - 4D + 4 = 0$$
$$r^2 - 4r + 4 = 0$$

Roots (by factoring and inspection)

$$(r-2)(r-2) = 0$$

or

$$r_1 = 2; \quad r_2 = 2 \quad \text{(repeated)}$$

General solution

$$y = C_1 e^{2x} + C_2 x e^{2x}$$

Substitution of conditions into y and y'

$$0 = C_1 + (0)$$
$$1 = 2C_1 + 2C_2(0)(1) + C_2$$

Integration constants

$$C_1 = 0; \quad C_2 = 1$$

Solution

$$y = x e^{2x}$$

EXAMPLE A1.6

Solution for a Homogeneous Model Having Complex, Nonrepeated, Characteristic Roots

Model

$$\frac{d^2 y}{dx^2} - \frac{dy}{dx} + 6.5y = 0; \qquad \begin{aligned} y(0) &= 0 \\ y'(0) &= 1 \end{aligned}$$

Characteristic equation

$$D^2 - D + 6.5 = 0$$
$$r^2 - r + 6.5 = 0$$

Roots (by applying the quadratic equation)

$$R_1 = 0.5 + 2.5i; \quad R_2 = 0.5 - 2.5i$$

or

$$r_1 = 0.5; \qquad r_2 = 2.5$$

General solution

$$y = C_1 e^{(0.5+2.5i)x} + C_2 e^{(0.5-2.5i)x}$$

$$y = e^{0.5x}(C_1 e^{2.5ix} + C_2 e^{-2.5ix})$$

$$y = e^{0.5x}(A_1 \cos 2.5x + A_2 \sin 2.5x)$$

Substitution of conditions into y and y'

$$0 = (1)[A_1(1) + C_2(0)]$$

$$1 = 0.5[A_1(1) + A_2(0)] + (1)[-2.5A_1(0) + 2.5A_2(1)]$$

Integration constants

$$A_1 = 0; \qquad A_2 = 0.4$$

Solution

$$y = 0.4e^{0.5x} \sin 2.5x$$

A1.3 THE UNDETERMINED COEFFICIENT METHOD

1. This method (abbreviated UC herein) is good for linear, ordinary, nonhomogeneous models with constant coefficients and certain types of driving functions. The model is assumed to be of the following form, where the a's are constants and $f(x)$ is the driving function.

$$\frac{d^n y}{dx^n} + a_1 \frac{d^{n-1}y}{dx^{n-1}} + \cdots a_n y = f(x); \qquad n \text{ conditions}$$

The solution consists of two parts: $y = y_c + y_p$. The first part, y_c, is called the "complementary solution" and is simply the homogeneous solution obtained by setting $f(x)$ equal to zero. The second part, y_p, is called the "particular solution" and accounts for the driving function effects.

2. In this method solutions usually can be obtained if $f(x)$ is a polynomial in x, a $\sin cx$ or $\cos cx$ expression, an e^{cx} expression, or combinations of the above. After the general complementary solution has been obtained, a particular solution is assumed and patterned after the general form of $f(x)$, as follows.
   (a) If $f(x) = A_0 x^n + A_1 x^{n-1} + \cdots A_n$, then assume that

   $$y_p = B_0 x^n + \cdots B_n$$

   (b) If $f(x) = C_1 \sin A_1 x$, or $C_2 \cos A_1 x$, or combinations of these, then assume that

   $$y_p = B_0 \cos A_1 x + B_1 \sin A_1 x$$

(c) If $f(x) = A_0 e^{A_1 x}$, then assume that

$$y_p = B_0 e^{A_1 x}$$

(d) If $f(x) = $ sum or product combinations of the above, then assume that

$$y_p = \text{combinations of the same form as } f(x)$$

In the above, the B's are the undetermined coefficients; the A's or C's are constants.

3. After y_p has been assumed, it must be inspected to make sure it does not contain any terms that duplicate terms already in existence in the y_c solution. If a replicate term is discovered, it can be multiplied by the independent variable, raised to an integer power just large enough to avoid duplication.

4. The assumed particular solution, y_p, is then differentiated n times and substituted (along with its derivatives) back into the original model. Coefficients of like terms are equated and the undetermined coefficients ($B_0, B_1, \ldots$) calculated.

5. The y_c and y_p solutions are combined into the complete general solution. The original conditions are then substituted to find the integration constants. Specifics of use are illustrated in Examples A1.7–A1.9.

EXAMPLE A1.7

UC Solution for a Polynomial Driving Function Model

Model
$$\frac{d^2 y}{dx^2} - 4y = 6x^2; \qquad \begin{aligned} y(0) &= 0 \\ y'(0) &= 1 \end{aligned}$$

Complementary Solution (from Example A1.4)

$$y_c = C_{1c} e^{2x} + C_{2c} e^{-2x}$$

Assumed Particular Solution

$$y_p = B_0 x^2 + B_1 x + B_2$$

Inspected for Duplicate Terms   None found
Differentiation

$$\frac{dy_p}{dx} = 2B_0 x + B_1; \qquad \frac{d^2 y_p}{dx^2} = 2B_0$$

Substitution

$$(2B_0) - 4(B_0 x^2 + B_1 x + B_2) = 6x^2$$

Equating Coefficients of Like Terms

x^2: $\qquad -4B_0 = 6; \qquad B_0 = -1.5$

x: $\qquad -4B_1 = 0; \qquad B_1 = 0$

Constant: $\quad 2B_0 - 4B_2 = 0; \qquad B_2 = -0.75$

Substituting and Combining

$$y = y_c + y_p = C_1 e^{2x} + C_2 e^{-2x} - 1.5x^2 - 0.75$$

Substituting conditions into y and y'

$$0 = C_1 + C_2 - 0.75$$
$$1 = 2C_1 - 2C_2$$

Integration constants

$$C_1 = 0.625; \qquad C_2 = 0.125$$

Solution

$$y = 0.625e^{2x} + 0.125e^{-2x} - 1.5x^2 - 0.75$$

EXAMPLE A1.8

UC Solution for a Trigometric Driving Function Model

Model

$$\frac{d^2y}{dx^2} - 4y = 6 \sin 2x; \qquad \begin{aligned} y(0) &= 0 \\ y'(0) &= 1 \end{aligned}$$

Complementary solution (from Example A1.4)

$$y_c = C_{1c} e^{2x} + C_{2c} e^{-2x}$$

Assumed particular solution

$$y_p = B_0 \cos 2x + B_1 \sin 2x$$

Inspected for duplicate terms   None found

Differentiation

$$\frac{dy_p}{dx} = -2B_0 \sin 2x + 2B_1 \cos 2x$$

$$\frac{d^2y_p}{dx^2} = -4B_0 \cos 2x - 4B_1 \sin 2x$$

Substitutions

$$-4B_0 \cos 2x - 4B_1 \sin 2x - 4B_0 \cos 2x + B_1 \sin 2x = 6 \sin 2x$$

Equating Coefficients of like terms

$$\cos 2x: \quad (-8B_0) = 0; \quad B_0 = 0$$
$$\sin 2x: \quad (-8B_1) = 6; \quad B_1 = -0.75$$

General solution

$$y = y_c + y_p = C_1 e^{2x} + C_2 e^{-2x} - 0.75 \sin 2x$$

Substituting conditions into y and y'

$$0 = C_1 + C_2 - 0; \qquad 1 = 2C_1 - 2C_2 - 1.5$$

Integration constants

$$C_1 = 0.625; \qquad C_2 = -0.625$$

Solution

$$y = 0.625 e^{2x} - 0.625 e^{-2x} - 0.75 \sin 2x$$

EXAMPLE A1.9

UC Solution for an Exponential Driving Function Model

Model

$$\frac{d^2 y}{dx^2} - 4y = 10 e^{2x}; \qquad \begin{aligned} y(0) &= 0 \\ y'(0) &= 1 \end{aligned}$$

Complementary solution (from Example A1.4)

$$y_c = C_{1c} e^{2x} + C_{2c} e^{-2x}$$

Assumed particular solution

$$y_p = B_0 e^{2x}$$

Inspected for duplicate terms: One found
Applying independent variable multiplier gives

$$y_p = B_0 x e^{2x}; \qquad \text{reinspect, no duplicate terms}$$

Differentiation

$$\frac{dy_p}{dx} = 2B_0 x e^{2x} + B_0 e^{2x}$$

$$\frac{d^2 y_p}{dx^2} = 2B_0 e^{2x} + 2B_0 e^{2x}(2x+1)$$

Substitution

$$[2B_0 e^{2x} + 2B_0 e^{2x}(2x+1)] - 4B_0 x e^{2x}) = 10e^{2x}$$

Equating coefficients of like terms

$$e^{2x}: \qquad 4B_0 = 10; \qquad B_0 = 2.5$$

General solution

$$y = y_c + y_p = C_1 e^{2x} + C_2 e^{-2x} + 2.5 x e^{2x}$$

Substituting conditions into y and y'

$$0 = C_1 + C_2$$
$$1 = 2C_1 - 2C_2 + 2.5$$

Integration constants

$$C_1 = -0.375; \qquad C_2 = 0.375$$

Solution

$$y = 0.375(e^{-2x} - e^{2x}) + 2.5 x e^{2x}$$

A1.4 THE VARIATION-OF-PARAMETERS METHOD

1. This method (abbreviated VP herein) is good for linear, ordinary, nonhomogeneous differential models with constant coefficients. The model is assumed to be of the following form, where the a's are constants.

$$\frac{d^n y}{dx^n} + a_1 \frac{d^{n-1} y}{dx^{n-1}} + \cdots a_n y = f(x); \qquad n \text{ conditions}$$

The solution consists of two parts: $y = y_c + y_p$. The first part, y_c, is called the "complementary solution," and is simply the homogeneous solution obtained by setting $f(x)$ equal to zero. The second part, y_p, is called the "particular solution," and accounts for the nonhomogeneous (driving) function, $f(x)$.

In this method the form of the particular solution, y_p, is chosen to be exactly the same as the previously determined general complementary solution, y_c, except that the integration constants are replaced by parameters of the independent variable. This particular solution is then differentiated n times but in a rather unique fashion. After *each* differentiation, all terms containing the derivatives of the parameters are grouped together and set to zero, except for the nth differentiation. For the nth differentiation, the derivative terms are set equal to the nonhomogeneous (driving) function, $f(x)$. The resulting set of linear algebraic equations has the parameter derivatives as the unknowns. Expressions for the derivatives are obtained by solving the set; the parameters are obtained by separating variables and integrating. Substitution of the parameter expressions back into the particular solution and combining with the complementary solution then gives the complete general solution. It should be observed that the constants of integration from the parameter integration can be merged with the constants of integration in the complementary solution at this stage. Finally, the constants of integration are evaluated by substituting the original conditions into the complete general solution (and its derivative equations).

2. A generalization of the procedure follows.

Complementary solution

$$y_c = C_1 e^{r_1 x} + C_2 e^{r_2 x} + \cdots C_{nc} e^{r_n x}$$

Particular solution

$$y_p = u_1 e^{r_1 x} + u_2 e^{r_2 x} + \cdots u_n e^{r_n x}$$

where
$$u_1 = u_1(x)$$
$$u_2 = u_2(x) \qquad \text{the parameters}$$
$$u_n = u_n(x)$$

Differentiation

$$\frac{dy_p}{dx} = u_1 r_1 e^{r_1 x} + u_2 r_2 e^{r_2 x} + \cdots u_n r_n e^{r_n x} + \underbrace{e^{r_1 x} \frac{du_1}{dx} + e^{r_2 x} \frac{du_2}{dx} + \cdots e^{r_n x} \frac{du_n}{dx}}_{= \text{zero}}$$

$$\frac{d^2 y_p}{dx^2} = u_1 r_1^2 e^{r_1 x} + u_2 r_2^2 e^{r_2 x} + \cdots u_n r_n^2 e^{r_n x} + \underbrace{r_1 e^{r_1 x} \frac{du_1}{dx} + r_2 e^{r_2 x} \frac{du_2}{dx} + \cdots r_n e^{r_n x} \frac{du_n}{dx}}_{= \text{zero}}$$

$$\frac{d^n y_p}{dx^n} = u_1 r_1^n e^{r_1 x} + u_2 r_2^n e^{r_2 x} + \cdots u_n r_n^n e^{r_n x}$$

$$\underbrace{+ r_1^{n-1} e^{r_1 x} \frac{du_1}{dx} + r_2^{n-1} e^{r_2 x} \frac{du_2}{dx} + \cdots r_n^{n-1} e^{r_n x} \frac{du_n}{dx}}_{= f(x)}$$

Algebraic set

$$(e^{r_1 x}) \frac{du_1}{dx} + (e^{r_2 x}) \frac{du_2}{dx} + \cdots (e^{r_n x}) \frac{du_n}{dx} = 0$$

$$(r_1 e^{r_1 x}) \frac{du_1}{dx} + (r_2 e^{r_2 x}) \frac{du_2}{dx} + \cdots (r_n e^{r_n x}) \frac{du_n}{dx} = 0$$

$$(r_1^{n-1} e^{r_1 x}) \frac{du_1}{dx} + (r_2^{n-1} e^{r_2 x}) \frac{du_2}{dx} + \cdots (r_n^{n-1} e^{r_n x}) \frac{du_n}{dx} = f(x)$$

Solution of set

$$\frac{du_1}{dx} = g_1(x) f(x)$$

$$\frac{du_2}{dx} = g_2(x) f(x)$$

$$\frac{du_n}{dx} = g_n(x) f(x)$$

Integration

$$u_1 = \int g_1(x) f(x)\, dx + C_{1P}$$

$$u_2 = \int g_2(x) f(x)\, dx + C_{2P}$$

$$u_n = \int g_n(x) f(x)\, dx + C_{nP}$$

Substitution into y_p

$$y_p = e^{r_1 x} \int g_1(x) f(x)\, dx + e^{r_2 x} \int g_2(x) f(x)\, dx + \cdots e^{r_n x} \int g_n(x) f(x)\, dx$$
$$+ C_{1P} e^{r_1 x} + C_{2P} e^{r_2 x} + \cdots C_{nP} e^{r_n x}$$

Combining to give complete general solution

$$y = y_c + y_p = C_1 e^{r_1 x} + C_2 e^{r_2 x} + \cdots C_n e^{r_n x} + e^{r_1 x} \int g_1(x) f(x)\, dx$$
$$+ e^{r_2 x} \int g_2(x) f(x)\, dx + \cdots e^{r_n x} \int g_n(x) f(x)\, dx$$

where

$$C_1 = C_{1C} + C_{1P}$$
$$C_2 = C_{2C} + C_{2P}$$
$$C_n = C_{nC} + C_{nP}$$

Here $C_1, C_2, \ldots, C_n$ are evaluated by substituting the original n conditions into y and its derivatives.

The use of the method for some specific problems is illustrated in Examples A1.10–A1.12.

EXAMPLE A1.10

VP Solution for an Exponential Driving Function Model

Model

$$\frac{d^2 y}{dx^2} - 4y = e^{2x}; \qquad \text{two conditions}$$

Complementary solution (from Example A1.4)

$$y_c = C_1 e^{2x} + C_2 e^{-2x}$$

Particular solution

$$y_p = u_1 e^{2x} + u_2 e^{-2x}$$

Differentiation

$$\frac{dy_p}{dx} = 2u_1 e^{2x} - 2u_2 e^{-2x} + \underbrace{e^{2x} \frac{du_1}{dx} + e^{-2x} \frac{du_2}{dx}}_{=0}$$

$$\frac{d^2 y_p}{dx^2} = 4u_1 e^{2x} + 4u_2 e^{-2x} + \underbrace{2e^{2x} \frac{du_1}{dx} - 2e^{-2x} \frac{du_2}{dy}}_{=e^{2x}}$$

Solution of algebraic set

$$\frac{du_1}{dx} = \frac{1}{4}; \qquad \frac{du_2}{dx} = -\frac{e^{4x}}{4}$$

Integration

$$u_1 = \frac{x}{4} + C_{1p}$$

$$u_2 = -\frac{1}{16} e^{4x} + C_{2p}$$

Substitution

$$y_p = \frac{x}{4} e^{2x} - \frac{1}{16} e^{2x} + C_{1p} e^{2x} + C_{2p} e^{-2x}$$

Combination

$$y = y_c + y_p = C_1 e^{2x} + C_2 e^{-2x} + \frac{1}{16} e^{2x}(4x - 1)$$

where

$$C_1 = C_{1c} + C_{1p}$$
$$C_2 = C_{2c} + C_{2p}$$

General solution (integration constants not evaluated)

$$y = \frac{e^{2x}(4x - 1)}{16} + C_1 e^{2x} + C_2 e^{-2x}$$

EXAMPLE A1.11

VP Solution For a Polynomial Driving Function Model

Model

$$\frac{d^2 y}{dx^2} - 4y = 6x^2; \qquad y(0) = 0; \qquad y'(0) = 1$$

Complementary solution (from Example A1.4)

$$y_c = C_{1c} e^{2x} + C_{2c} e^{-2x}$$

Particular solution

$$yp = u_1 e^{2x} + u_2 e^{-2x}$$

Differentiation

$$\frac{dy_p}{dx} = 2u_1 e^{2x} - 2u_2 e^{-2x} + \underbrace{e^{2x} \frac{du_1}{dx} + e^{-2x} \frac{du_2}{dx}}_{= 0}$$

$$\frac{d^2 y_p}{dx^2} = 4u_1 e^{2x} + 4u_2 e^{-2x} + \underbrace{2e^{2x} \frac{du_1}{dx} - 2e^{-2x} \frac{du_2}{dx}}_{= 6x^2}$$

Solution of algebraic set

$$\frac{du_1}{dx} = 1.5 e^{-2x} x^2; \qquad \frac{du_2}{dx} = -1.5 x^2 e^{2x}$$

Integration

$$u_1 = \int 1.5x^2 e^{-2x}\, dx + C_{1p}; \qquad u_2 = \int -1.5x^2 e^{2x}\, dx + C_{2p}$$

Using tables:

$$u_1 = 1.5e^{-2x}\frac{x^2}{-2} - \frac{2x}{4} - \frac{2}{8} + C_{1p}; \qquad u_2 = -1.5e^{2x}\frac{x^2}{2} - \frac{2x}{4} + \frac{2}{8} + C_{2p}$$

Substitution

$$y_p = 1.5\left(\frac{-x^2}{2} - \frac{x}{2} - \frac{1}{4}\right) - 1.5\left(\frac{x^2}{2} - \frac{x}{2} + \frac{1}{4}\right) + C_{1p}e^{2x} + C_{2p}e^{-2x}$$

or

$$y_p = -1.5x^2 - 0.75 + C_{1p}e^{2x} + C_{2p}e^{-2x}$$

Combination

$$y = y_c + y_p = C_1 e^{2x} + C_2 e^{-2x} - 1.5x^2 - 0.75$$

Integration constants   (see Example A1.7)

$$C_1 = 0.625, \qquad C_2 = 0.125$$

Solution

$$y = 0.625e^{2x} + 0.125e^{-2x} - 1.5x^2 - 0.75$$

EXAMPLE A1.12

VP Solution for a Polynomial Driving Function Model

Model

$$\frac{d^2y}{dx^2} - 2\frac{dy}{dx} + y = \frac{e^x}{x}; \qquad \text{two conditions}$$

Complementary solution

$$y_c = C_{1c}e^x + C_{2c}xe^x$$

Particular solution

$$y_p = u_1 e^x + u_2 xe^x$$

Differentiation

$$\frac{dy_p}{dx} = u_1 e^x + u_2(xe^x + e^x) + \underbrace{e^x\frac{du_1}{dx} + xe^x\frac{du_2}{dx}}_{=0}$$

$$\frac{d^2 y_p}{dx^2} = u_1 e^x + u_2(xe^x + 2e^x) + e^x \frac{du_1}{dx} + (xe^x + e^x) \frac{du_2}{dx}$$

$$= \frac{e^x}{x}$$

Solution of algebraic set

$$\frac{du_1}{dx} = -1; \qquad \frac{du_2}{dx} = \frac{1}{x}$$

Integration

$$u_1 = \int (-1) \, dx + C_{1p}; \qquad u_2 = \int \frac{1}{x} \, dx + C_{2p}$$

$$u_1 = -x + C_{1p}; \qquad u_2 = \ln x + C_{2p}$$

Substitution

$$y_p = -xe^x + x \ln(x)e^x + C_{1p}e^x + C_{2p}xe^x$$

Combination and general solution

$$y = y_c + y_p = C_1 e^x + C_2 xe^x + xe^x \ln x$$

where

$$C_2 = C_{2c} + C_{2p} - 1$$
$$C_1 = C_{1c} + C_{1p}$$

Appendix A2

Some General-Purpose
Numerical Methods

A2.1 DETERMINING ROOTS OF EQUATIONS

Any equation involving a single variable, x, can be expressed as $f(x) = 0$. Values of x that satisfy the foregoing are the roots sought. The first step is to get a general idea of how the function $f(x)$ behaves with respect to x. A rough graph of $f(x)$, the dependent variable, versus x, the independent variable, is recommended. Only a few values of x should be used. Wherever the $f(x)$ curve appears to cross the x axis is likely to be a real root location. If the $f(x)$ curve seems to only touch the x axis, and not cross it, then the tangent point probably represents replicate real roots. If the $f(x)$ curve does not cross the x axis but shows a minimum value, then the minimum point probably represents a complex root. It will be assumed here that if the appropriate real roots can be found, division, factoring, and the quadratic formula can be applied to find any complex roots. An approximate value of the sought-after real root is guessed at initially. This "starting" value is obtained from the rough graph and should be guessed on the low side. Then a computer program based on one of the following methods can be used to get the refined value.

A2.1.1 The Trial-and-Error Method

Among the simplest methods to use, many variations exist, for example, the "search" method, the "interval-halving" method, and so on. To start it, in addition to a starting value, a starting increment, Δx, an increment-reduction factor, DD, and a significant-digit test parameter, ε, must be chosen. The starting increment should be chosen reasonably large, based on the rough graph and the value chosen for DD. The increment is decreased by dividing it by DD every time the x axis is crossed. The magnitude of the increment is also compared with ε. If the increment becomes smaller than ε, then the last value of x is accepted as a root.

After the starting values and parameters have been chosen, values of $f(x)$ are computed for successive values of x, incremented by Δx, until $f(x)$ changes sign. Then the increment size is compared with ε. If it is smaller, the last x value is taken

313

as the root. If larger, the x value is backed up by one Δx, the Δx decreased by dividing it by DD, and the whole process repeated as many times as necessary. If an additional root is sought, new starting values have to be chosen, and the foregoing repeated.

An example program, coded in BASIC, is given as Figure A2.1.

A2.1.2 The Linear Interpolation Method

This method, once started, converges on the root much faster than the trial-and-error method. Some variations on this method are known as the "secant" method, the "regula/falsi" method, and the "method of false positions." Here two starting values, spanning the desired root, are required. These are usually obtained by applying the trial-and-error method initially, but they can also be obtained directly from the rough graph. Then linear interpolation between successive points, $f(x_1)$ and $f(x_2)$, is initiated and continued until successive values of the two

```
                          Main Program
010   REM--REF: INTRO. ENG'RNG. MODELING ......., RIEDER/BUSBY
100   REM--REAL ROOT BY TRIAL-AND-ERROR METHOD
110   REM--(INPUT) X0 = STARTER VALUE FROM ROUGH GRAPH
120   REM--(INPUT) DX = STARTER INCREMENT FOR X
130   REM--(INPUT) EP = SIGNIFICANT-DIGIT TEST PARAMETER
140   REM--(INPUT) DD = FACTOR (DIVISOR) FOR DECREASING DX
150   REM-- PLACE F(X) AS FX INTO SUBROUTINE 500
160   CLS: CLEAR10: DEFINT I-N
170   READ X0, DX, EP, DD
180      DATA 0.0, 0.1, 1E-4, 2.
190   X = X0
200   GOSUB 500
210   FF = FX
220   REM-- LOOP ALLOWS 100 TRIALS TO LOCATE ROOT ACCURATELY
230   REM-- IF ROOT NOT FOUND, CHECK FOR COMPLEX OR REPEATED ROOTS
240   FOR I = 1 TO 100
250      GOSUB 500
260      PRINT "X=";X;" F(X)=";FX
270      REM-- CHECKS FOR SIGN CHANGE IN F(X)
280      IF FF*FX = 0 GOTO 360
290      IF FF*FX > 0 THEN FF = FX: GOTO 330
300      IF DX <= EP GOTO 360
310      X = X + DX
320      DX = DX/DD
330      X = X + DX
340   NEXT I
350   LPRINT "ROOT NOT FOUND !!": END
360   LPRINT "ROOT IS: X =";X;"  DX =";DX
370   END
                          Subroutine
500   REM--SUBROUTINE TO DETERMINE F(X) VALUES
510   FX = X*X - 3*X + 1.3
520   RETURN
530   END
```

Figure A2.1   BASIC program for finding a real root by trial-and-error method.

x's differ by less than a significant-digit test parameter ε. The method, which requires that the $f(x)$ values used in the interpolation have opposite signs, is implemented by the following formula.

$$x_{new} = \frac{[x_1 f(x_2) - x_2 f(x_1)]}{[f(x_2) - f(x_1)]}$$

An example program, coded in BASIC, is given as Figure A2.2.

```
                              Main Program
010  REM--REF! INTRO. ENG'RNG. MODELING ...... RIEDER/BUSBY
100  REM--REAL ROOT BY LINEAR-INTERPOLATION METHOD
110  REM--USES TRIAL-AND-ERROR METHOD TO GET STARTED
120  REM--(INPUT) XO = STARTER VALUE FROM ROUGH GRAPH
130  REM--(INPUT) DX = STARTER INCREMENT FOR X
140  REM--(INPUT) EP = SIGNIFICANT-DIGIT TEST PARAMETER
150  REM-- PLACE F(X) AS FX IN SUBROUTINE 500
160  CLS! CLEAR10! DEFINT I-N
170  READ XO, DX, EP
180    DATA 0.0, 0.1, 1E-4
190  X = XO
200  GOSUB 500
210  F1 = FX
220  REM-- LOOP ALLOWS 20 TRIALS TO FIND ROUGH LOCATION OF ROOT
230  REM-- IF LOCATION NOT FOUND, CHECK FOR COMPLEX OR REPEATED ROOTS
240  FOR I = 1 TO 20
250    GOSUB 500
260    PRINT "X=";X;" F(X)=";FX
270    REM-- CHECKS FOR SIGN CHANGE IN F(X)
280    IF F1*FX = 0 THEN F3 = FX! GOTO 470
290    IF F1*FX < 0 GOTO 340
300    F1 = FX
310    X = X + DX
320  NEXT I
330  LPRINT "LOCATION OF ROOT NOT IDENTIFIED !!"! END
340    F2 = FX
350    X1 = X - DX
360    X2 = X
370    XX = X
380  X = (X1*F2 - X2*F1)/(F2 - F1)
390  GOSUB 500
400  PRINT "X=";X;" F(X)=";FX
410    F3 = FX
420    IF ABS(XX-X) <= EP GOTO 470
430    XX = X
440    IF F3*F1 = 0 GOTO 470
450    IF F3*F1 > 0 THEN X1 = X! F1 = F3! GOTO 380
460    X2 = X! F2 = F3! GOTO 380
470  LPRINT "ROOT IS! X =";X3;" F3=";F3
480  END
                              Subroutine
500  REM--SUBROUTINE TO DETERMINE F(X) VALUES
510  FX = X*X - 3*X + 1.3
520  RETURN
530  END
```

Figure A2.2 BASIC program for finding a real root by the linear interpolation method.

A2.1.3 The Newton–Raphson Method

This is the semianalytical version of the linear interpolation method and requires that the analytical first derivative of the function, $f'(x)$, be known. This method is the simplest version of a general technique known as "Newton's method."

A single starting value, relatively close to the desired root, must be available and is usually obtained from a preliminary application of the trial-and-error method, although it can also be approximated directly from the rough graph. Successive values of x are then obtained from the following formula.

$$x_{i+1} = x_i - \frac{f(x_i)}{f'(x_i)}$$

Main Program

```
010  REM--REF: INTRO. ENG'RNG. MODELING ,,,,,, RIEDER/BUSBY
100  REM--REAL ROOT BY NEWTON RAPHSON METHOD
110  REM--USES TRIAL-AND-ERROR METHOD TO GET STARTED
120  REM--(INPUT) XO = STARTER VALUE FROM ROUGH GRAPH
130  REM--(INPUT) DX = STARTER INCREMENT FOR X
140  REM--(INPUT) EP = SIGNIFICANT-DIGIT TEST PARAMETER
150  REM-- PLACE F(X) AS FX IN SUBROUTINE 500
160  REM-- PLACE DF(X)/DX AS FP IN SUBROUTINE 600
170  CLS: CLEAR10: DEFINT I-N
180  READ XO, DX, EP
190     DATA 0.0, 0.1, 1E-4
200  X = XO
210  GOSUB 500.
220  FF = FX
230  REM-- LOOP ALLOWS 20 TRIALS TO FIND ROUGH LOCATION OF ROOT
240  REM-- IF LOCATION NOT FOUND, CHECK FOR COMPLEX OR REPEATED ROOTS
250  FOR I = 1 TO 20
260     GOSUB 500
270     PRINT "X =";X;"  F(X) =";FX
280     IF FF*FX = 0 GOTO 380
290     IF FF*FX < 0 GOTO 340
300     FF = FX
310     X = X + DX
320  NEXT I
330  LPRINT "LOCATION OF ROOT NOT IDENTIFIED !!": END
340  GOSUB 600
350  X = X - FX/FP
360  PRINT "X =";X;"  F(X)/F'(X) =";FX/FP
370     IF FX/FP > EP THEN GOSUB 500: GOTO 340
380  LPRINT "ROOT IS: X =";X;"  F(X) =";FX
390  END
```
Subroutines
```
500  REM--SUBROUTINE TO DETERMINE F(X) VALUES
510  FX = X*X - 3*X + 1.3
520  RETURN
530  END
600  REM--SUBROUTINE TO DETERMINE F'(X) VALUES
610  FP = 2*X - 3
620  RETURN
630  END
```

Figure A2.3 BASIC program for finding a real root by the Newton–Raphson method.

If properly applied, successive approximations to the sought-after root get better until the procedure is terminated with satisfaction of the following condition.

$$\left|\frac{f(x_i)}{f'(x_i)}\right| \leqslant \varepsilon$$

Here ε is a significant-digit test parameter for the x value.

An example program, coded in BASIC, is given as Figure A2.3.

A2.2 SOLUTIONS OF LINEAR ALGEBRAIC EQUATION SETS

Two broad categories of methods exist for getting solutions to linear matrix equations. In the so-called exact methods, the number of operations to get a solution can be stated beforehand. Computer time is more or less predictable. However, significant roundoff inaccuracies can occur, even for 100 equation and smaller sets, and relatively large amounts of central processing storage are usually required. In the "iteration" methods, the number of operations to get a reasonably accurate solution cannot be stated beforehand. However, very large sets (thousands of equations) can be handled with minimal central processing storage, and roundoff inaccuracies are usually irrelevant.

A2.2.1 Exact Methods

A2.2.1.1 Gauss Elimination

In this method a set of n equations, in n unknowns, is reduced to an equivalent triangular set of equations, which in turn, is solved by "back substitution." This can be illustrated by considering the following general set of nonhomogeneous equations.

$$a_{11}x_1 + a_{12}x_2 + \cdots + a_{1n}x_n = c_1$$
$$a_{21}x_1 + a_{22}x_2 + \cdots + a_{2n}x_n = c_2$$
$$\vdots \qquad \vdots \qquad \qquad \vdots$$
$$a_{n1}x_1 + a_{n2}x_2 + \cdots + a_{nn}x_n = c_n$$

The reduction to a triangular set is started by dividing the first equation by a_{11}. The resulting normalized equation is then multiplied by a_{21} and subtracted from the second equation of the set to give a new second equation with x_1 missing. Next the normalized equation is multiplied by a_{31} and subtracted from the third equation to give a new third equation with x_1 missing. This process is repeated for the remaining equations; it is observed that all the x_1 values below the diagonal have been eliminated. Next, if the first equation is disregarded, it is noted that a new set with one less unknown exists. The normalization, multiplication, and subtraction process described above is then repeated for this new set to eliminate the x_2 values below the diagonal of the original set. If this is repeated for each unknown in turn, until only the last unknown, x_n, is left, the remaining set of equations is triangular.

$$a_{11}x_1 + a_{12}x_2 + \cdots + a_{1n}x_n = c_1$$
$$b_{22}x_2 + \cdots + b_{2n}x_n = d_2$$
$$\cdots \qquad \cdots \qquad \cdots$$
$$b_{nn}x_n = d_n$$

Main Program

```
010  REM--REF: INTRO, ENG'RNG, MODELING ......, RIEDER/BUSBY
100  REM--EQUATION-SET SOLUTION BY GAUSS-ELIMINATION METHOD
110  REM--(INPUT) N = NUMBER OF EQUATIONS (UNKNOWNS)
120  REM--(INPUT) A(I,J) = COEFFICIENTS OF UNKNOWNS
130  REM--------- I = ROW NO.; J = COLUMN NO.
140  REM--(INPUT) A(I,M) = NONHOMOGENEOUS CONSTANT FOR ROW I
150  REM-- PLACE COEFFICIENT DATA INTO SUBROUTINE 1000
160  CLS: CLEAR10: DEFINT I-N
170  READ N
180    DATA 4
190    M = N + 1
200    L = N - 1
210    DIM A(N,M), X(N)
220  GOSUB 1000
230  REM-- START OF REDUCTION TO TRIANGULAR FORM
240  FOR K = 1 TO L
250    K1 = K + 1
260    JJ = K
270    BG = ABS(A(K,K))
280    REM--START OF SEARCH FOR LARGEST PIVOT ELEMENT
290    FOR I = K1 TO N
300      AB = ABS(A(I,K))
310      IF BG > AB THEN BG = AB: JJ = I
320    NEXT I
330    IF JJ = K GOTO 410
340    REM--INTERCHANGES ROWS TO GET MAXIMUM PIVOT ELEMENT
350    FOR J = K TO M
360      TE = A(JJ,J)
370      A(JJ,J) = A(K,J)
380      A(K,J) = TE
390    NEXT J
400    REM--DETERMINES REDUCED ELEMENTS OF TRIANGULAR SET
410    FOR I = K1 TO N
420      Q = A(I,K)/A(K,K)
430      FOR J = K1 TO M
440        A(I,J) = A(I,J) - Q*A(K,J)
450      NEXT J
460    NEXT I
470    FOR I = K1 TO N
480      A(I,K) = 0
490    NEXT I
500  NEXT K
510  REM--BACKSUBSTITUTION FOR THE SOLUTIONS
520  X(N) = A(N,M)/A(N,N)
530  FOR NN = 1 TO L
540    SU = 0
550    I = N - NN
```

Figure A2.4*a*   BASIC program for solving a linear set of equations by the Gauss elimination method.

```
560    I1 = I + 1
570    FOR J = I1 TO N
580      SU = SU + A(I,J)*X(J)
590    NEXT J
600    X(I) = (A(I,M) - SU)/A(I,I)
610  NEXT NN
620  REM--OUTPUT OF THE SOLUTION VECTOR
630  FOR I = 1 TO N
640    LPRINT "X(";I;") =";X(I)
650  NEXT I
660  END
                          Subroutine
1000 REM--COEFFICIENT DATA INPUT; A(I,M) IS NONHOMO. CONSTANT
1010 FOR I = 1 TO N
1020   FOR J = 1 TO M
1030     READ A(I,J)
1040     PRINT "A(";I;J;") =";A(I,J)
1050   NEXT J
1060 NEXT I
1070 RETURN
1100 DATA 3E-5, -5, 47, 20, 18
1110 DATA 11, 16, 10, 41, 26
1120 DATA 56, 22, 11, -18, 34
1130 DATA 17, 66, -12, 7, 82
1140 END
```

Figure A2.4*b*   BASIC program for solving a linear set of equations by the Gauss elimination method *(continued)*.

It is evident that the "diagonal" is simply the group of terms that has the identical double subscripts on the coefficients. A back substitution process is then started by solving directly for x_n first ($x_n = d_n/b_{nn}$). The x_n value is then substituted back into the next-to-last equation, which is then solved directly for x_{n-1}. Both x_n and x_{n-1} are then substituted back into the next equation up, and x_{n-2} is solved for directly, and so on.

An example program, coded in BASIC, is shown as Figure A2.4. Since some sets may contain a zero (or very small coefficient) on the diagonal, and since it is undesirable to use such a coefficient as a divisor, an additional feature is illustrated in the program. This feature is called a "maximum pivot strategy," and is simply a process of interchanging rows in a set to always make sure that the largest-magnitude coefficient exists on the diagonal before each normalization. The normalized equation is often referred to as the "pivot equation"; the first coefficient (the divisor) from this equation is called the "pivot element."

A2.2.1.2 Gauss-Jordan Elimination

A number of simplifications become obvious to any user of the Gauss elimination method. The Jordan modification of the Gauss elimination method incorporates most of these simplifications. For example, it utilizes streamlined matrix notation and combines the process of backsubstitution with the reduction to give a very efficient matrix equation solver. In the Jordan method an augmented matrix of coefficients is the only input. The augmentation column consists of the nonhomogenous constants, $a_{im} = c_i$, as follows.

$$[A] = \begin{bmatrix} a_{11} & a_{12} & \cdots & a_{1n} & \vdots & a_{1m} \\ a_{21} & a_{22} & \cdots & a_{2n} & \vdots & a_{2m} \\ \cdots & \cdots & \cdots & \cdots & & \\ a_{n1} & a_{n2} & \cdots & a_{nn} & \vdots & a_{nm} \end{bmatrix}$$

The solution values (solution vector) are obtained by successively reducing the above matrix, giving a matrix with one less column at each step, until a single column matrix remains. This single-column matrix is the solution vector. The first reduced matrix is given by the $[B]$ matrix. It should be noted that the zeros are not kept.

$$[B] = \begin{bmatrix} b_{11} & b_{12} & \cdots & b_{1n} \\ \cdots & & \cdots & \cdots \\ b_{n1} & b_{n2} & \cdots & b_{nn} \end{bmatrix}$$

The $[B]$ elements are computed using the following formulas.

$$b_{i-1, j-1} = a_{ij} - \frac{a_{1j}a_{i1}}{a_{11}}$$

$$b_{n, j-1} = \frac{a_{1j}}{a_{11}}$$

$$1 < i \leqslant n$$

$$1 < j \leqslant m$$

Here the i and j refer to row and column number, respectively, from the $[A]$ matrix. The m value varies from $n+1$ to 2. A constraint in the use of these formulas is that the pivot element a_{11} must not be too small. Consequently, a row interchange scheme to always put a largest-magnitude element in the a_{ii} position is common in most algorithms. As in the Gauss elimination method, this scheme is known as a "maximum pivot strategy." Once the $[B]$ elements are all computed, they are stored back into the $[A]$ matrix, thus destroying the original matrix but saving storage. Repetition of the foregoing finally gives a single column matrix of $\{B\}$ elements. This is the $\{X\}$ solution vector.

$$\{X\} = \begin{Bmatrix} X_1 \\ X_2 \\ \cdots \\ X_n \end{Bmatrix} = \begin{Bmatrix} b_{11} \\ b_{21} \\ \cdots \\ b_{n1} \end{Bmatrix}$$

An example program, coded in BASIC, is given as Figure A2.5.

```
010  REM--REF: INTRO, ENG'RNG, MODELING ,,,,,, RIEDER/BUSBY
100  REM--EQUATION-SET SOLUTION BY GAUSS-JORDAN METHOD
110  REM--(INPUT) N = NUMBER OF EQUATIONS (UNKNOWNS)
120  REM--(INPUT) A(I,J) = COEFFICIENTS OF UNKNOWNS; THE MATRIX ELEMENTS
130  REM--------- I = ROW NO.; J = COLUMN NO.
140  REM--(INPUT) A(I,M) = NONHOMOGENEOUS CONSTANT FOR ROW I
150  REM-- PLACE COEFFICIENT DATA INTO SUBROUTINE 1000
160  CLS: CLEAR10: DEFINT I-N
170  READ N
180     DATA 4
190     M = N + 1
200     DIM A(N,M), B(N,M)
210  GOSUB 1000
220  REM-- START OF SEARCH FOR LARGEST PIVOT ELEMENT
230     BG = ABS(A(1,1))
240     K = M + 1
250     JJ = 1
260     FOR I = 2 TO K
270        IF BG >= ABS(A(I,1)) GOTO 300
280        BG = ABS(A(I,1))
290        JJ = I
300     NEXT I
310  REM-- TERMINATE IF ALL POTENTIAL PIVOT ELEMENTS ARE TOO SMALL
320     IF BG < 1E-15 THEN LPRINT "NO UNIQUE SOLUTION !!": END
330  REM-- INTERCHANGE ROWS TO GET LARGEST PIVOT ELEMENT
340     IF JJ = 1 GOTO 410
350     FOR J = 1 TO M
360        TE = A(JJ,J)
370        A(JJ,J) = A(1,J)
380        A(1,J) = TE
390     NEXT J
400  REM-- REDUCE MATRIX- COMPUTE NEW MATRIX ELEMENTS;
410  FOR J = 2 TO M
420     FOR I = 2 TO N
430        B(I-1,J-1) = A(I,J) - A(1,J)*A(I,1)/A(1,1)
440     NEXT I
450  NEXT J
460  FOR J = 2 TO M
470     B(N,J-1) = A(1,J)/A(1,1)
480  NEXT J
490  M = M - 1
500  REM-- STORE <B> ELEMENTS BACK INTO MATRIX <A>
510  FOR J = 1 TO M
520     FOR I = 1 TO N
530        A(I,J) = B(I,J)
540     NEXT I
550  NEXT J
560  IF M <> 1 GOTO 230
570  REM-- OUTPUT OF SOLUTION VECTOR
580  FOR I = 1 TO N
590     LPRINT "X(";I;") =";A(I,1)
600  NEXT I
```

(continued)

Figure A2.5*a* BASIC program for solving a linear set of equations by the Gauss—Jordan method.

```
610 END
                              Subroutine
1000 REM-- COEFFICIENT DATA INPUT; A(I,M) IS NONHOMO. CONSTANT
1010 FOR I = 1 TO N
1020    FOR J = 1 TO M
1030       READ A(I,J)
1040       PRINT "I=";I;" J=";J;" A(I,J) =";A(I,J)
1050 NEXT J,I
1060 RETURN
1100 DATA 3E-5, -5, 47, 20, 18
1110 DATA 11, 18, 10, 41, 26
1120 DATA 56, 22, 11, -18, 34
1130 DATA 17, 66, -12, 7, 82
1140 END
```

Figure A2.5b BASIC program for solving a linear set of equations by the Gauss–Jordan method *(continued)*.

A2.2.1.3 Inverse Matrix Method

In this method the coefficient matrix $[A]$ is inverted and multiplied times the nonhomogeneous-constants vector $\{C\}$ to give the solution vector $\{X\}$.

$$[A]\{X\}=\{C\}$$
$$\{X\}=[A]^{-1}\{C\}$$

This method is particularly good where a large number of different $\{C\}$ vectors must be treated parametrically, provided the matrix $[A]$ is not too large. (Much above about 50 equations is probably too large unless special processors are available.) The advantages of this method are associated to a large extent with the fact that extremely efficient algorithms and operations exist for matrix multiplication on a computer.

The Gauss–Jordan method is recommended for inverting the $[A]$ matrix. The starting matrix is augmented with the identity matrix $[I]$.

$$[A \mid I]$$

$$\begin{bmatrix} a_{11} & \cdots & a_{1n} & 1 & \cdots & 0 \\ \cdots & \cdots & & & \cdots & \\ a_{n1} & \cdots & a_{nn} & 0 & \cdots & 1 \end{bmatrix}$$

The Gauss–Jordan reduction scheme is then simply applied repetitively until only n columns remain. The remaining columns make up the inverse matrix. In accomplishing the reduction by computer, the identity matrix is not stored, since special statements can place the required ones and zeros.

An example matrix inversion program, coded in BASIC, is given as Figure A2.6. It should be noted that a maximum pivot strategy is not employed. The user should make sure that the starting matrix is appropriately "diagonalized," that is, the largest magnitude coefficients placed as the diagonal elements. Using the inverse matrix in an actual matrix multiplication is not illustrated because such programs are so widely available.

```
010   REM--REF: INTRO. ENG'RNG. MODELING ...... RIEDER/BUSBY
100   REM-- MATRIX INVERSION BY GAUSS-JORDAN METHOD
110   REM--(INPUT) N = NUMBER OF EQUATIONS (THE NO. OF ROWS IN MATRIX)
120   REM--(INPUT) A(I,J) = MATRIX ELEMENTS (THE COEFFICIENTS,.)
130   REM-- PLACE COEFFICIENT DATA INTO SUBROUTINE 1000, BUT ,.
140   REM----- WARNING !!! REARRANGE FOR MAXIMUM DIAGONAL FIRST !!
150   CLS: CLEAR10: DEFINT I-N
160   READ N
170     DATA 4
180     M = N + 1
190     DIM A(N,M), B(N,M)
200   GOSUB 1000
210   REM-- REDUCE MATRIX- COMPUTE NEW MATRIX ELEMENTS
220   FOR K = 1 TO N
230     A(1,M) = 1.
240       FOR J = 2 TO M
250         FOR I = 2 TO N
260           A(I,M) = 0
270             B(I-1,J-1) = A(I,J) -A(I,J)*A(1,1)/A(1,1)
280         NEXT I
290       NEXT J
300       FOR J = 2 TO M
310         B(N,J-1) = A(1,J)/A(1,1)
320       NEXT J
330   REM-- STORE <B> ELEMENTS BACK INTO MATRIX <A>
340       FOR I = 1 TO N
350         FOR J = 1 TO N
360           A(I,J) = B(I,J)
370         NEXT J
380       NEXT I
390   NEXT K
400   REM-- OUTPUT OF INVERSE MATRIX
410   FOR I = 1 TO N
420     FOR J = 1 TO N
430       LPRINT "I =";I;" J =";J;" IA(I,J) =";A(I,J)
440     NEXT J
450   NEXT I
460   END
```

```
1000 REM-- COEFFICIENT DATA INPUT
1010 FOR I = 1 TO N
1020   FOR J = 1 TO N
1030     READ A(I,J)
1040     PRINT "I =";I;" J =";J;" A(I,J) =";A(I,J)
1050 NEXT J,I
1060 RETURN
1100 DATA 56, 22, 11, -18
1120 DATA 17, 66, -12, 7
1130 DATA 3E-5, -5, 47, 20
1140 DATA 11, 16, 10, 41
1150 END
```

Figure A2.6   BASIC program for inverting a matrix by the Gauss–Jordan method.

323

A2.2.2 Iterative Methods

A2.2.2.1 Gauss-Seidel Method

In this method, the initial step is to choose starting values (estimates of unknowns) and a convergence parameter. The convergence parameter, ε, is simply a small number against which the absolute difference between two successive computations

```
                          Main Program
010  REM--REF: INTRO, EMG'RNG, MODELING ...... RIEDER/BUSBY
100  REM-- EQUATION-SET SOLUTION BY GAUSS-SEIDEL METHOD
110  REM--(INPUT) N = NUMBER OF EQUATIONS (UNKNOWNS)
120  REM--(INPUT) A(I,J) = COEFFICIENTS OF UNKNOWNS
130  REM---------- I = EQUATION NO,; J = UNKNOWN NO,
140  REM--(INPUT) A(I,M) = NONHOMOGENEOUS CONSTANT FOR EQUATION I
150  REM--- PLACE COEFFICIENT DATA INTO SUBROUTINE 1000, BUT ...
160  REM--- WARNING !! REARRANGE FOR MAXIMUM DIAGONAL FIRST !!!
170  REM--(INPUT) X(I) =STARTERS (THESE ARE ASTUTE ENGINEERING GUESSES)
180  REM--- DEFAULT VALUES ARE ZEROES, ALSO THE SOLUTION VECTOR ,
190  REM--(INPUT) MX = MAXIMUM NO, OF ITERATIONS DESIRED
200  REM--(INPUT) EP = SIGNIFICANT-DIGIT CONVERGENCE TEST PARAMETER
210  CLS: CLEAR10: DEFINT I-N'
220  READ N
230     DATA 4
240     M = N + 1
250     DIM A(N,M), X(N)
260  GOSUB 1000
270  REM-- INPUT STARTERS
280     FOR I = 1 TO N
290       X(I) = 0
300     NEXT I
310  REM-- START OF ITERATION LOOP
320  FOR IT = 1 TO MX
330     NF = 0
340  REM-- COMPUTES THE UNKNOWNS  X(1), X(2), ... X(N)
350     FOR J = 1 TO N
360       SA = X(J)
370       X(J) = A(J,M)
380       FOR K = 1 TO N
390         IF K = J GOTO 410
400           X(J) = X(J) - A(J,K)*X(K)
410         NEXT K
420       X(J) = X(J)/A(J,J)
430       DF = SA - X(J)
440       IF ABS(DF) > EP THEN NF = NF + 1
450     NEXT J
460     IF NF = 0 GOTO 510
470     PRINT "ITERATION NO, =";IT;" POINTS FAILING CONVERGENCE =";NF
480  NEXT IT
490  LPRINT "CONVERGENCE NOT ATTAINED IN";MX;" ITERATIONS !!"
500  REM-- OUTPUT OF SOLUTION VECTOR
510  FOR J = 1 TO N
520     LPRINT "X(";J;") =";X(J)
430  NEXT J
540  END
                          (continued)
```

Figure A2.7*a*   BASIC program for solving a linear set of equations by the Gauss—Seidel method.

```
                                Subroutine
1000 REM--COEFFICIENT DATA INPUT; A(I,M) IS NONHOMO, CONSTANT
1010 FOR I = 1 TO N
1020    FOR J = 1 TO M
1030       READ A(I,J)
1040       PRINT "I =";I;" J =";J;"  A(I,J) =";A(I,J)
1050 NEXT J,I
1060 RETURN
1100 DATA 56, 22, 11, -18, 34
1110 DATA 17, 66, -12, 7, 82
1120 DATA 3E-5, -5, 47, 20, 18
1130 DATA 11, 16, 10, 41, 26
1140 END
```

Figure A2.7*b* BASIC program for solving a linear set of equations by the Gauss–Seidel method *(continued)*.

(at successive iterations) of a given unknown can be compared. The better the starting estimates (i.e., the closer they are to the actual solution values), the faster the method produces a solution. The sequence of tasks, after the initial step, is relatively easy to implement but difficult to describe. If we assume a well-posed set of equations, the next step is to solve the first equation for the first unknown (x_1 using the set shown in A2.2.1.1), using the starting estimates for the unknowns needed. The second equation is then solved for the second unknown, x_2, using the just-computed value of x_1 and the estimates for the other unknowns. The process is repeated for the remaining equations, always using the latest computed values wherever possible. When all the equations have been solved, each for a different unknown, the first "iteration" has been completed. Then the whole procedure is repeated, using the latest calculated values wherever possible, to give the second iteration, and so on. However, after the first iteration, as each unknown is computed, it is subtracted from its immediately preceding value. The absolute value of this difference is then compared with ε. If less than ε, then the unknown is said to be "converged." Computations are continued until all unknowns satisfy the foregoing test at the same iteration. Then the latest computed values are accepted as a solution. The fewer the iterations required to obtain a solution, the faster the rate of convergence is said to be. The faster the rate of convergence, the greater will be the accuracy that can be specified through ε.

An example program, coded in BASIC, is given as Figure A2.7. The Gauss–Seidel method will always converge if, for a given set, each diagonal coefficient is larger in magnitude than the sum of the magnitudes of the other coefficients in the same equation. However, the method may also converge acceptably for other cases, including nonlinear sets. Two variations on this method are known as the "Jacobi" method and the "SOR" method.

A2.3 NUMERICAL INTEGRATION

Several different procedures exist for numerically evaluating the following integral.

$$g = \int_a^b f(x)\, dx$$

Here $f(x)$ may be either tabular or analytical. Only two of the better-known methods will be considered.

A2.3.1 Trapezoidal Method

Also known as the "trapezoidal rule," this method is among the easiest to apply. The following formula can be used for the case of nonuniform spacing between the function values.

$$g = 0.5 \sum_{i=1}^{n} (\Delta x)_i [f(x_i) + f(x_{i+1})] + \text{error}$$

Here n represents the number of function, $f(x_i)$, points less one. The spacing is accounted for by $(\Delta x)_i = x_{i+1} - x_i$. The "error" term may be significant but typically cannot be evaluated when $f(x)$ is a tabular function. For analytical $f(x)$'s, it is often easier to decrease the Δx's and compare results in an iterative fashion rather than evaluate the error term directly. Consequently, the error term, although shown, is not discussed further.

```
                           Main Program
010  REM--REF: INTRO, ENG'RNG, MODELING ,,,,,,, RIEDER/BUSBY
100  REM-- INTEGRATION OF F(X) BY TRAPEZOIDAL METHOD
110  REM--(INPUT) NP = NUMBER OF DISCRETE F(X) VALUES
120  REM--(INPUT) DX = INCREMENT OF X BETWEEN F(X) VALUES
130  REM--(INPUT) FX(I) = F(X) VALUES USED IN PROGRAM
140  REM--- PLACE F(X)'S (EITHER TABULAR OR ANALYTIC) IN SUBROUTINE 1000
150  CLS: CLEAR10: DEFINT I-N
160  READ NP, DX
170     DATA 7, .523599
180     DIM FX(NP)
190     GOSUB 1000
200     N = NP - 1
210     S = 0
220  REM-- LOOP ADDS F(X) VALUES
230  FOR I = 2 TO N
240     S = S + FX(I)
250  NEXT I
260     G = (DX/2)*(FX(1) + 2*S + FX(NP))
270  REM-- OUTPUT THE ANSWER
280  LPRINT "THE INTEGRAL OF F(X) IS";G
290  END
                           Subroutine
1000 REM-- DISCRETE F(X) VALUES INPUT; EITHER TABULAR OR ANALYTIC
1010 FOR I = 1 TO NP
1020    READ FX(I)
1030    PRINT "I =";I;" FX(I) =";FX(I)
1040 NEXT I
1050 RETURN
1100 DATA 0., .5, .866026, 1.0, .866026, .5, 0.,
1110 END
```

Figure A2.8 BASIC program for finding the integral of $F(x)$ by the trapezoidal rule.

For uniformly spaced $f(x)$ values, all Δx values are equal, and the following formula can be used.

$$g = \frac{\Delta x}{2}\{f(x_1) + 2[f(x_2) + f(x_3) + \cdots + f(x_n)] + f(x_{n+1})\} + \text{error}$$

An example program, coded in BASIC, for uniformly spaced $f(x)$ values, is shown as Figure A2.8. Occasionally, it is desired to evaluate an integral with a variable upper limit (i.e., b replaced by x). An example of a program to do this is shown as Figure A2.9.

A2.3.2 Simpson's 1/3 Rule

This method is more accurate than the trapezoidal method but requires three-point groups of $f(x)$ values, with the middle $f(x)$ value located exactly halfway between the other two $f(x)$'s. This restriction means that it is usually easier to require that

```
                        Main Program
010  REM--REF: INTRO, ENG'RNG, MODELING ....... RIEDER/BUSBY
100  REM-- TRAPEZOIDAL INTEGRATION OF  F(X)  TO UPPER LIMIT  X
110  REM--(INPUT) NP = NUMBER OF DISCRETE F(X)  VALUES
120  REM--(INPUT) DX = INCREMENT OF X BETWEEN F(X) VALUES
130  REM--(INPUT) FX(I) = VALUES OF F(X)
140  REM--(INPUT) XO = INITIAL VALUE OF X (THE LOWER LIMIT)
150  REM--- PLACE F(X) VALUES (TABULAR OR ANALYTIC) IN SUBROUTINE 1000
160  REM--------- G(X) = DISCRETE RESULTS FROM THE INTEGRATION
170  CLS: CLEAR10: DEFINT I-N
180  READ NP, DX, XO
190     DATA 7, .523599, 0.0
200     DIM FX(NP), GX(NP)
210     GOSUB 1000
220     X = XO
230     GX(1) = FX(1)
240  FOR I = 2 TO NP
250     GX(I) = GX(I-1) + (FX(I-1) + FX(I))*DX/2
260  NEXT I
270  REM-- OUTPUT OF THE SOLUTION VECTOR
280  FOR I = 1 TO NP
290     LPRINT "X(";I;")=";X;" F(X)=";FX(I);" G(X)=";GX(I)
300     X = XO + I*DX
310  NEXT I
320  END
                        Subroutine
1000 REM-- DISCRETE F(X) VALUES INPUT (TABULAR OR ANALYTIC)
1010 FOR I = 1 TO NP
1020    READ FX(I)
1030    PRINT "FX(";I;")=";FX(I)
1040 NEXT I
1050 RETURN
1100 DATA 0., .5, .866026, 1.0, .866026, .5, 0,
1110 END
```

Figure A2.9 BASIC program for finding the integral of $F(x)$ to upper limit x by the trapezoidal method.

all Δx's be the same, and that an even number of increments be used. If an even number is not available for a tabular $f(x)$, then it is common practice to treat the last increment by the trapezoidal method. The following formula can be used for the Simpson method with uniform Δx and an even number of Δx's.

$$g = \frac{\Delta x}{3} \{f(x_1) + 4[f(x_2) + \cdots f(x_{even})] + 2[f(x_3 + \cdots f(x_{odd})] + f(x_{n+1})\} + \text{error}$$

A popular iterative variation on this method, for analytical $f(x)$'s, is to start with two increments ($n = 2$) and successively apply the foregoing equation, doubling the number of increments (in effect halving the Δx) each time. When two successive g values differ by less than some arbitrary small value, ε, then the integration is accepted as reasonably accurate.

An example program, coded in BASIC, is given as Figure A2.10.

```
                              Main Program
010  REM--REF: INTRO. ENG'RNG. MODELING ....... RIEDER/BUSBY
100  REM-- INTEGRATION OF F(X) BY SIMPSON'S 1/3 RULE
110  REM--(INPUT) NP = NUMBER OF DISCRETE F(X) VALUES
120  REM----- MUST BE AN ODD NUMBER (GIVES EVEN NO. OF SUBINTERVALS)
130  REM--(INPUT) DX = INCREMENT OF X BETWEEN F(X) VALUES
140  REM--(INPUT) FX(I) = VALUES OF F(X)
150  REM--- PLACE F(X) VALUES (TABULAR OR ANALYTIC) IN SUBROUTINE 1000
160  CLS: CLEAR10: DEFINT I-N
170  READ NP, DX
180     DATA 7, .523599
190     DIM FX(NP)
200     GOSUB 1000
210     N = NP - 1
220     K = NP - 2
230     SE = 0
240     SO = 0
250  REM-- LOOPS ADD F(X)'S FOR EVEN AND ODD I'S
260  FOR I = 2 TO N STEP 2
270     SE = SE + FX(I)
280  NEXT I
290  FOR I = 3 TO K STEP 2
300     SO = SO + FX(I)
310  NEXT I
320     G = (DX/3)*(FX(1) + 4*SE + 2*SO + FX(NP))
330  REM-- OUTPUT OF THE ANSWER
340  LPRINT "THE INTEGRAL OF F(X) IS:";G
350  END
                              Subroutine
1000 REM--DISCRETE F(X) VALUES INPUT (TABULAR OR ANALYTIC)
1010 FOR I = 1 TO NP
1020    READ FX(I)
1030    PRINT I; " F(X) = "; FX(I)
1040 NEXT I
1050 RETURN
1100 DATA 0., .5, .866026, 1.0, .866026, .5, 0.
1110 END
```

Figure A2.10   BASIC program for finding the integral of $F(x)$ by Simpson's 1—3 rule.

A2.4 NUMERICAL DIFFERENTIATION

In this case the objective is to substitute $f(x)$ values to get approximations for the derivatives $d^k f(x)/dx^k$. One approach is to use truncated Taylor-series expansions to provide appropriate formulas. This leads to backward, forward, and central approximations at a given point. In general, the central approximation is the most accurate. If only the first few (typically two or three) terms are retained in the series, then the following well-known formulas are obtained. The abbreviations, f_i for $f(x_i)$, and Δx for $(x_{i+1} - x_i)$, are used in the following list.

$$\frac{df}{dx} = \frac{f_{i+1} - f_{i-1}}{2\,\Delta x}$$

First Central Difference Approximation for the First Derivative

$$\frac{df}{dx} = \frac{f_i - f_{i-1}}{\Delta x}$$

First Backward Difference Approximation for the Second Derivative

$$\frac{df}{dx} = \frac{f_{i+1} - f_i}{\Delta x}$$

First Forward Difference Approximation for the First Derivative

$$\frac{d^2 f}{dx^2} = \frac{f_{i+1} - 2f_i + f_{i-1}}{(\Delta x)^2}$$

First Central Difference Approximation for the Second Derivative

$$\frac{d^2 f}{dx^2} = \frac{f_i - 2f_{i-1} + f_{i-2}}{(\Delta x)^2}$$

First Backward Difference Approximation for the Second Derivative

$$\frac{d^2 f}{dx^2} = \frac{f_{i+2} - 2f_{i+1} + f_i}{(\Delta x)^2}$$

First Forward Difference Approximation for the Second Derivative

Since most applications of derivative approximations occur as part of other methods, separate programs to just evaluate derivatives are seldom useful. Consequently, no example program is shown.

A2.5 INTERPOLATION AND CURVE FITTING

Tabulations of discrete $f(x)$ values against x values are common and have been treated from the perspectives of integration and differentiation already. However, an analytical formula for $f(x)$, derived from the tabulated numbers, is often sought.

Two different approaches to generating such an analytic $f(x)$ can be used. If the data for $f(x)$ are so-called raw data, perhaps from an experiment, and show scatter, then some type of least-squares fitting is the best approach. If, on the other hand, the data are from published sources, and obviously have been manipulated and smoothed, then an exact fitting over a given data span with an "interpolating polynomial" is probably the best approach. In either case some independent knowledge of the theoretical behavior of $f(x)$ is usually desirable if fits over wide x spans are sought. In the following, the first two methods fall in the exact-fit approach, while the third is a least squares fit.

A2.5.1 Gregory Newton Interpolating Polynomial

In this method a polynomial in x is matched exactly to a group of $f(x)$ values. Differences between tabulated $f(x)$ values are used, and a uniform spacing, Δx, is assumed. The following formula is for the forward Gregory Newton polynomial.

$$f(x) \simeq P(x) = f_0 + s\,\Delta f_0 + s(s-1)\frac{\Delta^2 f_0}{2!} + s(s-1)(s-2)\frac{\Delta^3 f_0}{3!} + \cdots$$

where

$$s = \frac{(x-x_0)}{\Delta x}$$

$x_0 = x$ value at the first tabulated $f(x)$ in the span of interest

$f_0 = f(x_0)$

$\Delta f_0 = f(x_1) - f(x_0)$

$\Delta f_1 = f(x_2) - f(x_1)$

$\Delta f_2 = f(x_3) - f(x_2)$

$\Delta^2 f_0 = \Delta f_1 - \Delta f_0$

$\Delta^2 f_1 = \Delta f_2 - \Delta f_1$

$\Delta^3 f_0 = \Delta^2 f_1 - \Delta^2 f_0$

It should be observed that the number of data pairs required is the desired degree of the polynomial plus one. When interpolating with the foregoing polynomial, the span of the polynomial should be bisected by the interpolation point if possible. An example program, coded in BASIC, is given as Figure A2.11.

A2.5.2 Lagrange Interpolating Polynomial

In this method, uniform spacing of the data is not necessary. However, if widely different spacings are present, caution should be used in interpolating near the center of any large increment. The following formula can be used to get the poly-

```
010  REM--REF: INTRO, ENG'RNG, MODELING ....... RIEDER/BUSBY
100  REM-- INTERPOLATION BY FORWARD GREGORY NEWTON METHOD
110  REM--(INPUT) N = NUMBER OF F(X) VALUES IN DATA SET
120  REM--(INPUT) FX(I) = DISCRETE F(X) VALUES ,,
130  REM---- MUST SPAN THE INTERPOLATION POINT
140  REM--(INPUT) DX = INCREMENT IN X BETWEEN F(X) VALUES
150  REM--(INPUT) XI = VALUE OF X AT FIRST F(X) IN OVERALL DATA SET
160  REM--(INPUT) ID = DESIRED DEGREE OF POLYNOMIAL
170  REM--(INPUT) XU = VALUE OF X AT DESIRED INTERPOLATION POINT
180  REM--- PLACE F(X) VALUES IN SUBROUTINE 1000
190  CLS: CLEAR10: DEFINT I-N
200  READ N, DX, XI, ID, XU
210    DATA 7, .523599, 0, 3, 2.13
220    DIM FX(N), D(N)
230    GOSUB 1000
240    M = ID + 1
250    SM = M*DX/2
260    XF = (N-1)*DX
270  REM-- IDENTIFYING ACTIVE SPAN OF DATA CONSISTENT WITH ID
280    IF (XU-XI) <= SM THEN J = 1: GOTO 320
290    IF (XF-XU) <= SM THEN J = N - ID: GOTO 320
300    J = (XU-XI)/DX - M/2 + 2
310    REM-- FINDS FIRST X AND F(X) VALUES IN THE ACTIVE SPAN
320    X0 = XI + (J-1)*DX
330    F0 = FX(J)
340  REM--COMPUTES FINITE DIFFERENCES IN ACTIVE REGION
350  FOR I = 1 TO ID
360    D(I) = FX(J+1) - FX(J)
370    J = J + 1
380  NEXT I
390  FOR J = 2 TO ID
400    FOR I = J TO ID
410      K = ID - I + J
420      D(K) = D(K) - D(K-1)
430    NEXT I
440  NEXT J
450  REM--COMPUTES S
460    S = (XU-X0)/DX
470  REM--COMPUTES INTERPOLATED VALUE OF F(XU)
480  FU = F0
490  FN = S
500  DN = 1
510  FOR I = 1 TO ID
520    FU = FU + FN/DN*D(I)
530    FN = FN*(S-I)
540    DN = DN*(I+1)
550  NEXT I
560  REM-- OUTPUT ANSWER
570  LPRINT "F(X) AT X =";XU;" IS:";FU
580  END
```

(continued)

Figure A2.11*a* BASIC program for finding an interpolated value of *F(x)* by the forward Gregory—Newton method.

```
                              Subroutine
1000 REM--DISCRETE F(X) VALUES INPUT
1010 FOR I = 1 TO N
1020    READ FX(I)
1030    PRINT I;" F(X) =";FX(I)
1040 NEXT I
1050 RETURN
1100 DATA 0., .5, .866026, 1.0, .866026, .5, 0.
1110 END
```

Figure A2.11*b*   BASIC program for finding an interpolated value of *F(x)* by the forward Gregory–Newton method *(continued)*.

nomial.

$$f(x) \simeq P(x) = L_0(x)f_0 + L_1(x)f_1 + L_2(x)f_2 + \cdots$$

where
$$f_0 = f(x_0)$$
$$f_1 = f(x_1)$$
$$f_2 = f(x_2)$$
$$L_0(x) = (x - x_1)(x - x_2)(x_0 - x_1)^{-1}(x_0 - x_2)^{-1}$$
$$L_1(x) = (x - x_0)(x - x_2)(x_1 - x_0)^{-1}(x_1 - x_2)^{-1}$$
$$L_2(x) = (x - x_0)(x - x_1)(x_2 - x_0)^{-1}(x_2 - x_1)^{-1}$$

An example program, coded in BASIC, is given as Figure A2.12.

A2.5.3 Least-Squares Polynomial

In this method, the degree of the polynomial is not directly tied to the number of data pairs used. Instead, the polynomial coefficients are adjusted to minimize the sum of the squared differences between the curve and the actual $f(x_i)$ values. Generally, a low-degree (less than about degree five) polynomial is preferred to avoid inaccuracies in evaluating the coefficients. In many cases only a first-degree polynomial (linear least squares) is permitted. However, the following matrix equation can be solved for the appropriate coefficients for the following quadratic.

$$f(x) \simeq P(x) = a_0 + a_1 x + a_2 x^2$$

$$\begin{bmatrix} n & S(x) & S(x^2) \\ S(x) & S(x^2) & S(x^3) \\ S(x^2) & S(x^3) & S(x^4) \end{bmatrix} \begin{Bmatrix} a_0 \\ a_1 \\ a_2 \end{Bmatrix} = \begin{Bmatrix} S(f) \\ S(xf) \\ S(x^2 f) \end{Bmatrix}$$

Here, n represents the number of data pairs used, the S prefix represents a summation over all n of whatever follows in the parentheses, and f is $f(x_i)$.

No example program is given for this method, since the major effort is in evaluating the matrix equation. Such an evaluation has been illustrated already in Sections A2.2.1.1 and A2.2.1.2.

```
010  REM--REF: INTRO. ENG'RNG. MODELING ....... RIEDER/BUSBY
100  REM-- INTERPOLATION BY LAGRANGE POLYNOMIAL METHOD
110  REM--(INPUT) N = NUMBER OF F(X) VALUES IN DATA SET
120  REM--(INPUT) FX(I) = DISCRETE F(X) VALUE AT X(I)
130  REM--(INPUT) X(I) = DISCRETE X VALUE
140  REM--(INPUT) ID = DESIRED DEGREE OF POLYNOMIAL (KEEP SMALL)
150  REM--(INPUT) XU = VALUE OF X AT DESIRED INTERPOLATION POINT
160  REM--------- MUST FALL WITHIN SPAN OF AVAILABLE X(I) VALUES
170  REM--- PLACE F(X) AND X VALUES IN SUBROUTINE 1000
180  CLS: CLEAR10: DEFINT I-N
190  READ N, ID, XU
200     DATA 7, 3, 2.13
210     DIM FX(N), X(N)
220  REM--IDENTIFIES FIRST X AND F(X) VALUES IN ACTIVE SPAN
230  IF (ID+1) > N THEN LPRINT "NOT ENOUGH DATA FOR SPECIFIED ID !!": END
240  FOR I = 1 TO (N-1)
250     IF XU <= X(I) OR XU > X(I+1) THEN GOTO 300
260     IO = I - (ID+2)/2
270     IM = IO + ID + 1
280        IF IO < 1 THEN IO = 1: GOTO 340
290        IF IM > N THEN IO = N - (ID+1): GOTO 340
300     GOTO 340
310  NEXT I
320  LPRINT " XU IS OUT OF DATA-SET SPAN !!": END
330  REM-- COMPUTES LAGRANGE TERMS
340  FU = 0
350  FOR I = 1 TO (ID+1)
360     II = (I-1) + IO
370     AL = FX(II)
380     FOR J = 1 TO (ID+1)
390        JJ = (J-1) + IO
400        IF I = J GOTO 420
410        AL = AL*(XU-X(JJ))/(X(II)-X(JJ))
420     NEXT J
430     FU = FU + AL
440  NEXT I
450  REM-- OUTPUT ANSWER
460  LPRINT "F(X) AT X =";XU;" IS:";FU
470  END
```

```
1000 REM-- DISCRETE F(X) AND X VALUES INPUT
1010 FOR I = 1 TO N
1020    READ X(I), FX(I)
1030    PRINT I;" X =";X(I);"  F(X) =";FX(I)
1040 NEXT I
1050 RETURN
1100 DATA 0, 0,  .5, .479426
1110 DATA 1.3, .963558,  2., .909297
1120 DATA 2.6, .515501,  3., .141120
1130 DATA 3.14159, .000003
1140 END
```

Figure A2.12 BASIC program for finding an interpolated value of $F(x)$ by the Lagrange polynomial method.

Selected Answers

2.1 (b) $R = \ln(D_0/D_i)/(2\pi k1)$; $T = (T_h - T_c)$

2.3 (a) $R = 1/(A_1 F_{12})$; $F = J_1 - J_2$

2.5 (b) $R_{eq} = 78.8571$

2.7 (b) $R_{ys} = 5.0598$

2.17 $dh/dt + (0.019329\sqrt{h})/(1.5 - 0.2h)^2 - 0.15915[1 + \cos 0.05t]/(1.5 - 0.2h)^2 = 0.0$

2.19 $dh/dt - DA[1 + \cos Bt]/(hWL) = 0.0$

2.21 (b) $dP/dt + GT(P - P_e)^{(1-b)}/(Va) = GT\dot{m}_i/V$

2.23 $dT/dt + (mCR)^{-1}T = (mCR)^{-1}T_c$

2.25 $dI/dt + I/(L/R_1 + L/R_2) = (2/L)\sin 20t$

2.27 $dh_2/dt + \gamma(h_2 - h_1)(r_0 + 0.25h_2)^{-2}/(\pi C\sqrt{|h_2 - h_1|}) = 0.0$

2.29 $dN/dt - kN = P(t)$

2.31 $dN/dt + kN = 0$

2.33 $d^2T/dx^2 + [B/(A + BT)](dT/dx)^2 = 0$

2.35 $d^2\theta/dt^2 + (gC\sqrt{L}/W)\,d\theta/dt|d\theta/dt|^{0.5} + (g/L)\sin\theta = 0.0$

2.37 $d^2y/dt^2 + (CD^2g/W_0)[(V_x)^2 + (dy/dt)^2]^{0.25}\,dy/dt = g$

2.39 $dV/dt + (gC/W)|V_r|V_r = -gF_r/W$

2.41 $d^2s/dt^2 + (Cg/W)(ds/dt)^n - g\sin\theta s = (g/W)(F_p - F_r)$

2.43 $d^2x/dt^2 + (gC/W)(dx/dt)|dx/dt|^{(a-1)} = (g/W)(F_r - F_1)$

2.45 Same as 2.43 except $n = a$ and $y = x$.

2.47 $d^2\theta/dt^2 + [L_r F_r(x) - L_f F_f(x)]\sin\theta/J_0 = 0.0$

2.49 $d^2x/dt^2 + [gC/(LA\gamma)](dx/dt)|dx/dt|^{-0.3} + (0.68493g/L)x = Pg/(L\gamma)$

2.51 $d^2x_s/dt^2 + (gC_s/W_s)[d(x_s - x_c)/dt]|d(x_s + x_c)/dt|^{-0.7}$
$+ (gK_s/W_s)(x_s - x_c)|x_s - x_c|^{0.1} = 0.0$

2.53 $d^2x_2/dt^2 + (gC_2/W_2)d(x_2 - x_1)/dt + (gk_2/W_2)(x_2 - x_1) + (gk_3/W_2)x_2 = 0.0$

2.55 $d^2x/dt^2 - (gC/W)[(H - x)/R_2](dR_2/dt)|dR_2/dt|^{0.8}$
$+ (gk/W)[(H - x)/R_2](R_1 - R_2) = 0.0$, where $R_1 = \sqrt{A^2 + H^2}$ and
$R_2 = \sqrt{A^2 + (H - x)^2}$

2.57 $d^2X/dt^2 + (0.045714g)dX/dt + (0.45714g)dX/dt|dX/dt|^{0.55}$
$+ (0.14286g)X = 0.0$, where $X = x - 2[1 - \cos 2t]$

2.59 $d^2y/dt^2 + (gC/W)dy/dt|dy/dt|^{0.5} + [3gEI/(WL^3)]y = gF_w/W$

2.61 $d^2x/dt^2 - (gC/W)V|V|^{0.9} + (2gk/W)\sqrt{x^2 + L^2} = (2g/W)(kL - T_0) - gF_p/W$,
where $V = V_s - dx/dt$

2.63 $d^2y/dx^2 - (M/EI)[1 + (dy/dx)^2]^{1.5} = 0.0$, where $M = (L - x_w)(x/L)W$ for
$0 \leqslant x \leqslant x_w$; $M = (L - x)(x_w/L)G$ for $x_w < x \leqslant L$

2.65 $d^2y/dx^2 + [1 + (dy/dx)^2]^{1.5}(1.5x^3 - 60x^2 + 450x) = 0.0$

2.67 See 2.63.

3.1 Parabolic; $\partial^2 T/\partial y^2 = (1/\alpha)\partial T/\partial t$; $T(y, 0) = T_\infty$, $T(0, t) = a + bt$, $T(L, t) = T_\infty$

3.3 Elliptic; $\partial^2 T/\partial x^2 + \partial^2 T/\partial y^2 = 0.0$; $T(0, y) = 40 - 30y$ for $y \leqslant D$, $T(0, y) = 10$
for $y > D$; $T(x, 0) = 40 + 20x/W$ and $T(x, H) = 10 + 50x/W$ for all x's;
$T(W, y) = 60$ for all y's

3.5 Elliptic; $\partial^2 T/\partial x^2 + \partial^2 T/\partial y^2 = -(1/k)\partial k/\partial x \, \partial T/\partial x$; conditions as in 3.3

3.7 Elliptic; $\partial^2 P/\partial x^2 + \partial^2 P/\partial y^2 = 0.0$; $P(0, y) = 30.$, $P(L, y) = 10.$,
$dP/dy|(x, \text{up and down}) \simeq 0$

3.9 Same as 3.7

3.11 $d\rho/dt + d(\rho U)/dx + d(\rho V)/dy = 0.0$

3.13 $d^2T/dx^2 - [a_1/(a_0 + a_1 T)](dT/dx)^2 = (a_0 + a_1 T)|C/(x_e - x_i) \, dT/dt$;
$T(x, 0) = T_0$, $T(0, t) = T_1$, $T(L, t) = T_2$

3.15 $\partial^2 u/\partial x^2 = \gamma/(Eg)\partial^2 u/\partial t^2$; $u(0, t) = 0.0$, $u(x, 0) = 0.0$; $\partial u/\partial x|(L, t) = F_0(t)/(AE)$,
$\partial u/\partial x|(x, 0) = 0.0$

3.17 $\partial^2 z/\partial x^2 + \partial^2 z/\partial y^2 = (\gamma t/F_t) + [\gamma t/(gF_t)]\partial^2 z/\partial t^2$; with six conditions,
where $t = $ thickness, $\gamma = $ specific weight, and $g = $ gravitational acceleration

3.19 $\partial^2 S_x/\partial x^2 = \partial^2 S_x/\partial y^2$; $\partial^2 S_y/\partial x^2 = \partial^2 S_y/\partial y^2$; $R_x = [1 + (dz/dx)^2]^{1.5}/(\partial^2 z/\partial x^2)$;
$R_y = [1 + (dz/dy)^2]^{1.5}/(\partial^2 z/\partial y^2)$

3.21 Refer to Example 3.6; $S_{\max, x} = [0.5t/(1 - \mu^2)][\partial^2 w/\partial x^2 + \mu \, \partial^2 w/\partial y^2]$
$S_{\max, y} = [0.5t/(1 - \mu^2)][\partial^2 w/\partial y^2 + \mu \, \partial^2 w/\partial x^2]$

4.1 $d^2y/dx^2 + y/4 = 8.$; y, 2, L, NH, BC, constant, analytic, single equation

4.3 $dx/dt - x - z\sqrt{z} = 0.0$, $x(0) = 1.$; $dz/dt + z^2 - x - y = 0.0$, $z(0) = 2.$;
$dy/dt - x^2 - 1/y = 0.0$, $y(0) = 1.$; x, y, z, 1, NL, H, IC, constant, analytic,
three-equation set

4.5 $d^2y/dx^2 + (1/x) \, dy/dx - y = 0.0$; $y(0.1) = 1.0$, $y'(0.9) = 0.0$; y, 2, L, H, BC,
variable, analytic, single equation

4.7 $dx/dy \pm \sqrt{20/(3y) - x^2/(2y)} = 0.0$, with conditions; x, 1, NL, H, conditions
missing, variable, analytic, single equation

4.9 $d^3y/dt^3 + 3d^2y/dt^2 + 7dy/dt + 5y = 0.0$; $y(0) = 0.$, $y'(0) = 1.$, $y''(0) = 2.$;
y, 3, L, H, IC, constant, analytic, single equation

4.11 $d^2x/dy^2 - 0.32 \, dx/dy|dx/dy|^{0.8} + 45.2x = 10 \cos 5y$; $x(0) = 0.0$, $x'(0) = 5.$;
x, 2, NL, NH, IC, variable, analytic, single equation

4.13 $d^2\theta/dt^2 + (\frac{1}{93}) \sin \theta = (\frac{1}{93}) \cos 3t$; $\theta(0) = 0.0$, $\theta'(0) = 2.3$; θ, 2, NL, NH, IC, constant, analytic, single equation

4.15 $dy/dx = [g(x) + 14.8 \tan x]^{(1/3)}$; $y(0) = 6.$, $y'(0) = 2.$; y, 1, L, NH, IC, constant, tabular, single equation

4.17 Two-dimensional Laplace equation (elliptic); x, 2, L, H, BC, constant, analytic, single equation

4.19 Poisson equation (elliptic); driving function: 3P; z, 2, L, NH, BC, constant, analytic(?), single equation

4.21 $\partial^2 c/\partial b^2 = 6.3b\ \partial c/\partial a + 4.2b - 1.9$; conditions missing; (parabolic); c, 2, L, NH, BC and IC, variable, analytic, single equation

4.23 Poisson equation (elliptic); driving function: $-2.6 \sin 3x$; conditions missing; C, 2, L, NH, BC, constant, analytic, single equation

5.1 $T(0.1) = 70.833$; $T(0.4) = 73.231$

5.3 $x(0.1) = 13.000$; $x(0.3) = 22.003$

5.5 $x = 4.0 - 0.15 \sin 4y$

5.7 $y(1.5) = 12.183$; $y(2.0) = 20.633$

5.9 $x = 10.0 + 2.5y^2 - 0.1y^5$

5.11 $T(0.5) = 53.117$; $T(2.0) = 23.208$

5.13 $x(0.5) = 2.8000$; $x(1.0) = 5.8702$

5.15 $y = 0.5\ e^{-2x} \sin 2x$

5.17 $y(0.1) = 1.1107$; $y(0.2) = 1.2491$

5.19 $t(0.5) = 1.5$; $t(1.0) = 2.0$; $x(0.5) = 1.5$; $x(1.0) = 2.25$

5.21 $x(0.3) = 0.40778$; $x(0.4) = 0.64076$; $y(0.3) = 1.9208$; $y(0.4) = 1.8771$

5.23 $y_1(0.1) = 0.0$; $y_1(0.2) = 0.01$; $y_2(0.1) = 1.0$; $y_2(0.2) = 1.0$

5.25 $P(1.5) = 5.6958$; $P(2.0) = 2.4962$; $P'(1.5) = -5.6641$; $P'(2.0) = -6.9492$

5.27 $y(0.6) = 1.8452$; $y(0.8) = 2.5469$; $y'(0.6) = 2.9695$; $y'(0.8) = 4.0232$

5.29 Euler: $y(0.3) = 1.0148$; $y'(0.3) = 0.14274$; fourth-order Runge–Kutta: $y(0.3) = 1.0215$; $y'(0.3) = 0.13976$; third-order Adams predictor–corrector: $y(0.3) = 1.0215$; $y'(0.3) = 0.13975$

5.31 Exact: $y(1.0) = 0.79195$; $y'(1.0) = 0.55392$; fourth-order Runge–Kutta: $y(1.0) = 0.79354$; $y'(1.0) = 0.55021$

5.33 Exact: $x(0.4) = 2.7469$; $x'(0.4) = 3.0131$; fourth-order Runge–Kutta: $x(0.4) = 2.7383$; $x'(0.4) = 3.0367$

5.35 $y = \exp(2x)[C_1 + 0.25 \int \exp(x^2 - 2x)\ dx]$
$+ \exp(-2x)[C_2 - 0.25 \int \exp(x^2 + 2x)\ dx]$

5.37 Shoot no. 1: $y(1.5) = 1.875$; $y'(1.5) = 2.375$. Shoot no. 2: $y(1.5) = 3.5$; $y'(1.5) = 4.125$. Shoot no. 3: $y(1.5) = 2.6875$; $y'(1.5) = 3.25$

5.39 Shoot no. 1: $a(6.0) = 2.4676$; $a'(6.0) = 6.1635$. Shoot no. 2: $a(6.0) = 4.3771$; $a'(6.0) = 7.6170$. Shoot no. 3: $a(6.0) = 3.4314$; $a'(6.0) = 6.8770$

5.41 Shoot no. 3: $y(1.5) = 4.9977$; $y'(1.5) = 5.5793$. Shoot no. 4: $y(1.5) = 5.0000$; $y'(1.5) = 5.5819$

5.43
$$\begin{bmatrix} -0.6 & 1.6 & 0.0 \\ 0.0 & -1.0 & 2.0 \\ 0.0 & -0.4 & -1.4 \end{bmatrix} \begin{Bmatrix} y_2 \\ y_3 \\ y_4 \end{Bmatrix} = \begin{Bmatrix} 0.4 \\ 0.0 \\ 0.0 \end{Bmatrix}$$

5.45 $P(0.2) = 1.6751$; $P(0.4) = 2.7474$; $P(0.6) = 3.8022$ by tabular Gauss–Jordan elimination

5.47 $y(1.0) = 14.790$; $y(2.0) = 33.883$; $y(3.0) = 52.506$ by tabular Gauss–Jordan elimination

$y(1.0) = 13.601$; $y(2.0) = 32.523$; $y(3.0) = 51.727$ by tabular Gauss–Seidel elimination

5.49
```
LABEL EXERCISE 5.49
INITIAL
    RENAME TIME=X, FINTIM=XSPAN
    INCON YO=13.
    PARAMETER EPSI=0.001
DYNAMIC
    DYDX=0.5/(X+EPSI)-SIN(Y)*ABS(SIN(Y))**.5
    Y=INTGRL(YO,DYDX)
TERMINAL
    TIMER DELT=.2, OUTDEL=.2, XSPAN=100.
    PRTPLOT Y(DYDX)
    END
    STOP
ENDJOB
```

5.51
```
LABEL EXERCISE 5.51
RENAME TIME=X, DELT=DX, FINTIM=MAXX
INITIAL
    INCON ZO=20
    FUNCTION G=(-60,2),(-10,0),(100,3)
    FUNCTION F=(0,-2),(10,23),(500,1)
DYNAMIC
    DZDX=GF+FF
    GF=NLFGEN(G,Z)
    FF=NLFGEN(F,X)
    Z=INTGRL(ZO,DZDX)
TERMINAL
    METHOD RKSFX
    TIMER DX=.01, OUTDEL=.02, MAXX=5.
    PRTPLOT Z(GF,FF), DZDX
    END
    STOP
ENDJOB
```

5.53
```
LABEL EXERCISE 5.53
RENAME TIME=X, DELT=DX, FINTIM=MX
INITIAL
    CONSTANT YO=1., YPO=0.
DYNAMIC
    DYDX=INTGRL(YPO,D2YD2X)
```

```
            D2YDX2=5-Y**1.5+DYDX
            Y=INTEGRL(YO,DYDX)
        TERMINAL
          METHOD RKSFX
          TIMER DX=.01, OUTDEL=.02, MX=2.
          PRTPLOT X(DYDX,D2YDX2), DYDX(X)
          END
          STOP
        ENDJOB
```

5.55 LABEL EXERCISES 5.55
```
        INITIAL
            INCON YO=.1, YPO=0.0
            PARAMETER A=.9022,B=6.434,...
                      C=609.23, N=1.5, M= -2.32
        DYNAMIC
            FY=B*(1.+C*(1.-Y)**M)*Y
            D2YDT2=-A*DYDT**N-FY
            DYDT=INTGRL(YPO,D2YDT2)
            Y=INTGRL(YO,DYDT)
        TERMINAL
          METHOD RECT
          TIMER DELT=.002, OUTDEL= .004, FINTIM=.4
          PRTPLOT Y(DYDT), DYDT(Y)
          END
          STOP
        ENDJOB
```

5.57 LABEL EXERCISE 5.57
```
        INITIAL
            CONSTANT A=30, B=502, C=3, D=12,...
              E=7.9, F=4080., G=500.,AN=1.25, AM=1.1
            INCON XO=0., XPO=10., YO=0., YPO=10.
        DYNAMIC
            D2XDT2=-(A*(DXDT-DYDT)*ABS(DXDT-DYDT)**(AN-1)+B*(X-Y))
            DD=D*DYDT*ABS(DYDT)**(AM-1)
            EE=E*(DXDT-DYDT)*ABS(DXDT-DYDT)**(AN-1
            D2YDT2=-C*(DD-EE+F*Y-G*(X-Y))
            DXDT=INTGRL(XPO,D2XDT2)
            DYDT=INTGRL(YPO,D2YDT2)
            X=INTGRL(XO,DXDT)
            Y=INTGRL(YO,DYDT)
        TERMINAL
          METHOD RKSFX
          TIMER DEL=.001, OUTDEL=.002, FINTIM=.2
          PRTPLOT X(D2XDT2,Y), Y(D2YDT2,X)
          END
          STOP
        ENDJOB
```

5.59L $T(0.1)=62.16$; $T(0.5)=60.12$

5.61L $h(36.0)=1.3524$; $h(60.0)=0.94397$

5.63L $T_s(0900)=43.232$; $T_s(1600)=50.111$

5.65L $T(2.0)=319.42$; $T(5.0)=489.99$

5.67L $T(0.10) = 69.4$; $T(0.50) = 67.23$

5.69L $P(0.5) = 183.96$; $P(1.0) = 199.69$

5.71L $Z(1.0) = 0.53713$; $Z'(1.0) = -17.976$

5.73L $T(0.2) = 0.59361$; $T'(0.2) = 2.3215$

5.75L $\theta(7.0) = -9.1150$; $\theta'(7.0) = 1.1904$

5.77L $y(0.2) = 0.081743$; $y'(0.2) = 0.82881$

5.79L $y(4.0) = 0.80798$; $y'(4.0) = 0.44961$

5.81L $\theta(3.0) = 1434.1$; $\theta'(3.0) = 434.6$

5.83L $x(0.10) = -0.045783$; $x''(0.10) = 13.964$

5.85L $y(5.0) = -2.5698 \times 10^{-3}$; $y'(5.0) = -8.4653 \times 10^{-4}$

5.87L $x(8.0) = 1940.0$; $x'(8.0) = 32.136$

5.91L $T(0.10) = 298.75$; $T'(0.10) = 100.84$

5.93L $y(10.0) = -0.25652$; $y'(10.0) = 0.026246$

5.95L Convergence in four shoots gives: $F = 1756.04$

5.97L $T(9.1260) = 131.87$; $T'(9.1260) = 0.31$

6.1 $U_{2,2} = 225.20$, $U_{2,3} = 175.10$, $U_{3,2} = 225.10$, $U_{3,3} = 175.05$ (Gauss–Seidel)
$U_{2,2} = 225.00$, $U_{2,3} = 175.00$, $U_{3,2} = 225.00$, $U_{3,3} = 175.00$ (Gauss–Jordan)

6.3 $T_{2,2} = 296.44$, $T_{2,3} = 234.92$, $T_{3,2} = 325.03$, $T_{3,3} = 263.51$ (Gauss–Seidel)
$T_{2,2} = 296.48$, $T_{2,3} = 234.95$, $T_{3,2} = 325.06$, $T_{3,3} = 263.52$ (Gauss elimination)

6.5 $C_{2,2} = 4.4621$, $C_{3,2} = 5.2692$, $C_{2,3} = 3.5082$, $C_{3,3} = 4.4880$ (Gauss–Jordan)
$C_{2,2} = 4.4393$, $C_{3,2} = 5.2584$, $C_{2,3} = 3.4968$, $C_{3,3} = 4.4826$ (Gauss–Seidel)

6.7 $T_{2,2} = 107.78$, $T_{3,2} = 169.24$, $T_{4,2} = 269.24$, $T_{2,3} = 101.95$ (Gauss–Seidel)

6.9 $C(x, t)$; $C(2.0, 2.5) = 37.5$, $C(5.0, 2.5) = 12.5$, $C(9.0, 2.5) = 206.25$

6.11 At $t = 0.4$; $T_2 = 17.565$, $T_3 = 6.8964$, $T_4 = 3.1236$, $T_5 = 2.4745$; at $t = 0.8$:
$T_2 = 41.145$, $T_3 = 19.270$, $T_4 = 9.7695$, $T_5 = 6.9147$, using a tabular Gauss–Jordan solver

6.13 At $t = 10.417 \times 10^{-5}$: $T = -15.295$ at $x = 0.2$, $T = -19.165$ at $x = 0.4$

6.15 at $t = 0.4$: $T_2 = 13.633$, $T_3 = 3.7301$, $T_4 = 1.2880$, $T_5 = 1.4220$; at $t = 0.8$:
$T_2 = 37.712$, $T_3 = 14.718$, $T_4 = 6.2378$, $T_5 = 5.0815$, using a tabular Gauss–Jordan solver

6.17 at $t = 5.0$: $C = 29.375$ at $x = 2.0$, $C = 9.054$ at $x = 4.0$, $C = 17.751$ at $x = 6.0$

6.19 at $t = 8.0$: $T_2 = 0.170$, $T_3 = 12.692$, $T_4 = 19.362$, and $T_5 = 22.848$, using a tabular Gauss elimination solver

6.21 At $t = 4.0$: $T_2 = 0.88517$, $T_3 = 18.541$, $T_4 = 23.278$, $T_5 = 24.569$

6.23 $u(x, t)$: $u(x, 0.2) = 0.0$ except $u(2.0, 0.2) = 0.2$, $u(x, 0.4) = 0.0$ except $u(4.0, 0.4) = 0.2$

6.25L $T = 56.13$ at $x/W = 0.5$ and $y/H = 0.5$, $T = 50.49$ at $x/W = 0.91667$ and $y/H = 0.5$

6.27L $P(x, y)$: $P(0.5, 0.7) = 0.006$, $P(1.1, 0.7) = -0.024$, $P(1.3, 0.7) = -0.102$

6.29L $D(x/W, y/L)$: $D(0.25, 0.25) = 0.191$, $D(0.5, 0.25) = 0.221$, $D(0.5, 0.5) = 0.294$

6.31L $T(x/W, y/H)$: $T(0.25, 0.25) = 39.9$, $T(0.5, 0.5) = 30.4$, $T(0.75, 0.75) = 115.0$

6.33L $C(x/W, y/L)$: $C(0.875, 0.55556) = 14.1$, $C(0.5, 0.88889) = 5.0$

6.35L $w(x, y)$ with $w(0, 0)$ at center: $w(0.1, 0.0) = -0.533 \times 10^{-5}$; $w(0.7, 0.0) = -0.776 \times 10^{-5}$, $w(-0.7, 0.0) = 0.000$

6.39L $P(x)$ at $t = 1200$: $P(6.0) = 1.7950$, $P(18.0) = 1.3832$, $P(48.0) = 1.0041$

6.41L $y(x)$ at $t = 9.8 \times 10^{-3}$: $y(1.0) = -0.10046$, $y(2.3) = -0.21068$, $y(4.0) = -0.06031$

6.43L $C(x)$ at $t = 15$: $C(0.01) = 0.02175$, $C(0.03) = 0.00421$

6.45L $U(x)$ at $t = 10$: $U(0.02) = 33.23$, $U(0.06) = 0.24$

6.47L $v(y)$ at $t = 3.0 \times 10^{-3}$: $v(1.0) = 0.14998$, $v(2.5) = 0.25509$, $v(3.0) = 0.24261$

7.1 To make the stiffness nonsingular, one of the degrees of freedom will have to be eliminated, that is, apply a boundary condition and fix one end.

7.3 $f_1 = -(f_2 + f_3)$

INDEX